The Sonar of Dolphins

Whitlow W.L. Au

The Sonar
of Dolphins

With 237 illustrations

Springer-Verlag

New York Berlin Heidelberg London Paris
Tokyo Hong Kong Barcelona Budapest

Whitlow W.L. Au
Naval Ocean Systems Center
P.O. Box 997
Kailua, HI 96734 USA

Library of Congress Cataloging-in-Publication Data
Au, Whitlow, W.L.
  The sonar of dolphins / Whitlow W.L. Au.
    p.   cm.
  Includes bibliographical references and index.
  ISBN 0-387-97835-6.—ISBN 3-540-97835-6
  1. Dolphins—Physiology.   2. Sonar.   3. Echolocation (Physiology)
I. Title.
QL737.C432A92   1993
599.5′3—dc20   92-22696

Printed on acid-free paper.

Production managed by Karen Phillips; manufacturing supervised by Vincent Scelta.
Typeset by Asco Trade Typesetting Ltd., Hong Kong.
Printed and bound by Edwards Brothers, Inc., Ann Arbor, MI.
Printed in the United States of America.

9 8 7 6 5 4 3 2 1

ISBN 0-387-97835-6 Springer-Verlag New York Berlin Heidelberg
ISBN 3-540-97835-6 Springer-Verlag Berlin Heidelberg New York

I dedicate this book to my wife, Dorothy,
and my children: Wagner "Jim," Wailani,
Wesley and Wainani.

And God said, "Let the water teem with living creatures, and let birds fly above the earth across the expanse of the sky." So God created the great creatures of the sea and every living and moving thing with which the water teem, according to their kinds... And God saw that it was good.

Genesis 1: 20–21

Preface

Over the ages, humans have always been fascinated by dolphins. This fascination heightened in the 1950s when oceanariums and aquariums began to use dolphins as show performers to demonstrate their prowess and display how tractable and trainable they were. The television series "Flipper" brought considerable public awareness and, coupled with the growing sophistication and popularity of dolphin shows, helped to further heighten public interest in these intriguing marine mammals. Soon the alluring but unfounded myth began to surface that dolphins are the smartest of animals, with an intelligence approaching and perhaps surpassing that of humans. The popularity of dolphins has grown to the point where many feel that none of these ever smiling and lovable animals should be kept in captivity but that they should all be released back into the ocean. Such sincere but misplaced opinions threaten serious research on these animals. We have much to learn about dolphins and small whales, about their physiology, morphology, psychology, and sensory capabilities—including their auditory and sonar capabilities. Certain facets of dolphin biology can only be studied with captive animals, and increased knowledge will improve our ability to manage and preserve wild dolphins.

This book is about the sonar of dolphins. My goal is to synthesize the many research findings and current pool of knowledge on the auditory and sonar capabilities of dolphins in a cohesive manner and to present a comprehensive and organized treatise on the subject. In my opinion, such an effort is long overdue, since there is a paucity of books on this subject. I know of only one booklet and one book dealing with echolocation in dolphins. I have attempted to introduce as many concepts from physics as possible and also to create mathematical models as an aid to the quantification and understanding of biosonar capabilities. Topics are covered which range from auditory pathways and processes, to anatomy of the dolphin's head, to signal processing models, to a comparison of the sonar of bats and dolphins.

Research on the sonar of dolphins has always involved scientists of diverse academic background, experience and expertise. I have collaborated with biologists, psychologists, physiologists, physicists, electrical and mechanical engineers, veterinarians and a political scientist in my research. The political scientist was Bill Powell, my first division head (later the head of my department and directorate). The interdisciplinary nature of the field has made my work very interesting, informative, and enriching, but it did present a unique challenge in writing this book. It is my intention to present material that will be useful and informative to anyone interested in biosonar. This book is the result of many years of research on the sonar of dolphins as a civilian scientist at the Naval Command Control and Ocean Surveillance Center (NCCOS) [formerly the Naval Ocean Systems Center] located on Kaneohe Bay, Oahu, Hawaii. Most of the dolphin sonar and auditory research in the United States has had some affiliation with the U.S. Navy. An interesting book on the history of the Navy's role in dolphin research was written by Forrest Wood in 1973 under the title *Marine Mammals and Man: The Navy's Porpoises and Sea Lions*, published by Robert Luce Inc.

I am indebted to many collaborators who have provided valuable assistance in my research. These include animal behavioral scientists Deborah and Jeffery Pawloski, Patrick Moore, Ralph Penner, and Wayne Turl. Computer systems analyst Perry Keaney provided valuable assistance in configuring the software used for much of my backscatter work. I have spent many hours discussing various aspects of sonar signal processing with Robert Floyd. I have often used Bob as a sounding board, and our many discussions have enhanced and solidified my understanding of the subject. I would have been lost many times without the valuable assistance of John Tokunaga, an excellent and dedicated electronics technician. Finally, my various supervisors through the years, Mr. Richard Soule, Dr. James Fish, Dr. Michael Stallard, and Dr. Paul Nachtigall, have all been extremely supportive and cooperative in allowing me to pursue various avenues of research I considered important. Last but not least, I am very grateful to my wife Dorothy who has been extremely supportive of my career.

I wish to acknowledge Mr. William Friedl, Mr. Robert Floyd, and Dr. Paul Nachtigall (all of NCCOS) who reviewed the original manuscript and provided many helpful suggestions and comments. Special thanks are extended to Dr. Arthur Popper, Chairman of the Zoology Department, University of Maryland, for his helpful suggestions and comments and for his role in having my book published by Springer-Verlag New York.

Whitlow W.L. Au
Kailua, Hawaii, 1992

Contents

1

Introduction

Research in biosonar has always required a multidisciplinary approach involving scientists of widely varying academic backgrounds, experience and expertise. Scientists with backgrounds as diverse as biology, psychology, physiology, physics, engineering, and veterinary medicine have been involved in unlocking the mysteries of the dolphin sonar. The rich interdisciplinary nature of the field presents a special challenge in the writing of a technical, scientific book on the dolphin sonar system that would be meaningful to a broad audience. In order to approach this subject, this chapter will discuss some necessary and important background material which may be very basic and elementary to members of a particular academic discipline, yet not so obvious and easily understood for those coming from a different background. The chapter will begin with a brief history of biosonar research, followed by sections on underwater acoustics, Fourier analysis, psychophysics, and signal detection theory. The sections on underwater acoustics and Fourier analysis are included to provide helpful background information for those in the behavioral and biological sciences. Similarly, the section on psychophysics was written to assist those in the physical sciences in understanding the various behavioral experiments that will be discussed. Signal detection theory is introduced to emphasize the statistical nature of auditory sensory perception. Throughout the book, excursions will be taken at times to

assist those not familiar with specific topics. The goal is to make this book useful, interesting, and informative to all who are interested in the sonar of dolphins.

My intent is to present a comprehensive, organized, and up-to-date treatise on the dolphin sonar system. Unfortunately, there are only very few books addressing this topic outside the former Soviet Union. However, there have been three excellent international NATO Advanced Study Institute symposia on animal sonar systems, involving bats, dolphins, and birds, with the proceedings published as books. The first symposium was held in Frascati, Italy, in 1966 (Busnel 1967); the second, on the island of Jersey in the English Channel in 1979 (Busnel and Fish 1980); and the third in Helsingor, Denmark, in 1986 (Nachtigall and Moore 1988). One purpose of these symposia was to present and discuss material that represented the "state of the art" knowledge of animal sonar systems. This book will synthesize into a unified treatise much of the relevant materials on the dolphin sonar system presented at these symposia, along with other research efforts, especially my own and those of my colleagues.

1.1 Historical Perspective

The term *SONAR*, coined late in World War II as a counterpart of the then famous word *RADAR*, is an acronym for *SO*und *NA*vigation and *R*anging. It originally referred to the principle of detecting and localizing underwater objects, such as submarines and mines, by projecting pulses of sound and detecting the echoes from the objects. Another word that was coined in analogy to the terminology of radar systems used in World War II is the term *echolocation*. Griffin (1944) originally applied this term to animal orientation based on the transmission of ultrasonic pulses and the reception of echoes from objects. Sonar, echolocation, and the term *biological sonar* or *biosonar* are all used interchangeably in referring to the concept of object detection, localization, discrimination, recognition, and orientation or navigation by animals emitting acoustic energy and receiving echoes.

1.1.1 Discovery of Sonar in Bats

Research on animal sonar can be traced back to the work of the Italian scientist Lazzaro Spallanzani in the 1770s. In 1773 Spallazani observed that bats could fly freely in a dark room where owls were helpless. This observation served as a starting point for a number of ingenious experiments in which Spallanzani studied the orientation capabilities of bats. He was amazed to observe that bats that had been blinded could fly and avoid obstacles as well as those that could see. Spallanzani's experiments inspired Charles Jurine, a Swiss scientist, to conduct a number of experiments in which the ears of bats were plugged with wax. Jurine (1798) found that the bats became helpless and collided with obstacles. Spallanzani repeated and improved on the experiments of Jurine and obtained similar results. He concluded that the bats' hearing was an important component of the bats' orientation and obstacle avoidance capabilities, but he could not explain how or why, because the bats appeared to be totally silent. His findings were ridiculed, rejected, and forgotten for more than a hundred years.

It was not until 1912 that the notion of bats using sounds inaudible to humans to detect objects in darkness was suggested, by an engineer named Maxim (1912). He thought that bats used the reflection of very low frequency subsonic sounds created by flapping of the wings. In the wake of the *Titanic's* tragic sinking, Maxim proposed a subsonic device that could warn a ship of an approaching iceberg, based on his erroneous view of bat echolocation.

The first to propose that the paradox of bats "seeing with their ears" could be explained by echolocation involving the use of ultrasonic sounds above the frequency range of human hearing was an English neurophysiologist named Hartridge (1920). He observed several bats that happened to fly into his Cambridge office one night. After closing the window, he was amazed to find that the bats could fly rapidly from room to room in the dark. He then experimented by varying the gap of the door from narrow to wide and found that the bats did not attempt to fly through when the gap was narrow. However, Hartridge did not conduct further experiments to confirm his hypothesis, and his notion was not well known.

It was not until 1938 that the "Spallanzani bat problem" was experimentally solved at Harvard University. Pierce and Griffin (1938), using an ultrasonic detector consisting of a piezoelectric crystal microphone and a superheterodyne receiver, were able to detect and tape-record the sonar signals of bats (*Myotis lucifugus* and *Eptesicus fuscus*). They found that the sounds had frequency components between 30 and 70 kHz. They also discovered that the bats emitted discrete pulses of 1–2 ms duration. Galambos (1941, 1942), working with Griffin at Harvard, further demonstrated that bats could not only emit but also hear ultrasonic sounds. By measuring cochlear potentials, they found that bats could hear sounds of frequencies from 30 Hz to 90 kHz (the limit of their apparatus). These and other experiments at Harvard removed all doubts concerning the use of sonar by bats for orientation in flight. Further details on the discovery of bat sonar and on the beginnings of modern research into bat sonar can be found in the excellent book *Listening in the Dark* by Donald Griffin (1958).

1.1.2 Discovery of Sonar in Dolphins

The research of Griffin and Galambos on bat echolocation paved the way for the discovery of echolocation in dolphins. In 1947 Arthur McBride, the first curator of Marine Studios (later Marineland) in Florida, presented evidence in his personal notes that the Atlantic bottlenose dolphin (*Tursiops truncatus*) may detect objects underwater by echolocation (McBride 1956). During the course of capture operations performed at night in the turbid waters of Florida's brackish inland waterways, McBride observed that *Tursiops* could avoid fine mesh nets although they readily charged large (10-sq.in.) mesh nets, and could go over nets where the cork lines were pulled below the water. They also seemed to detect any opening in the net from distances beyond visual range. The dolphins' behavior reminded McBride of the "sonic sending and receiving apparatus which enables the bat to avoid obstacles in the dark." He also considered the enormous development of the dolphins' cerebral cortex and the importance of acoustic perception to these animals. McBride died in 1950 and it was only later that his notes were discovered and published with an introductory note by William Schevill (McBride 1956).

During the early 1950s several investigators began to suspect echolocation in dolphins. Kellogg and Kohler (1952) were the first to publicly hypothesize on the possibility of dolphin echolocation. They performed a crude sound avoidance experiment and found that dolphins could hear ultrasonic frequencies up to 50 kHz. They concluded that "it follows as a unique possibility ... that they may also produce or emit ultrasonic vibrations. The inference seems inescapable that the porpoise, like the bat, may orient itself with respect to objects in its environment by echolocation...." A year later, Kellogg, Kohler, and Morris (1953) reported that *Tursiops* could hear ultrasonic frequencies up to 80 kHz. During this same time period, Forrest Wood, the new curator of Marineland, Florida, began to listen to and record the sounds produced by *Tursiops* and the spotted dolphin, *Stenella plagiodon*, housed in one of the oceanarium tanks. Whenever the transducer was introduced into the tank, the dolphins would inspect it, swimming by and emitting rasping and grating sounds (Wood 1952, 1953). This suggested that they were "echo-investigating" the transducer. The same response was noted when other objects were introduced into the tank. Schevill and Lawrence (1953a,b), using a call tone between 3 and 5 seconds, found that a *Tursiops* could hear frequencies as high as 120 kHz. Although the animal swam through "more than usually opaque water," Schevill and Lawrence listened for but could not detect any sounds related to the dolphin's navigation or food getting.

Schevill and Lawrence (1956) continued to search for evidence of echolocation in dolphins by conducting a simple food-finding experiment. By splashing the water, they summoned a captive dolphin to the boat in which they sat, holding a fish in the water. A net was stretched 2.5 m away from and perpendicular to the side of the boat. The dolphin had to decide on which side of the net the fish was lowered. The fish was lowered in a random fashion on one side of the net or the other. In 75% of the trials, the dolphin selected the correct side of the net. The use of vision by the dolphin was eliminated by conducting the experiment in turbid water and during dark nights. During this experiment, Schevill and Lawrence did hear the dolphin emit pulses as it swam toward the fish. Their results provided strong evidence of possible echolocation by dolphins.

Kellogg (1958) also conducted a series of experiments that provided strong evidence of echolocation by dolphins. He suspended 36 poles spaced 2.4 m apart in a 17 m × 21 m pool and observed two dolphins as they swam through the obstacles. In the first 20-minute session, a total of four collisions for both animals was observed. However, it appeared that the horizontal tail flukes were touching the obstacles as the animals swam past. All subsequent sessions were perfect, showing no collisions even on dark, moonless nights. (In one series of observations made during a dark moonless night, the dolphins swam through the obstacles without collisions.) Kellogg then performed a fish food discrimination experiment with one of the dolphins, which learned to discriminate between a preferred fish, spot (*Leiostomus xanthurus*), and a nonpreferred fish, mullet (*Mugil cephalus*). An experimenter located behind a plywood barrier with its bottom edge 2.5 to 5 cm below the water surface simultaneously inserted the two fish into the water, holding their tails. The dolphin was required to swim directly to and touch or take the preferred spot, which was randomly inserted to the right or left. The dolphin improved its performance progressively, and during the final 140 trials, some in near-total darkness at night, made no errors. In another experiment, Kellogg presented two spot fish, one behind a glass barrier and the other in a clear aperture. The glass window could be slid from side to side, and the fish were presented while the experimenter was behind the plywood barrier. One of the dolphins was trained to swim and take the fish presented in the clear aperture, ignoring the fish behind the glass. In 202 trials, the dolphin did not make a single error. In his final experiment, Kellogg required the two dolphins to swim through one of two openings in a net barrier. A Plexiglas barrier was randomly placed across one of the openings. In 50 trials each, the dolphins performed at 98% accuracy. Kellogg concluded that "the porpoises avoided the solid but invisible Plexiglas door by means of echo ranging."

Although the experiments of Schevill and Lawrence (1953a,b) and Kellogg (1958) offered interesting evidence that dolphins may have been echolocating, these experiments were not absolutely conclusive since vision was not entirely eliminated. It was not until 1960 that Kenneth Norris and his colleagues performed the first unequivocal demonstration of echolocation in dolphins (Norris et al. 1961). They placed rubber suction cups over the eyes of a *Tursiops* to eliminate its use of vision. The dolphin swam normally, emitting ultrasonic pulses and avoiding obstacles, including pipes suspended vertically to form a maze. The dolphin was able to retrieve fish tossed into the water as they drifted downward. Norris and his colleagues also noted that when the fish drifted below the level of it's melon, the dolphin did not retrieve it. From this they speculated that the sonar sounds were directional and were projected from the melon. During the same time period, Forrest Wood also used the same suction cup blindfold technique on a recently caught *Tursiops* (Wood and Evans 1980). The untrained dolphin swam about the tank with no indications that it lacked vision, and was able to retrieve fish tossed into the water. A *Tursiops truncatus* wearing suction cup blindfolds is shown in Figure 1.1.

Since the vivid demonstration of echolocation in *Tursiops* by Norris and his colleagues, the list of odontocetes that have been shown to echolocate has grown. The echolocation sound field of the rough-tooth porpoise (*Steno bredanensis*) was mapped by Norris and Evans (1966). They required the animal to swim between two taut lines while echolocating on a fish target. Busnel and Dziedzic (1967) trained a blindfolded harbor porpoise (*Phocoena phocoena*) to swim through a maze of vertically hanging wire. Norris (1968) demonstrated that a swimming blindfolded Pacific whitesided dolphin, *Lagenorhynchus obliquidens*, could avoid obstacles. The common dolphin, *Delphinus delphis*, was involved in a sonar experiment to examine it's ability to discriminate differences in complex geometric figures (Bel'kovich et al. 1969; Gurevich 1969). Penner and Murchison (1970) trained an Amazon River dolphin (*Inia geofrensis*) to respond to the difference in the diameter of thin wires presented as pairs behind a visually opaque but acoustically transparent screen. A Pacific pilot whale (*Globicephala melaena*) was trained to wear eye cups and retrieve rings tossed into the water (Wood and Evans 1980). A killer whale (*Orcinus orca*) was also trained to retrieve rings while blindfolded, during a study to measure its hearing sensitivity (Hall and Johnson 1971). Gurevich and Evans (1976) repeated the test for discrimination of three-dimensional figures by echolocation with a blindfolded Beluga (*Delphinapterus leucas*). The false killer whale (*Pseudorca crassidens*) was used by Thomas et al. (1988) in a sonar target detection experiment, with the target located behind a visually opaque but acoustically transparent screen. Finally Hatakeyama and Soeda (1990) observed Dall's porpoises (*Phocoenoides dalli*) avoid gill nets and swim through a 1.5 m × 1 m hole in a gill net. They concluded that the Dall's porpoises possess a high resolution sonar capability.

1.2 Some Underwater Acoustics

In this section, a brief discussion of some basic underwater acoustic principles and ideas will be offered, primarily for those who are not familiar with underwater acoustics. Readers well versed in underwater acoustics can skip to Section 1.3.

Acoustic energy propagates in water more efficiently than almost any form of energy so that

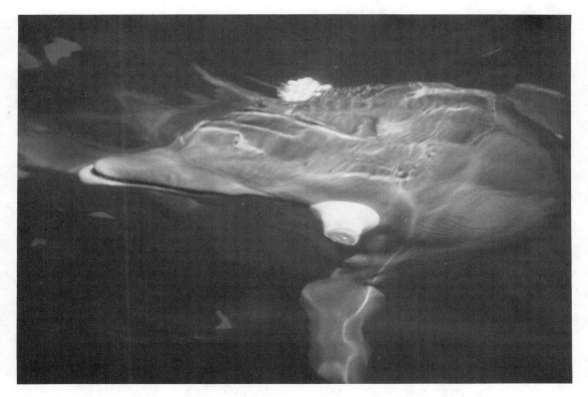

Figure 1.1. *Tursiops truncatus* wearing suction cup blindfolds.

the use of sonar and passive acoustics (listening) by dolphins is ideal. Electromagnetic, thermal, light, and other forms of energy are severely attenuated in water. Therefore, the most effective method for an animal to probe an underwater environment for the purposes of navigation, obstacle and predator avoidance, and prey detection is by sonar. Acoustic energy in water consists of molecular vibrations that travel at the speed of sound. The vibrations in water and other fluids are along the direction of propagation and are therefore referred to as longitudinal waves. In inelastic materials, acoustic vibrations can also occur in a direction perpendicular to the direction of propagation; such vibrations are referred to as transverse waves. Although fluids will only support longitudinal waves, transverse waves can be generated within targets in the scattering and reflection processes. However, any internal transverse waves will be transformed back into longitudinal waves when exiting the target and propagating in the fluid.

1.2.1 The Decibel and Sound Pressure Level

The decibel system has traditionally been used to discribe the intensity and pressure of acoustic waves. Decibel units provide a convenient way of expressing large changes in pressure. They also permit quantities to be multiplied or divided simply by adding or subtracting their decibel equivalents, respectively. Finally, decibels are a convenient measure of ratios: in underwater acoustics, the primary interest is often in ratios rather than in absolute quantities. The original and traditional use of decibels involved power ratios and is given by the equation

$$\text{power ratio in dB} = 10\log(P_1/P_2) \quad (1\text{-}1)$$

Where P_1/P_2 is the ratio of two powers. Throughout this book *log* refers to common logarithm and *ln* to natural logarithm. Acoustic energy is transmitted as an acoustic wave propagates in a medium, and the amount of energy per second (power) crossing a unit area is referred to as the intensity of the wave. Intensity ratios are defined similarly to power ratios:

$$\text{intensity ratio in dB} = 10 \log(I_1/I_2) \quad (1\text{-}2)$$

The unit of intensity in underwater acoustics is defined as the intensity of a plane wave having a pressure p of 1 micropascal (μPa). Since the relationship between acoustic pressure and intensity is

$$I = \frac{p^2}{\rho c} \quad (1\text{-}3)$$

where ρ is the density of water and c is the speed of sound, a pressure ratio in decibels can be expressed as

$$\text{pressure ratio in dB} = 20 \log(p_1/p_2) \quad (1\text{-}4)$$

Voltage and current ratios can also be expressed in decibels, in a similar manner as in (1-4). Strictly speaking, however, decibels refer to a ratio of intensities rather than pressures in acoustics and a ratio of powers instead of voltages or currents in electrical engineering. It is generally assumed that measured intensity is being referenced to the intensity of a plane wave of pressure equal to 1 μPa.

In acoustics, the basic measurement is one of pressure and not of intensity. Most hydrophones are sensitive to pressure, particle velocity, or pressure gradient. The sound pressure is defined in terms of sound pressure level (SPL), expressed as

$$\text{SPL} = 20 \log(p/p_0) \quad (1\text{-}5)$$

where p_0 is a reference pressure, usually 1 μPa. The reference of 1 μPa was adopted as the American National Standard in 1968. Previously, the common reference was 1 μbar (1 dyne/cm^2) for underwater acoustics and 0.000204 μbar for airborne acoustics. In order to convert SPL expressed in dB re 1 μbar into SPL in dB re 1 μPa, (re is a short hand notation for referenced) add 100 dB to the SPL expressed in dB re 1 μbar. Similarly, a voltage level can be expressed in dB by letting the reference voltage be 1 volt, so that the voltage is expressed as dBV, which reads as "decibels re 1 volt."

The conventional method of representing the magnitude of pressure and voltage signals is by

their root-mean-square, or rms, value, which is defined for an instantaneous pressure $p(t)$ as

$$p_{\text{rms}} = \sqrt{\frac{1}{T} \int_0^T p^2(t)\, dt} \quad (1\text{-}6)$$

where T is the duration of the signal. If the pressure signal is a sine wave of peak amplitude A, frequency f, and period T, then the rms pressure for N periods will be

$$\begin{aligned}
p &= \sqrt{\frac{A^2}{NT} \int_0^{NT} \sin^2 2\pi f t \, dt} \\
&= \sqrt{\frac{A^2}{NT} \int_0^{NT} \frac{1 - \cos 4\pi f t}{2} \, dt} \\
&= \sqrt{\frac{A^2}{NT} \left[\frac{NT}{2} - \frac{\sin 4\pi N}{8\pi f} \right]} \quad (1\text{-}7)
\end{aligned}$$

Since N is an integer, $\sin 4\pi N = 0$, so that

$$p = \frac{A}{\sqrt{2}} = 0.707\, A \quad (1\text{-}8)$$

Therefore, the rms value for a continuous sine wave or a sine pulse is simply 0.707 times the peak amplitude or 3 dB less than of the peak pressure expressed in dB. Furthermore, since the peak-to-peak amplitude is twice the peak amplitude, the rms pressure in dB is approximately 9 dB less than the peak-to-peak pressure expressed in dB. Most AC (alternating current) meters and other electronic measuring instruments are calibrated in rms voltage or current.

Unfortunately, dolphin sonar signals do not readily lend themselves to the rms convention. The signals are transient-(like) with nonuniform, exponentially decaying amplitudes, as will be seen in Chapter 5. Therefore, the simple expression given in (1-8) is not applicable and the integral expression (1-6) must be used. However, since the signals decay exponentially, it is difficult to select an appropriate duration T in (1-6). To simplify the description of dolphin sonar signals, the peak-to-peak amplitude has been adopted by Au et al. (1974). Peak-to-peak amplitude can be easily read off an oscilloscope trace or a strip chart or be measured with a peak/hold semiconductor device. The integration time of most AC meters is generally several orders of magnitude too long to be applicable for dolphin sonar signals.

1.2.2 Transmission Loss

Transmission loss refers to the decay of acoustic intensity as a signal travels from a source. If the intensity of the acoustic wave at a point 1 m from the source is I_0, and I_1 is the intensity at a distant point, then the transmission loss (TL) in dB is

$$TL = 10 \log I_0/I_1 = 20 \log p_0/p_1 \quad (1\text{-}9)$$

where p_0 and p_1 are the corresponding sound pressure levels at 1 m and at the distant point. Transmission loss may be considered to be the sum of loss due to *spreading* and loss due to *absorption* or *attenuation*.

The simplest kind of spreading loss is spherical spreading loss. This type of spreading loss applies to dolphin sonar since it occurs at relatively short distances, usually a few hundred meters. An acoustic signal will not normally experience considerable diffraction or ray bending over such distances unless there are extremely large variations in sound velocity along the path of propagation. Furthermore, sonar signals are relatively short (approximately 70 μs and less), so that components of the signal reflecting off various boundaries (water surface, bottom, or other boundaries) will not normally interfere with the direct component. We can examine spherical spreading loss by placing a sound source at the center of an imaginary sphere of radius 1 m and of another larger sphere of radius R. If for the moment we assume that there is no absorption loss, the power crossing both spheres must be the same. Since power equals intensity times area, we obtain

$$4\pi R_0^2 I_0 = 4\pi R_1^2 I_1 \quad (1\text{-}10)$$

Since R_0 is equal to 1 m, the transmission loss due to spreading can be derived from (1-9) and (1-10) as

$$TL = 10 \log R_1^2 = 20 \log R_1 \quad (1\text{-}11)$$

The spherical spreading loss is also referred to as *inverse square* loss, since the SPL decreases with square range so that transmission loss increases with square range.

Sound is also lost in the conversion of acoustic energy into heat. Absorption losses in seawater are caused by the effects of shear and volume viscosity, by ionic relaxation of magnesium sul-

fate ($MgSO_4$) molecules, and by a complicated boric acid ionization process. In the frequency range between 10 kHz and 200 kHz, the dominant absorption is due to ionic relaxation of $MgSO_4$. The attenuation rate in this frequency range was derived by Fisher and Simmons (1977) as

$$\alpha = \frac{A_2 f_2 f^2}{f_2^2 + f^2} \, dB/m \quad (1\text{-}12)$$

where

$A_2 = (48.83 \times 10^{-8} + 65.34 \times 10^{-10}\, T)$ sec/m
$f_2 = 1.55 \times 10^7 (T + 273.1)\exp[-3052/(T + 273.1)]$ Hz
f = frequency in Hz
T = temperature in degrees Celsius

Examples of typical acoustic absorption coefficients are plotted as a function of frequency for several water temperatures in Figure 1.2; the curves are for shallow depths (< 300 m). The effect of depth on absorption is relatively small: absorption decreases by about 2% for every increase of 300 m in depth. The transmission loss can be expressed as

$$TL = 20 \log R + \alpha R \quad (1\text{-}13)$$

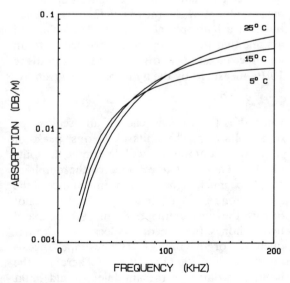

Figure 1.2. Absorption coefficient as a function of frequency for several different temperatures.

The absorption term can normally be ignored for ranges less than about 10 to 20 m where it does not significantly contribute to transmission loss.

1.2.3 Measurement and Production of Acoustic Signals

Underwater sounds are measured with hydrophones. These are usually piezoelectric transducers which convert sound pressure into proportional electrical voltages. For the measurement of dolphin sonar signals, a hydrophone should have a relatively flat frequency response up to approximately 150 kHz. It should also be as small as practical to minimize distortion of the sound field and reflection of the animal's signal off the hydrophone. One of the best hydrophones for dolphin sonar research is the Brüel & Kjaer 8103 miniature hydrophone: it is small, rugged, broadband, and built with precision, minimizing response variations from hydrophone to hydrophone. A schematic drawing of the B&K 8103 hydrophone, along with a typical response curve, is shown in Figure 1.3. The 8103 receiving response is relatively flat ($\pm$ 2 dB) up to a frequency of 160 kHz.

The sound pressure level measured by a hydrophone can be determined from the hydrophone free field sensitivity, if known, the gain in the measurement system, and the amount of voltage measured. Let M_x be the free field voltage sensitivity of a hydrophone in dB re 1 V/μPa, and assume that the hydrophone is connected to an amplifier of gain G in dB; then the SPL in dB re 1 μPa measured by the hydrophone is given by

$$\text{SPL} = |M_x| - G + 20 \log V \qquad (1\text{-}14)$$

where V is the voltage read off an oscilloscope or a voltmeter. The units of V, rms, peak, or peak-to-peak amplitude will determine the units of SPL. The input impedance of the amplifier should be much higher than the impedance of the hydrophone so that "loading" effects will not distort the measurement. In most broadband hydrophones, the impedance decreases with frequency. For the B&K 8103, the impedance is approximately 500 Ω at 10 kHz. Therefore, the input impedance of the amplifier should be at least 5 kΩ for measurements of acoustic signals with frequencies equal to or greater than 10 kHz.

If the measurement was conducted at a range R and the source level (SL), defined as the SPL at a reference range of 1 m, is desired, the transmission loss must be taken into accounted by adding it to (1-14):

$$\text{SL} = \text{SPL} + 20 \log R + \alpha R \qquad (1\text{-}15)$$

Other miniature hydrophones that can be used to measure dolphin sonar signals include the NRL (Naval Research Laboratory) F-42D, Edo-Western 6166, Bruel & Kjaer 8105, and Celesco LC-10.

Sounds can be produced in water by driving a transducer with a signal generator and a power amplifier. The transmit sensitivity of a transducer is generally given in terms of its source level per volt or per ampere of drive. Let M_T be the transmit sensitivity of a transducer in dB re 1 μPa/volt; the source level in dB re 1 μPa will be

$$\text{SL} = M_T + 20 \log V_{\text{in}} \qquad (1\text{-}16)$$

where V_{in} is the voltage at the input of the transducer cable. The SPL at any distance from the transducer can be calculated by subtracting the transmission loss from (1-16).

1.2.4 Noise Measurement

Noise, whether generated internally or externally, can have a tremendous, if not a dominating, influence on the detection range of any sonar system (biological or man-made). Noise can be defined as any unwanted acoustic signal that will interfere with the effectiveness of a sonar system. Ambient noise is the noise in the surrounding environment that is generated by various sources. One source is the interaction of the sea with its surroundings: noise is produced by tidal and wind-generated wave motions, the flow of currents, eddies and local turbulence, seismic activities, rain, and cavitation or collapse of air bubbles formed by turbulent wave action in the air-saturated near-surface waters. Sounds produced by biological organisms such as shrimp, fish, and marine mammals are another source of ambient noise. Ship traffic and other man-made machinery sounds can also contribute to the ambient noise condition of a location.

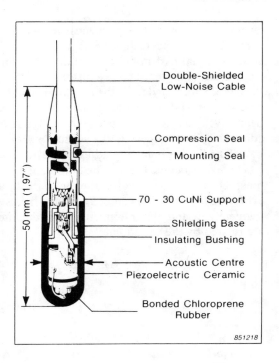

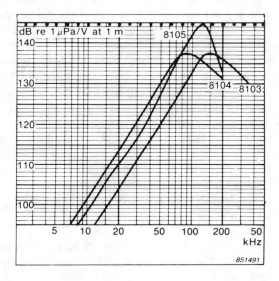

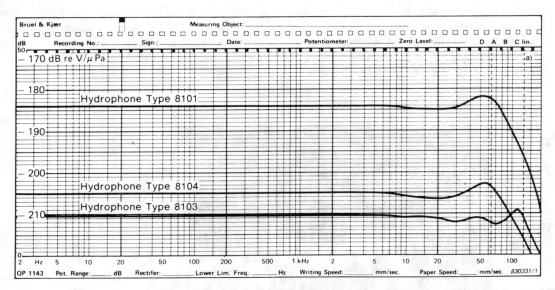

Figure 1.3. Schematic drawing of the B&J 8103 hydrophone along with typical transmitting and receiving sensitivity curves for the 8101, 8103, 8104, and 8105 hydrophones.

Ambient noise generally fluctuates unpredictably in a random fashion and has a continuous distribution of frequencies, or a continuous spectrum. Therefore, the amount of noise measured will depend on the bandwidth of the measurement system. The broader the bandwidth, the greater the amount of noise measured. If the noise is "white" (same mean amplitude at all frequencies), the amount of noise measured will be directly proportional to the bandwidth. In order to generalize noise measurements and make them independent of the bandwidth of the measurement system, researchers have adopted a standard convention of measuring noise on a per bandwidth basis by dividing the noise SPL by the system's bandwidth. In the decibel system, this is equivalent to subtracting out the bandwidth:

$$NL = 20 \log N_J - 10 \log \Delta f \qquad (1\text{-}17)$$

where N_J is the SPL of the noise and Δf is the bandwidth in Hertz. NL is referred to as the *noise spectrum level*, and is expressed either in dB re $1\ \mu Pa^2/Hz$ or in dB re $1\ \mu Pa/\sqrt{Hz}$. It is the intensity of a plane wave having an rms pressure of $1\ \mu Pa$ in a 1-Hz frequency band.

Ambient noise should be measured with a nondirectional hydrophone to ensure that the entire noise field is being measured. There are several common methods to measure ambient noise. One simple technique is to connect a nondirectional hydrophone to an amplifier and a spectrum analyzer. The resolution bandwidth should be subtracted from the amplitude scale as described in (1-17) in order to obtain the spectrum level. Instead of a spectrum analyzer, an adjustable filter connected to a voltmeter is often used. The filter is generally adjusted in $\frac{1}{3}$-octave, $\frac{1}{2}$-octave or 1-octave bands. The filter bandwidth must be subtracted out of the noise SPL as expressed in (1-17). The center frequency f_c, lower frequency f_1, and upper frequency f_u, of the filter are related by the equations

$$f_c = \sqrt{f_1 f_u} \qquad (1\text{-}18)$$

and

$$f_u/f_1 = 2^n \qquad (1\text{-}19)$$

where $n = \frac{1}{3}$, $\frac{1}{2}$, or 1 for $\frac{1}{3}$-octave, $\frac{1}{2}$-octave, or 1-octave band measurements, respectively. An increasingly common technique in this era of digital electronics is to digitize the hydrophone output after appropriate amplification and perform a Fourier transform on the data to determine the noise spectrum level. This technique will be discussed in Section 1.3.

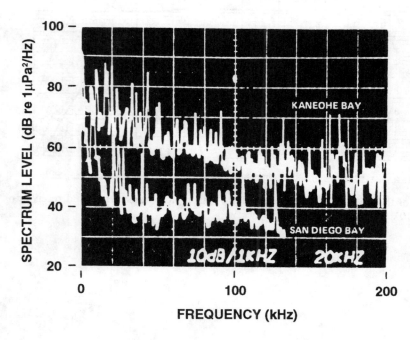

Figure 1.4. Ambient noise in Kaneohe Bay and San Diego Bay, measured with a H-52 hydrophone and a spectrum analyzer.

Figure 1.5. Ambient noise in Kaneohe Bay and San Diego Bay, measured with a H-52 hydrophone, 1/3-octave bandfilters, and an rms voltmeter.

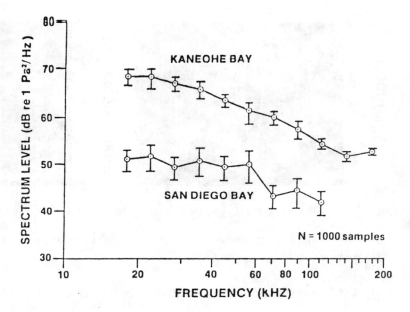

Examples of ambient noise measurements performed at the Naval Ocean Systems Center (NOSC) facilities in Kaneohe Bay, Oahu, Hawaii, and in San Diego, California, are shown in Figures 1.4 and 1.5. The noise in Figure 1.4 was measured with an NRL H-52 hydrophone and a spectrum analyzer. The noise in Figure 1.5 was measured with the same H-52 hydrophone connected to a filter and an rms voltmeter. The measurement was made in $\frac{1}{3}$-octave bands. The ambient noise of both bays is dominated by the presence of snapping shrimp; Kaneohe Bay has one of the highest levels of snapping shrimp noise in the world (Albers 1965).

1.3 The Time and Frequency Domains

1.3.1 Waveform and Spectrum

Any signal can be represented either in the time domain, with its amplitude displayed as a function of time, or in the frequency domain, with its amplitude displayed as a function of frequency. The time domain representation is usually referred to as the *waveform*, or *waveshape*, of the signal. Oscilloscopes are often used to observe the waveforms of signals. The frequency repre-

sentation is usually referred to as the *frequency spectrum* (or just *spectrum*) of the signal. Spectrum analyzers are often used to observe the spectral characteristics of continuous or long duration (on the order of several seconds) signals. An example of a 1-v rms, 120-kHz sinusoidal signal observed with an oscilloscope and a sweep frequency spectrum analyzer is shown in Figure 1.6. The frequency resolution of the spectrum analyzer used to obtain the display was 1 kHz. The frequency analyzer display is the output of a variable filter of 1-kHz bandwidth as the frequency of the filter is electronically tuned from 0 to 200 kHz. Since there is energy only at 120 kHz, the output of the filter should be zero for all frequencies except 120 kHz. In actuality, a self-noise voltage of -76 dBV appears in Figure 1.6. Another popular type of spectrum analyzer often used is the real-time spectrum analyzer, which consists of a bank of many contiguous constant-bandwidth filters all tuned to different frequencies. The outputs of the filters can be observed simultaneously, so that the spectrum of a signal can be observed in real time. However, real-time spectrum analyzers are usually three to four times more expensive than sweep frequency spectrum analyzers. They also tend to be larger and are limited in the upper frequency to about 100 kHz.

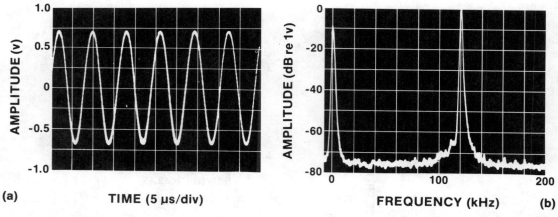

Figure 1.6. Example of a sinusoidal signal (1-v rms, 120 kHz) measured with (*A*) an oscilloscope, and (*B*) a spectrum analyzer.

1.3.2 Fourier Transformation

The frequency domain representation of a signal can be obtained by applying the Fourier Transform to the time domain signal. The time function is multiplied by $e^{-j2\pi ft}$ and integrated from $-\infty$ to $+\infty$ to transform it into the frequency domain, where $j = \sqrt{-1}$. Let $s(t)$ be the time domain waveform and $S(f)$ be the frequency domain representation of the signal; then the Fourier transform of $s(t)$ is

$$S(f) = \int_{-\infty}^{\infty} s(t)e^{-j2\pi ft}\,dt \qquad (1\text{-}20)$$

The kernel of the Fourier transform can be expressed as

$$e^{\pm j2\pi ft} = \cos(2\pi ft) \pm j\sin(2\pi ft) \quad (1\text{-}21)$$

Although $s(t)$ is a real function, its Fourier transform $S(f)$ is complex, having a real and an imaginary part. The spectrum of the signal is considered to be the absolute value of $S(f)$

$$|S(f)| = \sqrt{Re[S(f)]^2 + Im[S(f)]^2} \quad (1\text{-}22)$$

where $Re[S(f)]$ is the real part and $Im[S(f)]$ is the imaginary part of $S(f)$. Mathematically speaking, certain conditions must be satisfied for the Fourier transform of (1-20) to exist. However, the question of existence of the transform can be ignored for all signals that occur in real life. The signal $S(f)$ in the frequency domain can be transformed into the time domain by taking the

inverse Fourier transform of $S(f)$. The inverse Fourier transform is defined as

$$s(t) = \int_{-\infty}^{\infty} S(f)e^{j2\pi ft}\,dt \qquad (1\text{-}23)$$

If a signal can be expressed analytically, the integrals in (1-20) and (1-23) may be evaluated to obtain an analytical expression of the signal in both domains. However, this is rarely done except for the simplest types of signals (i.e., sine, square, rectangular, triangular, etc., signals). Instead, the waveforms of signals are usually digitized, and the Fourier transform is evaluated numerically with a computer.

Assume that the input waveform $s(t)$ is sampled and digitized N times at a fixed time interval of Δt, and $s(n)$ stands for the sampled values of $s(t)$, where $n = 0, 1, 2, \ldots, N-1$. Equations (1-20) and (1-23) can now be expressed as

$$S(k) = \sum_{n=0}^{N-1} s(n)e^{-2\pi nk/N} \qquad (1\text{-}24)$$

and

$$s(n) = \frac{1}{N}\sum_{k=0}^{N-1} S(k)e^{j2\pi nk/N} \qquad (1\text{-}25)$$

Equations (1-24) and (1-25) are the *discrete Fourier Transform* (DFT) pair. $S(k)$, where $k = 0, 1, 2, \ldots, N-1$, is the frequency component evaluated at

$$\Delta f = \frac{1}{N\Delta t} = \frac{1}{T} \qquad (1\text{-}26)$$

The *fast Fourier transform* (FFT) is a special, highly efficient and fast algorithm invented by Cooley and Tukey (1965) to evaluate the DFT of (1-24) and (1-25). The number of points in a sample had to be an integer power of 2, i.e., 2^i, where i is an integer. Cooley and Tukey's FFT algorithm requires approximately $N \log_2 N$ operations, whereas a brute force evaluation of the DFT requires about N^2 operations. For example, if $N = 1024 = 2^{10}$, the FFT algorithm would require approximately 10,240 operations, compared with the 1,048,576 operations for the DFT a computational saving of 102 times. Because of its efficiency in evaluating the summations in (1-24) and (1-25), the FFT algorithm has revolutionized the spectral analysis field. FFT programs have been written in every computer language known to man. Many digital oscilloscopes have the capability to Fourier transform the waveform data using the FFT configured in hardware, firmware or software. There have also been various modifications of the basic algorithm of Cooley and Tukey (1965) for special situations. Some good and easily read books on the FFT have been written by Bringham (1974), Ramirez (1985), and Burrus and Parks (1985).

1.3.3 Sampling Interval

The time interval Δt that is used in sampling and digitizing a waveform depends on the maximum frequency one is interested in. Since $S(k)$ in (1-24) is complex, with an equal number of real and imaginary parts, the maximum frequency that can be represented is

$$F_{max} = \frac{N}{2} \Delta f \qquad (1.27)$$

Substituting (1-27) into (1-26) we obtain

$$\Delta t = \frac{1}{N \Delta f} = \frac{1}{2F_{max}} \qquad (1.28)$$

If F_{max} is the maximum frequency of interest, then the sampling interval must be smaller than or equal to Δt of (1-28). Once the sampling interval is determined, caution must be taken to ensure that the input data do not contain frequencies above F_{max}. If higher frequencies are present in the data, they will fold back onto the lower frequencies producing false low-frequncy results.

This effect is referred to as "aliasing." Therefore, the input data should be passed through an anti-aliasing low-pass filter with an upper frequency limit equal to or less than F_{max}.

1.4 Experimental Psychological Methodology

This section consists of a brief discussion of experimental psychological methodology used in echolocation research with dolphins, intended to assist the nonbehavioralist with understanding some of the fundamental notions, ideas, and procedures associated with dolphin echolocation and passive auditory experiments that will be discussed in latter chapters. For an in-depth discussion of behavioral methodology applied to dolphin echolocation research, the reader is referred to the excellent review article by Schusterman (1980).

1.4.1 Stimulus Control and Operant Conditioning

One of the major considerations in performing any sensory sensitivity experiments with dolphins (or any subject) is the need to establish and maintain *stimulus control* of the subject's behavior. Establishing stimulus control implies that a stimulus exerts control over an animal's behavior and that introducing any changes in the characteristics of the stimulus will result in some measurable change in the animal's behavior, usually its performance accuracy. For example, in a sonar detection experiment the presence of a target (stimulus) will evoke a certain measurable behavioral response by the dolphin while the absence of the target will evoke a different response. Dolphins in the United States are generally trained with *operant conditioning* techniques (Skinner 1961) to respond to the controlling stimulus in a particular manner; they receive positive reinforcement, usually a fish reward. Often a secondary reinforcement (bridge tone) in the form of an underwater tone or an airborne whistle sound precedes the primary fish reinforcement. Improper responses to the controlling stimulus generally result in absence of any reinforcement, or in a time out. Once stimulus control has been established, an experimenter can test a dolphin's acuity in detecting, discriminating, recognizing, and classifying various kinds of stimuli.

The classical, or *Pavlovian*, procedure, in which an electric shock is paired with an acoustic stimulus, has been employed by Russian scientists to study sound reception in dolphins (Schusterman 1980). After repeated pairing of shock and stimulus, a highly stereotyped response —changes in galvanic skin condition, respiratory rate, or heart rate, or some other conditioned reflexes—is elicited (Supin and Sukhoruchenko 1970; Sukhoruchenko 1971). All of these experiments seem to involve passive hearing in dolphins and not echolocation. Electrophysiological techniques measuring evoked potentials from the midbrain generated by acoustic stimuli have also been used to study dolphin hearing, both in the United States (Bullock et al. 1968; Ridgway 1980) and in the Soviet Union (Supin et al. 1978). Electrophysiological techniques measuring cochlear potentials have been used in a single study of sound conduction in dolphins (McCormick et al. 1970, 1980). However, it seems that only operant conditioning procedures have been used in echolocation research (Schusterman 1980).

1.4.2 Two-Response Forced Choice Paradigm

The two-response forced choice paradigm, in which the subject dolphin has to make one of two specific responses on each trial, has been very popular. Variations on this technique depend on the simultaneous or successive presentation of the stimulus. In a simultaneous presentation experiment, two stimuli, a standard and a comparison, are positioned on either side of an imaginary center line or an actual dividing barrier, and the dolphin is required to examine the stimuli and indicate the position of the standard. Several different types of responses are commonly used. In a two-response paddle experiment the dolphin is required to touch the paddle that is on the same side of the center line as the standard stimulus. In another scenario, the dolphin may be required to swim and touch

the standard stimulus, or may be required to swim to that side of a dividing barrier where the standard is located.

Another experimental technique involving simultaneous presentation of stimuli involves a same–difference discrimination. Two identical stimuli or two different stimuli are positioned on either side of a center line. The dolphin, after examining the stimuli, is required to strike a specific paddle if the stimuli are identical or another paddle if the stimuli are different.

Experiments involving successive stimulus presentation and a two-response forced choice procedure are slightly simpler than experiments involving simultaneous stimulus presentation. Only one stimulus is presented per trial, and the dolphin, after examining the stimulus, is required to touch a specific paddle if the standard stimulus is presented or another paddle if a comparison stimulus, or no stimulus, is presented. In the simultaneous presentation method, the stimuli and associated response paddles have approximately the same relative spatial location (two stimuli from different locations along with spatially separated paddle locations). In the successive presentation method, however, the dolphin must make a bilateral or spatially different response to different stimuli originating from the same location in space. The two-response forced choice technique involving successive presentation is typically referred to as the "Yes/No" procedure in human psychoacoustic research (Green and Swet 1966).

1.4.3 Go/No-Go Response Paradigm

The go/no-go procedure is perhaps the most simple and straightforward procedure to study acoustic sensory perception in dolphins (Schusterman 1980). The procedure involves successive stimulus presentation and a dolphin that is trained to swim to a particular location or station. Some common stationing devices include a hoop through which a dolphin can extend its head, a bite plate for it to position on, and a chin cup or support bar on which it can rest its lower jaw. A trial begins with the subject swimming to its stationing and orienting itself toward the direction of the stimulus. Upon presentation of the stimulus, the dolphin leaves the station and touches a response paddle (go response) if it perceives the stimulus in a detection experiment or, if the stimulus is the standard, in a discrimination task. Upon striking the response paddle, a bridge tone is presented conveying to the dolphin that it has made a correct detection and will be receiving a fish reward. The subject has a definite time period to indicate a go response. Any response after that time period is considered a no-go response. If the stimulus is absent or is not the standard stimulus, the dolphin remains at its station (no-go response) for a short fixed time period until the experimenter presents a bridge or secondary reinforcement tone or sound indicating a correct response. The bridge tone signals to the dolphin that it has made a correct rejection and that it can now leave its station and collect a fish reward.

1.4.4 Psychophysical Procedures

After reliable stimulus control is achieved, some sort of psychophysical testing procedure is used to estimate the dolphin's absolute or differential sensitivity. The two most widely used techniques for studying acoustic perception and echolocation in dolphins have been the *method of constant stimuli* and the *up/down staircase*, or titration, method. The method of constant stimuli is a non-adaptive method in which the stimulus values are selected before testing begins. A session is divided into several blocks with an equal number of trials per block. The stimulus values are preselected and usually assigned to each block on a random basis. An alternative to a multiblock session is one that consists of one single block; here the stimulus value is kept constant throughout the session and only changes from session to session. In either case the stimulus presentation schedule (stimulus present or absent in a detection task, standard stimulus on the right or left in a simultaneous discrimination task, and standard or comparison stimulus present in a successive stimulus discrimination task) is randomized. An experimental session usually starts off with a block in which the animal can easily perform the detection or discrimination task at a high level of accuracy. The detection or discrimination task is

gradually made more difficult by manipulating the stimuli (i.e., decreasing the signal-to-noise ratio in a detection task) in blocks. Behavioral performance accuracy is recorded as a function of the stimulus value; the threshold is usually taken as some interpolated stimulus value associated with an arbitrary level of performance, typically 75% or 70% correct responses for all trials or 50% correct detection involving only stimulus-present trials in a detection task. This method yields threshold estimates with the same variability as and less bias than the up/down staircase method (Simpson 1988).

The up/down staircase method is an adaptive method in which the stimulus values are chosen while testing is in progress. The stimulus value is progressively made weaker or more difficult to discriminate as the animal correctly detects or identifies the stimulus. The process continues until the animal makes an error; then the stimulus value is progressively stepped up or made easier to discriminate, until the animal once again responds correctly to the stimulus. An example of how the signal-to-noise ratio may be varied (by using an attenuator to control the signal level) in a detection experiment using the staircase testing procedure is shown in Figure 1.7. Note that Figure 1.7 only displays the results of stimulus-present trials. Usually an equal number of stimulus-present and -absent trials, randomly distributed, is used during a session. The stimulus-absent trials or "catch" trials are inserted to keep the animal from prospecting or guessing but do not affect the level of the stimulus. The catch trials are also used to calculate the false alarm level (animal reporting the stimulus as present when it is in fact absent). Often the results of a session may be discarded if the false alarm level exceeds an arbitrary, predefined level. Having a relatively constant false alarm level is important if results are compared over many sessions, since the thresholds can vary significantly if false alarm levels vary.

An estimate of the animal's sensitivity is obtained by averaging the stimulus values at all the reversal points which are the local maxima and minima, or peaks and valleys, in a curve such as the one plotted in Figure 1.6. In the example of Figure 1.7, the threshold signal-to-noise ratio is

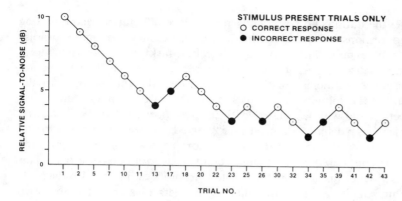

Figure 1.7. Variation in the signal-to-noise ratio in an experiment using the staircase testing procedure. Only stimulus-present trials are shown since the signal-to-noise ratio is changed only after stimulus-present trials.

threshold S/N

$$= \frac{6 + 4 + 7 + 6 + 7 + 6 + 8 + 6 + 8 + 7}{10}$$

$$= 6.5 \pm 1.2 \text{ dB} \tag{1-29}$$

The threshold obtained with the up/down staircase method is equivalent to the 50% correct detection threshold obtained with the method of constant stimuli (Levitt 1970). The standard deviation given in (1-29) is understood to be the standard deviation associated with the variation of the value of a signal or noise attenuator. Strictly speaking, the signal-to-noise ratio is a ratio of power, so that its standard deviation is not symmetrical about its mean when it is expressed in dB, but is symmetrical about its mean when expressed as a dimensionless ratio. However, for the sake of convenience, this point is often ignored or overlooked. A staircase threshold which is equivalent to the 70% correct detection threshold can be obtained by modifying the staircase procedure slightly. Instead of decreasing the stimulus value after each correct detection, the stimulus value is changed after two consecutive correct detection responses.

1.4.5 Binary Decision Matrix for Detection Experiments

In experiments involving the detection of a stimulus, the subject has four possible responses, two associated with correct responses and two associated with wrong responses. A correct detection ("hit") is one in which the subject responds correctly to the presence of the stimulus. A correct rejection occurs when the subject responds correctly to the absence of the stimulus. A miss occurs when the subject fails to report the presence of a stimulus, and a false alarm occurs when the subject indicates that the stimulus is present when it is not. When considering an experiment involving many trials, we can refer to the dolphin's performance in terms of conditional response probabilities. $P(Y/sn)$ is the conditional probability of a correct detection, that is, the probability of responding "yes" given the presence of the stimulus (sn). It is often referred to as the "hit rate." $P(Y/n)$ is the conditional probability of a false alarm, that is, the probability of responding "yes" given that the stimulus is absent and only noise (n) is present. $P(N/sn)$ is the conditional probability of a miss, the probability of responding "no" given the presence of the stimulus. $P(N/n)$ is the conditional probability of a correct rejection, the probability of responding "no" given that the stimulus is absent and only noise is present. The binary decision matrix relating the stimulus condition to possible responses is depicted in Figure 1.8. The different conditional probabilities are related according to the two equations

$$P(Y/sn) + P(N/sn) = 1 \tag{1-30}$$

and

$$P(Y/n) + P(N/n) = 1 \tag{1-31}$$

If we define the probability of the stimulus being present as $P(sn)$ and the probability of the stimulus being absent as $P(n)$, then the probability of a correct response can be expressed as

		RESPONSE	
		Yes	**No**
STIMULUS	**Present**	P(Y/sn)	P(N/sn)
	Absent	P(Y/n)	P(N/n)

Figure 1.8. Binary decision matrix relating the stimulus condition to possible responses.

$$P(C) = P(Y/sn)P(sn) + [1 - P(Y/n)]P(n)$$
$$(1\text{-}32)$$

In most experiments there are equal numbers of stimulus-present and stimulus-absent trials so that $P(sn) = P(n) = 0.5$. Thus (1-32) can be reduced to

$$P(C) = 0.5\{P(Y/sn) + [1 - P(Y/n)]\} \quad (1\text{-}33)$$

Often the four conditional probabilities are expressed as $P(D)$, $P(FA)$, $P(CR)$, $P(M)$ to denote the probability of detection, false alarm, correct rejection, and miss, respectively.

1.5 Signal Detection Theory

In this section a brief introduction to Signal Detection Theory (SDT) will be presented, motivated by the fact that some researchers have applied the ideas of SDT to dolphin bioacoustics and it is therefore important to have a basic understanding of SDT. For more extensive discussion, interested readers should refer to some excellent books on the subject written by Green and Swets (1966), Swets (1964), and Egan (1975).

The discussion of detection and discrimination threshold determination presented in Section 1.4

did not take into account nonsensory biasing factors such as motivation and expectation. Yet, response bias can affect the measurement of a subject's sensitivity in a detection or discriminating task. The response criterion or bias adopted by a subject can encompass a continuum that is affected by such factors as the cost of different types of errors (misses or false alarms), expectation of stimulus condition, and motivation. Toward one end of this continuum a subject can adopt a conservative criterion and only respond to the presence of a stimulus when he or she is absolutely sure that the stimulus is present. A subject adopting such cautious or conservative response bias will commit few false alarm errors but will also have a low probability of detection score. On the other end of this continuum, a subject can adopt a very liberal or lax criterion and respond as if a stimulus is present with only a weak notion or hint that it is. Such liberal response bias will result in many false alarm errors but also in a high probability of detection. The above example shows that the same subject having the same sensitivity to a stimulus can have quite different probability of detection scores depending on his or her response criterion.

The Theory of Signal Detectability (TSD), originally devised by Peterson, Birdsall, and Fox (1954) to study the detection of electronic signals in noise, provides a way to separate sensitivity and response bias, the two aspects of a subject's performance. Since the pioneering work of Peterson, Birdsall, and Fox in the field of radar signal detection, many have applied TSD to psychophysics, among the first being Tanner and Swet (1954), and the TSD nomenclature was gradually changed to Signal Detection Theory (SDT).

SDT holds that there is no true sensory threshold of the type discussed in Section 1.4. Rather, sensation is a continuously varying quantity and is zero only when the stimulus intensity is zero. Therefore, detection or discrimination of a stimulus in the presence of interference and noise is essentially a decision process and the threshold lies in the subject's decision-making process. Since noise is a random process, the problem is statistical in nature; there is always a finite probability of making an error. In the decision process only the two conditional probabilities,

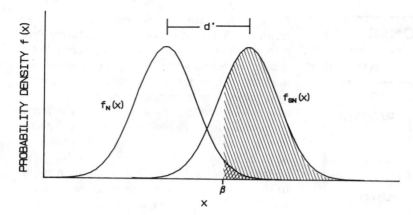

Figure 1.9. Probability density function of the noise (N) and signal plus noise (SN) in the decision process.

$P(Y/sn)$ and $P(Y/n)$ are of interest, since $P(N/sn)$ and $P(N/n)$ can be determined from (1-30) and (1-31). If we assume that the interfering noise has a normal or Gaussian distribution and that the summation of signal plus noise is also normal, with the same variance as the noise, the decision process can be described by the diagram in Figure 1.9, showing the probability density functions of noise and the signal plus noise. Let β be an arbitrary threshold level, so that whenever $x \geq \beta$ the subject responds that the stimulus is present and whenever $x < \beta$ the subject responds that the stimulus is absent. The crosshatched area to the right of β, under the signal-plus-noise probability density function $f_{SN}(x)$, equals the probability of detection. The crosshatched area to the right of β, under the noise-only probability density function $f_N(x)$, equals the probability of false alarm. If β is shifted toward the right (subject becoming more conservative), $P(Y/sn)$ and $P(Y/n)$ will both decrease in value. If β is shifted toward the left (subject becoming more liberal), $P(Y/sn)$ and $P(Y/n)$ will both increase in value.

Peterson, Birdsall, and Fox (1954) introduced a function called the *likelihood ratio*, which is defined as the ratio $f_{SN}(x)/f_N(x)$. They concluded from their analysis that a receiver which calculates the likelihood ratio for each receiver input is the optimum or ideal receiver for detecting signals in noise. Using the likelihood ratio, they found that for a situation in which the signal to be detected is completely known, the sensitivity parameter d' is related to the signal-to-noise ratio, according to the relationship

$$d' = \sqrt{2E/N} \qquad (1\text{-}34)$$

where E is the energy in the signal and N is the noise spectral density (rms noise in a 1-Hz band or rms noise per Hz). Since d' in (1-34) is the sensitivity associated with an optimum receiver, it represents the best sensitivity possible. Any other receiver will have a lower sensitivity than the optimum receiver. From Figure 1.9, we can see that d' is also the difference between the means of the two distributions. Our discussion of Figure 1.9 illustrates the point that a subject's sensitivity can be constant yet performance can vary depending on his or her response criterion. The location of β in Figure 1.9 is an index of the subject's response bias. Both d' and β can be calculated from the probability pair $P(Y/sn)$ and $P(Y/n)$, as shown by Snodgrass (1972) using z-scores. Elliott (1964) has also tabulated d' values as a function of $P(Y/sn)$ and $P(Y/n)$.

In SDT, the performance of a subject is usually plotted on a graph with $P(Y/sn)$ as the ordinate and $P(Y/n)$ as the abscissa. Figure 1.10 illustrates this format. The major diagonal is the chance line, since $P(Y/sn)$ is equal to $P(Y/n)$ on this line. The major diagonal represents the poorest a subject can perform short of making deliberate error. Points to the right of the major diagonal represent situations in which a subject makes deliberate errors. On the minor diagonal of Figure 1.10, $P(Y/sn) = 1 - P(Y/n)$. Now consider the situation depicted in Figure 1.9, with d' being constant and β allowed to vary along the horizontal axis. As β moves towards the right

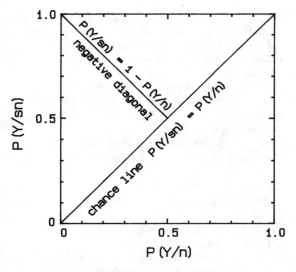

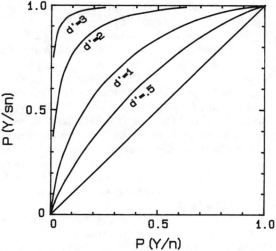

Figure 1.10. Format for plotting performance data, with the probability of detection plotted against the probability of false alarm.

Figure 1.11. Theoretical ROC curves for different d' values.

(subject becoming more conservative), the areas under both the f_{SN} and f_N curves will decrease, causing $P(Y/sn)$ and $P(Y/n)$ to also decrease in value. As β moves toward the left (subject becoming more liberal), the areas under both the f_{SN} and f_N curves will increase, causing $P(Y/sn)$ and $P(Y/n)$ to also increase in value. As β is varied, the probability pair will take on different values, and if we plot these values using the format of Figure 1.10, a curve called the *Receiver Operating Characteristic* (ROC) curve will be formed. A family of ROC curves can be generated by varying β for different d' values, as shown in Figure 1.11. Here, each ROC curve represents a constant sensitivity, with different values of $P(Y/sn)$ and $P(Y/n)$ generated by varying the response bias, β. Points on the left side of the minor diagonal in Figure 1.11 represent conservative response biases and points toward the right stand for liberal response biases. A neutral or moderate bias occurs when $\beta = 1$ and indicates a lack of response bias toward a "Yes" or "No" response. Points for a β of 1 will fall on the negative diagonal of Figure 1.10.

Most of the bioacoustics research performed on marine mammals has used the traditional psychological testing and threshold determina-

tion procedures discussed in Section 1.4. As mentioned, however, there is a scattering of research results published from the SDT prospective. Some publications contain results obtained from traditional psychoacoustic threshold determination procedures but with results plotted as ROC curves along with calculated values of d' and β.

References

Albers, V.M. (1965). *Underwater Acoustics Handbook II*. University Park: Pennsylvania State University Press.

Au, W.W.L., Floyd, R.W., Penner, R.H., and Murchison, A.E. (1974). Measurement of echolocation signals of the Atlantic bottlenose Dolphin, *Tursiops truncatus* Montagu, in open waters. J. Acoust. Soc. Am. 56: 1280–1290.

Bel'kovich, V.M., Borisov, V.I., Gurevich, V.S., and Krushinskaya, N.L. (1969). Echolocation capabilities of the common dolphin (*D. delphis*). Zoologicheskii Zhurnal 48: 876–883.

Brigham, E.O. (1974). *The Fast Fourier Transform*. Englewood Cliffs, N.J.: Prentice-Hall.

Bullock, T.H., Grinnel, A.S., Ikezono, E., Kamedo, K., Katsuki, Y., Nomoto, M., Sato, O., Suga, N., and Yanagisawa, K. (1968). Electrophysiological studies of central auditory mechanism in cetaceans. Zeitschrift für Vergleichende Phys. 59: 117–

Burrus, C.S., and Parks, T.W. (1985). *DFT/FFT and Convolution Algorithms*. New York: John Wiley and Sons.

Busnel, R.-G., ed. (1967). *Animal Sonar Systems: Biology and Bionics, Vol. 1*. Laboratoire de Physiologie Acoustique, Jouy-en-Josas, France.

Busnel, R.-G., and Dziedzic, A. (1967). Resultats metrologiques experimentaux de l'echolocation chez le *Phocaena phocaena* et leur comparison avec ceus de certaines chauves—souris. In R.-G. Busnel, ed., *Animal Sonar Systems. Biology and Bionics, Vol. 1*. Laboratoire de Physiologie Acoustique, Jouy-en-Josas, France, pp. 307–335.

Busnel, R.-G., and Fish, J.F., eds. (1980). *Animal Sonar Systems*. New York: Plenum Press.

Cooley, P.M., and Tukey, J.W. (1965). An Algorithm for the machine computation of complex Fourier series. Mathematics of Computation 19: 297–301.

Egan, J.P. (1975). *Signal Detection Theory and ROC Analysis*. New York: Academic Press.

Elliott, P.B. (1964). Appendix 1: Tables of d'. In: J. Swets, ed., *Signal Detection and Recognition by Human Observers*. New York: John Wiley and Sons, pp. 651–678.

Fisher, F.H., and Simmons, V.P. (1977). Sound absorption in sea water. J. Acoust. Soc. Am. 62. 558–564.

Galambos, R. (1941). Cochlear potentials from the bat. Science 93: 215.

Galambos, R. (1942). Cochlear potentials elicited from bats by supersonic sounds. J. Acoust. Soc. Am. 14: 41–49.

Green, D., and Swets, J.A. (1966). *Signal Detection Theory and Psychophysics*. Huntington, N.Y: Krieger Pub.

Griffin, D.R. (1944). Echolocation in blind men, bats and radar. Science 100: 589–590.

Griffin, D.R. (1958). *Listening in the Dark*. New Haven: Yale University Press.

Gurevich, B.S. (1969). Echolocation discrimination of geometric figures in the dolphin, *Delphinus delphis*. Vestnik Moskovskoga Universiteta, Biologiya, Pochovedeniye 3: 109–112 (English translation JPRS 49281).

Gurevich, B.S., and Evans, W.E. (1976). Echolocation discrimination of complex planar targets by the beluga whale (*Delphinapterus leucas*). J. Acoust. Soc. Am. 60: 5–6.

Hall, J.D., and Johnson, C.S. (1971). Auditory thresholds of a killer whale *Orcinus orca* Linnaeus. J. Acoust. Soc. Am. 51: 515–517.

Hartridge, H. (1920). The avoidance of objects by bats in their flight. J. Physiol. 54: 54–57.

Hatakeyama, Y., and Soeda, H. (1990). Studies on echolocation of porpoises taken in salmon gillnet fisheries. In: J.A. Thomas and R. Kastelein eds., *Sensory Abilities of Cetaceans*, New York: Plenum Press, pp. 269–281.

Jurine, L. (1798). Experiments on Bats Deprived of Sight by M. de Jurine. Philos. Mag. 1: 136–140.

Kellogg, W.N. (1958). Echo ranging in the porpoise. Science 128: 982–988.

Kellogg, W.N., and Kohler, R. (1952). Responses of the porpoise to ultrasonic frequencies. Science 116: 250–252.

Kellogg, W.N., Kohler, R., and Morris, H.N. (1953). Porpoise Sounds as Sonar Signals. Science 117: 239–243.

Levitt, H. (1970). Transformed up-down methods in psychoacoustics. J. Acoust. Soc. Am. 49: 467–477.

Maxim, H. (1912). The sixth sense of the bat. Sir Hiram's contention. The possible prevention of sea collisions. Sci. Amer. 7: 148–150.

McBride, A.F. (1956). Evidence for echolocation by cetaceans. Deep-Sea Research 3: 153–154.

McCormick, J.G., Wever, E.G., Palin, J., and Ridgway, S.H. (1970). Sound conduction in the dolphin ear. J. Acoust. Soc. Am. 48: 1418–1428.

McCormick, J.G., Wever, E.G., Ridgway, S.H., and Palin, J. (1980). Sound reception in the porpoise as it relates to echolocation. In: R.-G Busnel and J.R. Fish, eds., *Animal Sonar Systems*. New York: Plenum Press, pp. 449–467.

Nachtigall, P.E., and Moore, P.W.B. (1988). *Animal Sonar: Processes and Performance*. New York: Plenum Press.

Norris, K.S. (1968). The echolocation of marine mammals. In: H.T. Anderson, ed., *The Biology of Marine Mammals*, New York: Academic Press, pp. 391–423.

Norris, K.S., and Evans, W.E. (1966). Directionality of echolocation clicks in the rough-tooth porpoise, *Steno bredanensis* (Lesson). In: W.N. Tavolga, ed., *Marine Bio-Acoustics*. New York: Pergamon Press, pp. 305–316.

Norris, K.S., Prescott, J.H., Asa-Dorian, P.V., and Perkins, P. (1961). An experimental demonstration of echolocation behavior in the porpoise, *Tursiops truncatus* (Montagu). Biol. Bull. 120: 163–176.

Penner, R., and Murchison, A.E. (1970). Experimentally demonstrated echolocation in the Amazon river porpoise, *Inia geofrensis* (Blainville). In: T. Poulter, ed., *Proc. 7th Ann. Conf. Bio. Sonar and Diving Mammals*. Menlo Park, Cal.: Stanford Research Institute, pp. 17–38.

Peterson, W.W., Birdsall, T.G., and Fox, W.C. (1954). The theory of signal detectability. Inst. Radio Engr. PGIT 4: 171–212.

Pierce, G.W., and Griffin, D.R. (1938). Experimental determination of supersonic notes emitted by bats. J. Mamm. 19: 454–455.

Ramirez, R.W. (1985). *The FFT Fundamentals and Concepts*. Englewood Cliffs, N.J.: Prentice-Hall.

Ridgway, S.H. (1980). Electrophysiological experiments on hearing in odontocetes. In R.-G Busnel and J.F. Fish, eds., *Animal Sonar Systems*. New York: Plenum Press, pp. 483–493.

Schevill, W.E., and Lawrence, B. (1953a). "Auditory response of a bottle-nosed porpoise, *Tursiops truncatus*, to frequencies above 100 kc. J. Exper. Zool. 124: 147–165.

Schevill, W.E., and Lawrence, B. (1953b). High-frequency auditory response of a bottlenosed dolphin, *Tursiops truncatus* (Montagu). J. Acoust. Soc. Am. 24: 1016–1017.

Schevill, W.E., and Lawrence, B. (1956). Food-finding by a captive porpoise (*Tursiops truncatus*). Breviora (Mus. Comp. Zool., Harvard) 53: 1–15.

Schusterman, R.J. (1980). Behavioral methodology in echolocation by marine mammals. In: R.G. Busnel and J.F. Fish, eds., *Animal Sonar Systems*. New York: Plenum, pp. 11–41.

Simpson, W.A. (1988). The method of constant stimuli is efficient. Percep. & Psych. 44: 433–436.

Skinner, B.F. (1961). *Cumulative Record*. New York: Appleton-Century-Crofts.

Snodgrass, J.G. (1972). *Theory and Experimentation in Signal Detection*. Baldwin, N.Y.: Life Science Assoc.

Sukhoruchenko, M.N. (1971). The maximum hearing frequency range in the dolphin. Tr. Akust. Inst. (Moscow). 17: 54–59.

Supin, A.Y., and Sukhoruchenko, M.N. (1970). Determination of delphinid auditory thresholds by the method of conditioned galvanic skin reaction. Tr. Akust. Inst. (Moscow) 12: 194–199.

Supin, A.Y., Mukhametov, L.M., Ladygina, R.F., Popov, V.V., Mass, A.M., and Sukhoruchenko, M.N. (1978). Electrophysiological studies of the dolphin's brain. V.E. Sokolov, ed. Moscow: Izdatel'stvo Nauka.

Swets, J.A. (1964). *Signal Detection and Recognition by Human Observers*. New York: John Wiley and Sons.

Tanner, W.P., Jr. and Swets, J.A. (1954). A decision-making theory of visual detection. Psych. Rev. 61: 401–409.

Thomas, J., Stoermer, M., Bowers, C., Anderson, L., and Garver, A. (1988). Detection abilities and signal characteristics of echolocating false killer whale (*Pseudorca crassidens*). In P. Nachtigall, ed., *Animal Sonar Systems II*. New York: Plenum Press, pp. 323–328.

Wood, F.G., Jr. (1952). Porpoise sounds. Underwater sounds made by *Tursiops truncatus* and *Stenella plagiodon*. (phonograph record). Marineland Research Lab., Florida.

Wood, F.G., Jr. (1953). Underwater sound production and concurrent behavior of captive porpoises *Tursiops truncatus* and *Stenella plagiodon*. Bull. Marine Sci. of the Gulf and Caribbean 3: 120–133.

Wood, F.G., Jr., and Evans, W.E. (1980). Adaptiveness and ecology of echolocation in toothed whales. In R.G. Busnel and J.F. Fish, eds., *Animal Sonar Systems*. New York: Plenum Press, pp. 381–425.

2

The Receiving System

A man-made sonar system can be subdivided into three major subsystems, the *receiving, transmitting*, and *signal processing/decision* subsystems. The capabilities of a sonar are dependent on the integrated, interdependent functioning of the three major subsystems. Each subsystem contributes in a specific, but different, manner to the functioning of the total unit. There is no priority of importance assigned to any subsystem; only the capability of the total sonar system can be evaluated. The dolphin sonar system can also be subdivided into the same three major components, and here as well the understanding of each subsystem is equally important in understanding the total unit. In this chapter the receiving system will be studied and its characteristics discussed.

2.1 The Outer Ears

The receiving system of a dolphin's sonar is nothing other than its auditory or hearing system, consisting of its outer, middle and inner ears. A drawing of the dolphin's ear showing the external auditory meatus, the middle and inner ears is presented in Figure 2.1. Middle and inner ear are encased in a bony structure (tympanic bulla) which does not have a bony connection to the skull but is connected by cartilage, connective tissue, and fat (McCormick et al. 1970). Interested readers should refer to some excellent anatomical descriptions of the dolphin ear presented by Wever et al. (1971a, 1971b, 1972), Ridgway et al. (1974), and Fleischer (1976a, 1976b).

Unlike other mammals, dolphins and other cetaceans do not have pinnae and their external auditory meatus is not readily visible. The meatus in most dolphins is like a pinhole with parts of it (nearest the skin surface) composed of fibrous tissue. The external meatus with its narrow cross section and fibrous structure hardly seems capable of being an acoustic pathway to the middle and inner ears via the tympanic ligament. It is not clear how sound can enter into the meatus and be conducted along its interior to the middle ear. Nevertheless, there have been some who have argued in favor of the external auditory meatus being the primary pathway for sound transmission (Fraser and Purves, 1954, 1959, 1960). Since the auditory meatus in cetaceans is so different from that in all terrestrial mammals, one could justly ask, if sound is not conducted in the external auditory meatus, then how does it enter into the inner ears of dolphins. An alternate theory of sound conduction in dolphins suggested that the auditory meatus is nonfunctional and that sound enters the dolphin's head through the thinned posterior portion of the mandible and is transmitted via a fat-filled canal to the tympano-periotic bone which contains the middle and inner ears (Norris 1964, 1968a). A schematic diagram of a dolphin's head showing the mandibular window and the fat

Figure 2.1. Drawing of a dolphin's skull showing (*top*) the external auditory meatus and (*bottom*) the tympanic bulla. (From Johnson 1986.)

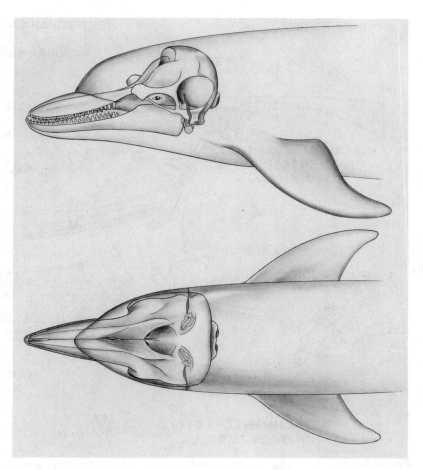

channel is displayed in Figure 2.2. The region of the pan bone or "acoustic window" of the mandible is very thin, varying from 0.5 mm to 3.0 mm in different species of dolphins (Norris 1968b). The fatty tissue of the oval acoustic window is almost completely free of muscle strands (Norris 1980). The interior portion of the pan bone region is filled with a translucent fat rich in oil that extends from the pan bone to the bulla. The mandibular fat body is a well-defined subcylindrical mass of fat flaring posteriorly. According to the theory, once sound enters into this fat body through the pan bone, it may be transmitted directly to the tympanic bulla to which the mandibular fat is attached. Varanasi and Malins (1971, 1972) found that this lipid material had a very low sound absorption characteristic, leading them to coin the term "acoustic fat." Interest-

ingly, acoustic fat is found only in the mandibular channel and the melon, and nowhere else in odontocete (Varanasi and Malins 1971; Varanasi et al. 1975).

Bullock et al. (1968) were the first to employ electrophysiological techniques in the study of dolphin hearing and were able to shed some important light on the viability of the two competing theories of sound conduction in dolphins. They measured auditory evoked potentials from the midbrain auditory structure as various types of sound stimuli were played to the dolphins. Using a hydrophone pressed against the head surface of their subjects, they obtained the largest potentials when the source was placed in the vicinity of the lower jaw acoustic window identified by Norris (1968a) and on the melon. The pattern of sensitivity of an individual *Stenella* is

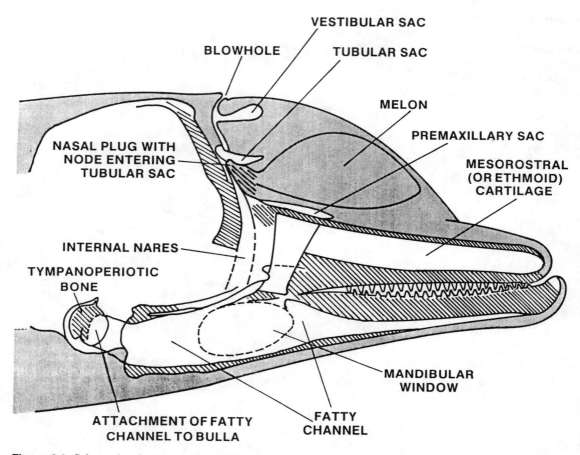

Figure 2.2. Schematic of a dolphin's head showing various structures associated with sound reception and production. (After Norris 1968a © Yale University Press.)

shown in Figure 2.3. They also found that when a distant sound source was used and an acoustic shield (10 × 12 cm piece of foam rubber) was held over the sensitive part of the lower jaw and melon, the evoked potential response decreased noticeably. The findings of Bullock et al. (1968) strongly supported Norris's theory that the pathway of sound to the cochlea is via the lower jaw.

McCormick et al. (1970, 1980) used a different electrophysiological technique to study sound conduction in the dolphin ear, the cochlear potential method, which measures the potential generated by hair cells in the organ of Corti (see Section 2.3). Using a vibrator in contact with the dolphin's skin, they mapped the level of the cochlear potential with the location of the vibrator on the dolphin's body. Their results with four *Tursiops truncatus* are summarized in the sche-

matic of Figure 2.4 showing the area of greatest sensitivity to sounds. Contrary to the results of Bullock et al. (1968), they found that the upper jaw and most of the skull seem to be acoustically isolated from the ear. When the external auditory meatus was stimulated, the cochlear potentials were 6 to 14 times less intense than when the most sensitive part of the lower jaw was stimulated. They also found that cochlear potentials were not affected after the auditory meatus was severed. They concluded that the meatus and tympanic membrane were acoustically nonfunctional and also, that sound enters the dolphin through the lower jaw.

Although the electrophysiological measurements of Bullock et al. (1968) and McCormick et al. (1970) seemed irrefutable, their conclusions concerning the lower jaw pathway have not been

Figure 2.3. Pattern of sensitivity of an individual *Stenella* to sound produced at 30 and 60 kHz by a hydrophone pressed against the head surface at the points shown. The numerical values represent attenuation at threshold; therefore, the largest numbers represent greatest sensitivity. Contour lines are drawn at intervals of 5 dB in sensitivity. Recording was from the inferior colliculus. (From Bullock et al. 1968.)

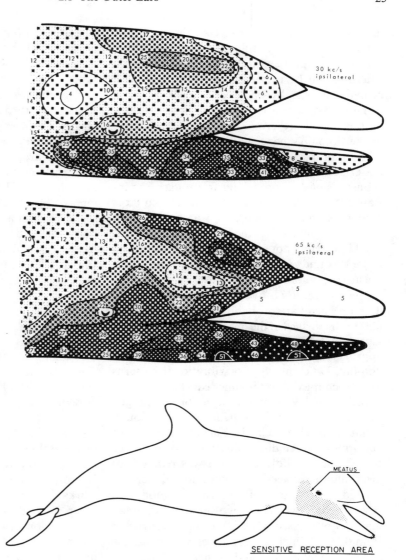

Figure 2.4. The three regions of greatest sensitivity to sound for a *Tursiops truncatus*. (From McCormick et al. 1970.)

totally accepted. Purves and Pilleri (1973) discounted the results of McCormick et al. (1970) on the basis that their stimulus levels, which were about 50 dB above the behavioral threshold measured by Johnson (1967), were too high, confounding the use of normal auditory pathways. Purves and Pilleri (1973) also argued that the surgical procedure damaged the auditory system. They continued to argue in favor of the external auditory meatus but without giving any experimental evidence (Purves and Pilleri 1983). Johnson (1986) correctly emphasized that the measurements of both Bullock et al. (1968) and

McCormick et al. (1970) need to be interpreted with a degree of caution. In both studies, the animal's mouth was open and its larynx connected to a gas-filled tube so that unnatural pockets of gas may have been present to alter the transmission of sound in the dolphin's head. These gas-filled pockets would not be present in a natural state. There are other reasons for caution in examining the results of these two electrophysiological studies. The acoustic conditions for both experiments were less than ideal: the subjects were confined to small holding tanks and their heads held near the surface, keeping incisions out of the water so that the electrodes could be electrically isolated from each other. The acoustic propagation conditions for such a situation can be extremely variable, with SPL changing drastically, on the order of 10 to 20 dB, if the sound source is moved as little as a few centimeters. Furthermore, exciting the surface of the dolphin's head with a hydrophone or a vibrator may set up a totally different vibrational pattern on the skin of the animal than would an acoustic signal approaching from afar. For instance, McCormick et al. (1970) measured large potentials when the vibrator was held against one of the front teeth of the lower jaw. That tooth would probably not experience similar vibration from a sound wave propagating in the water toward the dolphin. Finally, vibrational displacements induced by a hydrophone held against a dolphin's head are probably directed perpendicular to the surface of the head, thus differing from vibrational displacements caused by a sound approaching from a direction in front of the dolphin. Different types of vibrational patterns may be coupled into the inner ear differently.

Brill et al. (1988) used a behavioral approach to investigate the role of the lower jaw in sound conduction. They trained a *Tursiops truncatus* to perform a target discrimination task using its sonar. The blindfolded animal was required to discriminate between an aluminum cylinder and a sand-filled ring while wearing a hood that covered the lower jaw. Two hoods of different material were used. One hood was constructed from 0.16-cm thick gasless neoprene having low acoustic attenuation, and the other hood was constructed from 0.48-cm thick closed-cell neoprene having high acoustic attenuation. The acoustic attenuation of both materials was measured with simulated dolphin echolocation signals. The gasless neoprene had attenuation values of 2.2 dB and 1.2 dB for signals with peak frequencies of 35 and 110 kHz, respectively. The closed cell neoprene had attenuation values of 39 dB and 36 dB for signals with peak frequencies of 55 and 115 kHz. The difference between the animal's performance without hood and with the low attenuation hood was insignificant, but the difference between its performance without hood and with the high attenuation hood was significant. The behavioral results of Brill et al. (1988) are consistent with Norris's theory that the pathway of sound to the cochlea is via the lower jaw.

Taken together, the results of these three experiments seem to provide overwhelming evidence in support of Norris's theory of sound conduction through the lower jaw. Norris and Harvey (1974) made acoustic measurements on the head of a recently dead *Tursiops* by inserting a small hydrophone into different portions of the head to receive sounds from a projector located away from it. They found a two-fold increase in intensity as the sound traveled from the anterior to the posterior portion of the mandibular fat body, suggesting a structural intensification of received sound similar to the manner in which sound is intensified in the external ear canal of man.

2.2 The Middle Ear

The anatomy of the middle and inner ears of *Tursiops* and other dolphins was the topic of some studies by Ernest Wever and his colleagues at Princeton University (Wever et al. 1971a, 1971b, 1972 McCormick et al. 1970). Fleischer (1976a,b; 1980) has also studied the structure of the dolphin ear and has created descriptive analogs of the auditory processes in the dolphin. Interested readers should refer to these excellent works. Only a gross description of the middle and inner ear will be given here in order to foster appreciation of the dolphin ear's suitability for the reception of high frequency, low intensity underwater sounds.

A drawing of the right auditory bulla of *Tursiops truncatus* showing the relationships among its various parts is shown in Figure 2.5. The ossicular chain consisting of the malleus, incus, and stapes is housed within the bulla. A drawing of the ossicular chain is shown in Figure 2.6. There is no direct connection between the tympanic membrane and the malleus. The tympanic ligament, which is firmly attached to the bony edge of the dorsal opening in the tympanic bone, serves mainly as a suspension system for the malleus (McCormick et al. 1970). The incus is connected to the malleus by a saddle-shaped joint. The stapes is connected to the incus, and its head surface articulates with the major pro-

Figure 2.5. The right auditory bulla of *Tursiops truncatus*, opened on its lateral side to show its contents. (From McCormick et al. 1970.)

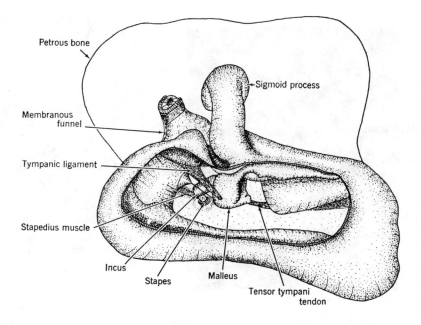

Figure 2.6. The ossicles, rotated about 80° from their position in the preceding figure to show the articulation between stapes and incus and the union of incus and malleus. (From McCormick et al. 1970.)

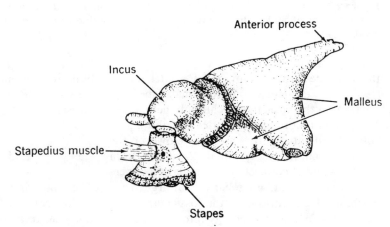

cess of the incus. The footplate of the stapes lies on the oval window connecting the middle ear to the cochlea and is attached by a narrow annular ligament as shown in Figure 2.7. After sound arrives at the bulla, propagating through the oval window into the mandibular fat channel, it is not clear how it is coupled into the bulla. Norris (1980) suggested that acoustic energy could possibly propagate through the tin-walled portion of a water- (or blood-) filled bulla into the middle ear without significant losses, or perhaps flexural waves could propagate around the bulla and reach the ossicular chain (malleus, incus, and stapes) through the processus gracilis of the malleus. McCormick et al. (1980) argued for translational bone conduction, and Johnson (1986) favored compressional bone conduction stimulated by pressure waves and not by mechanical bone stimulation. Translational bone conduction would exploit the same stapes–oval window "piston-cylinder" system which functions in aerial hearing (McCormick et al. 1980). In compressional bone conduction, compressional waves compress the bulla from all sides and cause cochlear fluid to move, exciting the cochlear hair cells as the fluid is pressed out

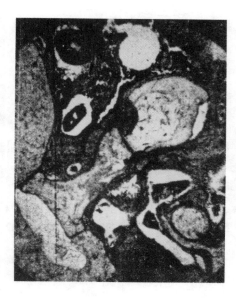 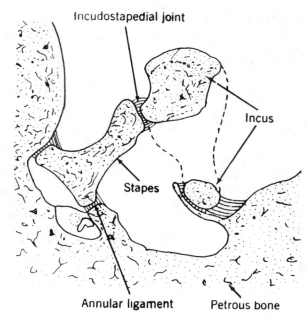

Figure 2.7. (*Left*) Photomicrograph showing the oval window with the footplate of the stapes held by its narrow annular ligament, and portions of the incus; (*right*) identification of the various parts. (From McCormick et al. 1970.)

through the oval window. McCormick et al. (1980) argued against compressional bone conduction on the basis that it can only be realized if the bulla is contained in a denser solid and is compressed equally from all sides. However, the bulla is completely separated physically from the skull and is connected by cartilage, connective tissue, and fat. Fleischer (1980) argued for mechanical activation of the ossicular chain which couples acoustic energy into the cochlear capsule. In the electrophysiological experiments conducted by McCormick et al. (1970, 1980), it was found that the cochlear potential was only slightly affected (4 dB loss) by removal of the malleus after it was severed from the incus. Although the malleus did not have an acoustic role, they concluded that the innermost portion of the ossicular chain was essential for the reception of sounds by the cochlea. An 18-dB drop was observed in the cochlear potential when the ossicular chain was damped during high frequency stimulation. McCormick et al. (1980) concluded that the ossicular chain must function as a rigid spring mounted inside the tympanic cavity, and that when acoustic energy sets the

bulla in motion, the inertia and stiffness of the ossicular chain probably cause the stapes to move relative to the oval window.

2.3 The Inner Ear

The oval window is the boundary between the middle ear and the cochlea, a spiral tube of the inner ear resembling a snail shell and containing the auditory nerve endings. A drawing showing the mode of coiling in the cochlea and an internal cross-sectional view of the cochlea for an odontocete are given in Figure 2.8. The cochlea is divided along its length by two very thin membranes (basilar membrane and spiral laminae) into two large chambers and is filled with fluid. It begins at the basal end where the oval and round windows are located and coils to a blind end at the apex. Acoustic energy enters the oval window as a compressional wave and causes the basilar membrane to vibrate, deforming the acoustic receptor hair cells located on the basilar membrane. The hair cells and their supporting structure are referred to as the organ of Corti.

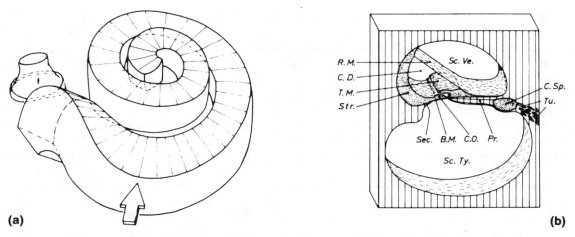

(a)   **(b)**

Figure 2.8. (*A*) Mode of coiling of the right cochlea and the location of the stapes in an odonotocete; (*B*) cross-sectional drawing of a cetacean cochlea. Abbreviations: R.M., Reissner's membrane; C.D., cochlear duct; T.M., tectorial membrane; Str., stria vascularis plus spiral ligament; C.O., Corti's organ; B.M., basilar membrane; Tu., tubuli for the nerve fibers; Sc.Ty., scala tympani; Sc.Ve., scala vestibuli; Sec., secondary spiral lamina; Pr., primary spiral lamina; C.sp., spiral canal. (From Fleischer 1976a.)

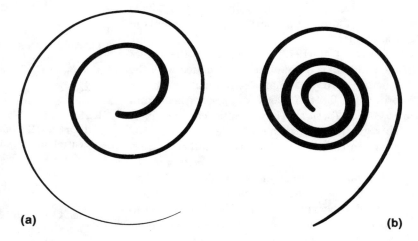

Figure 2.9. (*A*) Right basilar membrane of *Tursiops truncatus*; (*B*) left basilar membrane of a human. (From Wever et al. 1971b.)

(a)   **(b)**

The basilar membrane is suspended between bony supports along both edges, as shown in Figure 2.8B.

Wever et al. (1971b) measured the width of the basilar membrane of four *Tursiops truncatus* and found it to vary greatly along the cochlea in each individual. The basilar membrane is narrow, around 25 μm in width, at the basal end and increases in width at a uniform rate over the first 20 mm, and then at a greater rate over the rest of the cochlea. The maximum width is around 350 μm near the apical end. A 14-fold increase in width is typical for *Tursiops*. This is approximately twice as much variation as shown by the human basilar membrane, which has a 6.25-fold increase. Scale drawings of the form and changing width of the dolphin and human basilar membranes are shown in Figure 2.9. The number of cochlear turns for *Tursiops* is slightly more than two, compared to nearly three for humans. The number of cochlear turns in *Lagenorhynchus obliquidens* is about 1.75, and the variation in width of its basilar membrane is about 11 (Wever et al. 1972). Wever et al. (1971b) argued that the most important condition for frequency differentiation in the cochlea is the variation in stiffness of the vibrating structure. Therefore, the variation in width of the basilar membrane and the rigidity of the suspension should contribute to an excellent capability of pitch discrimination for *Tursiops*. They also argued that the narrow width of the basilar membrane near the basal end of the cochlea (25 μm compared to 80 μm for humans) is consonant with the ability of *Tursiops* to hear very high frequency sounds.

The structure of the hair cells along the basilar membrane of *Lagenorhynchus* and *Tursiops* was studied by Wever et al. (1971b, 1972), who also estimated the number of hair cells. The number of inner hair cells was approximately 3,451 for *Tursiops* and 3,275 for *Lagenorhynchus*. The number of outer hair cells was approximately 13,933 for *Tursiops* and 12,899 for *Lagenorhynchus*. By comparison, the human ear typically has 3,475 inner hair cells and 11,500 outer hair cells. Therefore, the two species dolphin and man have similar amount of primary receptor elements in the cochlea.

Wever et al. (1972) also estimated the size of the ganglion cell population associated with the hair cells. They estimated a population on the order of 60,000 to 70,000 for *Lagenorhynchus* and 95,000 for *Tursiops*. These numbers are considerably greater than man's complement of 30,500 ganglion cells. Comparing the number of ganglion cells to that of hair cells, *Lagenorhynchus* has about four times more ganglion cells than hair cells, and *Tursiops*, about five times more. These numbers are again high in comparison to those for the human cochlea, which has about twice as many ganglion cells as hair cells. Wever et al. (1972) suggested that the high ratio of ganglion cells to hair cells in the dolphin may aid in the representation of high frequency acoustic information and of fine details of cochlear events to higher centers of the auditory nervous system.

2.4 Summary

The auditory system of dolphins and other cetaceans seems well adapted to an aquatic environment. The absence of any protruding parts associated with the external ears, such as pinnas,

is useful in providing low hydrodynamic drag. The presence of a relatively thin region on each side of the mandible (pan bone region) that is in intimate contact with a fatty (lipid) material of excellent acoustic properties extending to the auditory bullae suggests an alternate acoustic pathway contrasting with the auditory meatus pathway in other mammals. Thus the dolphin can have low hydrodynamic drag but yet have a way for sound to enter into its auditory system. Having the bullae physically isolated from the skull and therefore isolated from each other allows it to localize sounds received by bone conduction. Finally, the variations in stiffness of the basilar membrane along with the large population of ganglion cells (two to three times greater than in humans) suggest an excellent capability for pitch discrimination and the perception of high frequency sounds.

References

Brill, R.L., Sevenich, M.L., Sullivan, T.J., Sustman, J.D., and Witt, R.E. (1988). Behavioral evidence for hearing through the lower jaw by an echolocating dolphin (*Tursiops truncatus*). Mar. Mamm. Science 4: 223–230.

Bullock, T.H., Grinnell, A.D., Ikezono, E., Kameda, K., Katsuki, Y., Nomoto, M., Sato, O., Suga, N., and Yanagisawa, K. (1968). Electrophysiological studies of central auditory mechanisms in cetaceans. Zeitschrift für Vergleichende Phys. 59: 117–316.

Fleischer, G. (1976a). Hearing in extinct cetaceans as determined by cochlear structrue. J. Paleontology 50: 133–152.

Fleischer, G. (1976b). On bony microstructures in the dolphin cochlea, related to hearing. N. Jb. Geol. Palaont. Abhl. 151: 161–191.

Fleischer, G. (1980). Morphological adaptations of the sound conducting apparatus in echolocating mammals. In: R.G. Busnel and J.F. Fish, eds. *Animal Sonar Systems*. New York: Plenum Press, pp. 895–898.

Fraser, F.C., and Purves, P.E. (1954). Hearing in cetaceans. Bull. British Mus. (Nat. Hist.) 2: 103–116.

Fraser, F.C., and Purves, P.E. (1959). Hearing in whales. Endeavour 18: 93–98.

Fraser, F.C., and Purves, P.E. (1960). Hearing in cetaceans: evolution of the accessory air sacs and the structure and function of the outer and middle ear in recent cetaceans. Bull. British Mus. (Nat. Hist.) 7: 1–140.

Johnson, S.C. (1967). Sound detection thresholds in marine mammals. In: W. Tavolga, ed., *Marine Bioacoustics*, New York: Pergamon Press, pp. 247–260.

Johnson, S.C. (1986). Dolphin audition and echolocation capacities In: R.J. Schusterman, J.A. Thomas and F.G. Wood, eds., *Dolphin cognition and behavior: a comparative approach.* Itillsdale, N.J.: Lawrence Erlbaum Associates, pp. 115–136.

McCormick, J.G., Wever, E.G., Palin, J., and Ridgway, S.H. (1970). Sound conduction in the dolphin ear. J. Acoust. Soc. Am. 48: 1418–1428.

McCormick, J.G., Weaver, E.G., Ridgway, S.H., and Palin, J. (1980). Sound reception in the porpoise as it relates to echolocation. In: R.G. Busnel and J.F. Fish, eds., *Animal Sonar Systems.* New York: Plenum Press, pp. 449–467.

Norris, K.S. (1964). Some problems of echolocation in cetaceans. In: W.N. Tavolga, ed., *Marine Bioacoustics.* New York: Pergamon Press, pp. 316–336.

Norris, K.S. (1968a). The evolution of acoustic mechanisms in odontocete cetaceans. In: E.T. Drake, ed., *Evolution and Environment.* New Haven: Yale University Press, pp. 297–324.

Norris, K.S. (1968b). The echolocation of marine mammals. In: H.T. Andersen, ed., *The Biology of Marine Mammals.* New York: Academic Press, pp. 391–423.

Norris, K.S. (1980). Peripheral sound processing in odontocetes. In: R.G. Busnel and J.F. Fish, eds., *Animal Sonar Systems.* New York: Plenum Press, pp. 495–509.

Norris, K.S., and Harvey, G.W. (1974). Sound transmission in the porpoise head. J. Acoust. Soc. Am. 56: 659–664.

Purves, P.E., and Pilleri, G.E. (1983). *Echolocation in Whales and Dolphins*, New York: Academic Press.

Purves, P.E., and Pilleri, G.E. (1973). Observations on the ear, nose, throat, and eye of *Platanista indi.* Invest. on Cetacea 5: 13–57.

Ridgway, S.H., McCormick, J.G., and Wever, E.G. (1974). Surgical approach to the dolphin's ear. J. Exp. Zoo. 188: 265–276.

Varanasi, U., and Malins, D.C. (1971). Unique lipids of the porpoise (*Tursiops gilli*): differences in triacylglycerols and wax esters of acoustic (Mandibular and melon) and blubber tissues. Biochim. Biophys. Acta 231: 415–418.

Varanasi, U., and Malins, D.C. (1972). Triacylglycerols characteristic of porpoise acoustic tissues: molecular structures of diisovaleroylglycerides. Science 176: 926–928.

Varanasi, U., Feldman, H.R., and Malins, D.C. (1975). Molecular basis for formation of lipid sound lens in echolocating cetaceans. Nature 255: 340–343.

Wever, E.G., McCormick, J.G., Palin, J., and Ridgway, S.H. (1971a). The cochlea of the dolphin, *Tursiops truncatus*: general morphology. Proc. Nat. Acad. Sci. USA 68: 2381–2385.

Wever, E.G., McCormick, J.G., Palin, J., and Ridgway, S.H. (1971b). The cochlea of the dolphin, *Tursiops truncatus*: the basilar membrane. Proc. Nat. Acad. Sci. USA 68: 2708–2711.

Wever, E.G., McCormick, J.G., Palin, J., and Ridgway, S.H. (1972). "Cochlear structure in the dolphin, *Lagenorhynchus obliquidens.* Proc. Nat. Acad. Sci. USA 69: 657–661.

3

Characteristics of the Receiving System for Simple Signals

The dolphin must have a fairly general and flexible sonar system in order to maximize its survival. Since acoustic energy propagates in water more efficiently than almost any other form of energy, a good passive and active sonar system is ideal for the aquatic environment. In order to have good passive and active sonar capabilities, the receiving system of the dolphin should have certain important characteristics. The receiver should be very sensitive over a wide frequency range to detect minute echoes and externally generated sounds. It should also be sensitive in both quiet and noisy environments and should be able to detect short and long duration sounds. A good spectral analysis capability, including frequency discrimination, is important in identifying and recognizing predators, prey, and other objects in the environment. Other important characteristics of a good sonar receiver include the capability to (a) spatially resolve and localize sounds, (b) reject externally generated interferences, and (c) recognize temporal and spectral patterns of sounds. In this chapter, these characteristics of the dolphin's receiving system will be examined.

3.1 Hearing Sensitivity

3.1.1 Sensitivity to Continuous Tones

In Section 1.1 it was mentioned that Kellogg et al. (1953) and Schevill and Lawrence (1953) were the first to obtain behavioral evidence that a bottlenose dolphin could hear sounds above 100 kHz. It was not until 1967, however, that a carefully performed hearing experiment was conducted to study the range and sensitivity of the dolphin auditory system. In this pioneering study, Johnson (1967) performed a carefully controlled psychophysical experiment to measure the underwater auditory sensitivity of a bottlenose dolphin to continuous wave (cw) tones as a function of frequency. A cw tone is a signal with a long enough duration to produce a steady-state response by the subject; i.e., the subject's sensitivity would not improve if the signal duration was longer. The dolphin was required to station in a sound stall and respond to the presence or absence of a two-second tone signal. A go/no-go response paradigm was used in conjunction with the up/down staircase psychophysical testing procedure. The dolphin's audiogram measured by Johnson is shown in Figure 3.1, along with a human audiogram measured by Sivian and White (1933). Johnson's results indicate that a bottlenose dolphin can hear over a wide frequency range between 75 Hz and 150 kHz, and has good sensitivity (within 10 dB of the maximum sensitivity) between approximately 15 kHz and 110 kHz. The upper frequency limit of hearing cuts off very sharply, at about 495 dB per octave, which is far greater than any electronic filter (Johnson 1986).

The only meaningful way thresholds in air and

Figure 3.1. Auditory sensitivity of a bottlenose dolphin and human subjects. The solid curve is the human audiogram of Sivian and White (1933) plotted against the right ordinate in Watts/m². Left ordinate: sound pressure level in dB (re 1 μPa) in water. Types of hydrophones used are shown in the legion. (Adapted from Johnson 1967 © Pergamon Press, Ltd.)

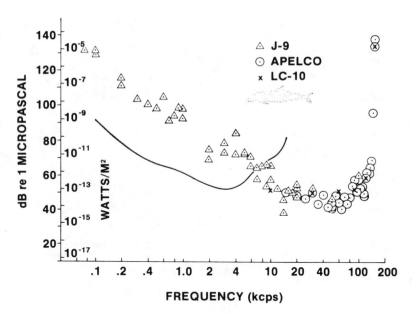

water can be compared is on the basis of intensity; hence the human audiogram was plotted in watts per cm² by Johnson (converted to watts/m² in Figure 3.1). The shapes of the dolphin and human audiograms are similar, with the dolphin's shifted to higher frequencies by a factor of 10. The data also indicate that the maximum hearing sensitivity of the dolphin under water and of the human in air are similar.

Johnson's auditory sensitivity results probably represent a very conservative estimate of the bottlenose dolphin's capability. The animal was conditioned against committing false alarm errors. A 90-second "time out" which included the removal of the response levers from the tank, was taken every time the dolphin committed a false alarm error. This type of experimental procedure probably caused the dolphin to adopt a relatively conservative response pattern in which it would choose the go response only if it was fairly sure that a signal was present.

Since the experiment of Johnson (1967) with *Tursiops truncatus*, the auditory sensitivity of many other small cetacean species has been measured. Audiograms have been measured for the harbor porpoise, *Phocoena phocoena* (Andersen 1970), killer whale, *Orcinus orca* (Hall and Johnson 1972; Bain and Dahlheim, pers. comm.), Amazon River dolphin, *Inia geoffrensis* (Jacobs

and Hall 1971), beluga or white whale, *Delphinapterus leucas* (White et al. 1978), Pacific bottlenose dolphin, *Tursiops gilli* (Ljungblad et al. 1982), and the false killer whale, *Pseudorca crassidens* (Thomas et al. 1988). The results of auditory sensitivity measurements with different cetacean species are depicted in Figure 3.2. In order to present the audiograms in a single figure with minimum confusion, the original results were curve fitted so that each audiogram would be a smooth rather than a ragged curve. The curve for *Orcinus Orca* is from Bain and Dalheim (pers. comm.) and is the average from two animals, one having a high frequency cutoff at 104 kHz and the other at 120 kHz. Previous data from Hall and Johnson (1971) indicated a high frequency limit of 35 kHz for the killer whale. Apparently, that particular whale experienced high frequency hearing losses. In all of these studies, with the exception of Andersen's, the up/down staircase psychophysical procedure was used to measure the animals' threshold. Andersen used the method of constant stimuli to test the *Phocoena*'s hearing sensitivity. Thomas et al. (1988) and possibly Hall and Johnson (1971) were the only investigators that did not observe a "time out" when the animal committed a false alarm error. Ljungblad et al. (1982) used a rather unusual modification to the staircase procedure

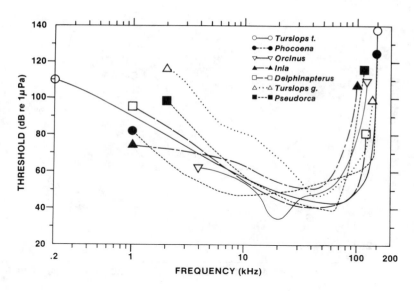

Figure 3.2. Audiograms for different cetacean species.

Table 3.1. Some important properties of the audiograms plotted in Figure 3.2

Species	Maximum Sensitivity (dB re 1 μPa)	Frequency of Best Hearing (kHz)	Upper Frequency Limit (kHz)
Tursiops truncatus	42	15–110	150
Phocoena phocoena	47	3–70	150
Orcinus orca	34	15–30	120
Inia geoffrensis	51	12–64	100
Delphinapterus leucas	40	11–105	120
Tursiops gilli	47	30–80	135
Pseudorca crassidens	39	17–74	115

in measuring the audiogram of the *Tursiops gilli*. After each correct signal-present trial, the signal level was reduced by 5 dB until the animal missed the signal. The signal level was then increased by 15 dB for the next signal-present trial. It seems that with this procedure the animal would not be motivated to detect a difficult signal, since the signal level would be raised (by 15 dB) if it missed the stimulus, making the task considerably easier. The relatively low sensitivity of the *Tursiops gilli* for low frequencies may be a reflection of the unusual testing procedure rather than the actual sensitivity of the dolphin.

The audiograms depicted in Figure 3.2 can be divided into three frequency regions. The first region is the low frequency region in which sensitivity gradually improves with frequency at

a rate of approximately 10 to 15 dB per octave (with the exception of the *Tursiops gilli*). The second region is the mid-frequency range where the sensitivity forms a U-shaped trough, or dip, in which the maximum sensitivity is reached. The third region is the high frequency cutoff region where the sensitivity decreases rapidly with frequency (at a minimum rate of 100 dB per octave, depending on the specific animal).

A summary of some important properties of the different audiograms of Figure 3.2 is given in Table 3.1. In the table, the frequency of best hearing is arbitrarily defined as the frequency region in which the auditory sensitivity is within 10 dB of the maximum sensitivity depicted in each audiogram of Figure 3.2. With the exception of the *Orcinus* and the *Inia*, the maximum sensi-

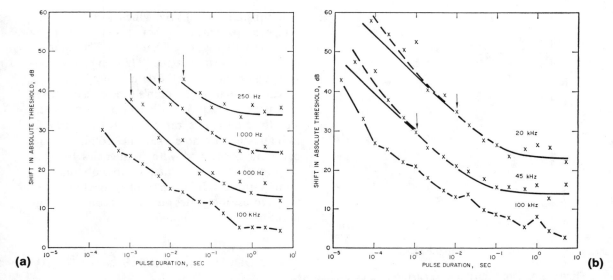

Figure 3.3. Absolute thresholds versus pulse duration for the data obtained (*A*) with a J-9 projector and (*B*) with an LC-10 hydrophone. The arrows indicate the shortest pulses used in least-square fit of the data to eq. (3.1). (From Johnson 1968a.)

tivities of the species represented are very similar within experimental uncertainties, especially for audiograms obtained with the staircase procedure using relatively large step sizes of 5 dB or greater. At the frequency of best hearing, the threshold for *Orcinus* is much lower than those for the other animals. It is not clear whether this keen sensitivity is a reflection of a real difference or a result of some experimental artifact. The data of Table 2.1 also indicate that *Tursiops truncatus* and *Delphinapterus leucas* seem to have the widest auditory bandwidth. All of the animals tested could hear high frequency sounds beyond 100 kHz, considerable higher than what most mammals can hear.

3.1.2 Sensitivity to Pulse Tones

The phenomenon of temporal auditory summation, in which the auditory sensitivity to a tone increases as the duration of a tone pulse increases, is well known in human audition (Plomp and Bouman 1959; Zwislocki 1960). The ear behaves like an integrator with an "integration time constant" (Zwislocki 1960). Energy is summed over the duration of a tone pulse until the pulse is longer than the integration time constant.

Therefore, as the duration of a pulse tone is made longer, more energy is contained within the pulse, and the ear is better able to perceive the tone. For tone pulses with durations shorter than the integration time, the amount of energy increases by 3 dB for each doubling of the duration. Hughes (1946) found that the change in the absolute hearing threshold as a function of the duration of a single tone pulse could be describe by the equation

$$I/I_\infty = 1 + (\tau/t) \qquad (3\text{-}1)$$

where I = threshold intensity of a tone pulse of duration t; I_∞ = threshold intensity at $t = \infty$; and τ is a constant. For signals longer than τ, the threshold is essentially constant and equal to I_∞.

Johnson (1968a) studied temporal auditory summation in a dolphin, using the same animal, experimental configuration, and procedure as in his cw audiogram measurement (Johnson 1967). He found a phenomenon similar to that in humans of increasing auditory sensitivity with increasing duration of the pulse tone. The results of Johnson's measurements are shown in Figures 3.3A,B for tone frequencies of 0.25, 1, 4, 20, 40, and 100 kHz. The dolphin's threshold shifted downward as the pulse duration increased and

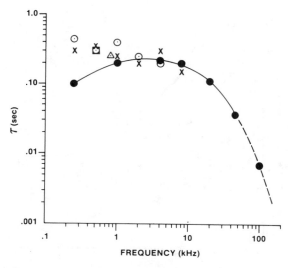

Figure 3.4. Time constants versus frequency for the dolphin and for humans. Open circles and crosses: data from Plomp and Bouman (1959); square: Blodgett et al. (1958); triangle: Hamilton (1957); closed circles: data for the dolphin, all other data for humans. (After Johnson 1968a.)

gradually leveled off as the duration approached and exceeded the integration time constant. The dashed curves represent equation (3-1) fitted to the data with a minimum least square error for each frequency except 100 kHz. The data of Figure 3.3 vary in a manner consistent with (3-1). The dolphin's threshold shifted at approximately 3 dB per doubling of duration in the time region of steepest descent.

The time constants used in the curve fit are plotted in Figure 3.4, along with time constants obtained in experiments with humans. Johnson (1968a) plotted the time constant as a function of frequency on a semi-log plot. Here, a log-log graph was used instead so that the time constant at 100 kHz could be included on the graph. Johnson (1968a) did not include the time constant at 100 kHz because he found that the transient response of the J-9 transducer used to obtain the data of Figure 2.8A caused the signal to have a broad spectrum. The LC-10 had a better transient response but still introduced unwanted frequency spreading, particularly above 100 kHz. Nevertheless, the 100-kHz data obtained with the LC-10 are included in Figure 3.4 in order to get an estimate of the dolphin's

time constant at 100 kHz, which is close to typical peak frequencies of sonar signals used by *Tursiops truncatus* (see Chapter 7). The estimate of the time constant at 100 kHz seems to be consistent with the rest of the data. From Figure 3.4, we can see that the dolphin's time constant rises to a peak at frequencies between 1 and 4 kHz, and then decreases rapidly with frequency. It is interesting to note that the dolphin's time constants associated with frequencies between 0.5 and 10 kHz closely match those obtained for human subjects. Johnson (1968a) concluded that the porpoise integrates pure-tone acoustic energy in the same way as humans.

3.2 Spectral Analysis Sensitivity

One of the fundamental characteristics of the human auditory system is its ability to derive spectral information from sounds. Our ability to hear one sound in the presence of other sounds and to discern individual harmonics within complex tones is an example of the ear's frequency resolution capability. The idea that our auditory frequency analyzer could be thought of as a bank of overlapping linear bandpass filters was suggested as early as 1863 by Helmholtz (Patterson and Moore 1986). But it was not until 1940 that Fletcher, in his famous article, suggested several ways in which the width of the bandpass filters could be estimated (Fletcher 1940). The first technique is commonly referred to as the *critical ratio* technique and the second one as the *critical bandwidth* technique. These techniques, developed to study the human auditory system, can also be applied to study the auditory system of dolphins.

3.2.1 Critical Ratio

In the critical ratio technique, the auditory sensitivity of subjects to tones masked by broadband white noise is measured. Fletcher hypothesized that (a) a pure tone would be masked only by a narrow band of noise surrounding the frequency of the tone and (b) at the threshold of hearing, the noise power would be equal to the signal power. Given the second hypothesis, the following equation would be applicable at the masked threshold:

$$I_{th} = N_0 \times \Delta f \qquad (3\text{-}2)$$

where: I_{th} = intensity of the tone at threshold in μPa^2

N_0 = noise spectral density in $\mu Pa^2/Hz$

Δf = bandwidth of the auditory filter at the tone frequency

Therefore, the bandwidth of the auditory filters can be indirectly estimated by performing masked-threshold measurements. The bandwidth is the ratio of threshold intensity to noise spectral density, and the bandwidths derived from this technique are referred to as critical ratios. The shape of the filters is assumed to be rectangular, with infinitely steep low and high pass cutoffs. The critical ratio can be expressed in dB as

$$CR = 10\log(\Delta f) \qquad (3\text{-}3)$$

Johnson (1968b) used the critical ratio technique of Fletcher (1940) to measure the filter bandwidth of the auditory system of *Tursiops truncatus*. He measured the masked thresholds for 15 different frequencies at six different noise levels, using the same animal, experimental configuration, and procedure as for his cw audiogram (Johnson 1967). The critical ratio results obtained by Johnson are shown in Figure 3.5, along with human monaural and binaural re-

sults. The dolphin's critical ratio seems like an extension of the human critical ratio to higher frequencies. The dolphin's critical ratio increases almost proportionately with frequency, suggesting that the dolphin's auditory system may be modeled by a bank of constant-Q filters. The Q, or quality, factor of a bandpass filter is defined as the ratio of center frequency to bandwidth as expressed by the equation

$$Q = f_0/\Delta f \qquad (3\text{-}4)$$

where f_0 is the center frequency of the filter. The dashed curve in Figure 3.5 is the constant-Q curve ($Q = 14.4$) which best fitted the dolphin data in a least mean square error manner. The constant-Q curve is consistent with the measured critical ratios.

The effectiveness of noise in masking pure tones can be determined by plotting the shift in threshold (difference between threshold in quiet and threshold in noise) as a function of the effective level (difference between the threshold in quiet and total noise energy in one critical band). Johnson (1968b) compared the effectiveness of noise in masking tones for *Tursiops* with data obtained in human experiments as illustrated in Figure 3.6. The human data are from the studies of French and Steinberg (1947) and Hawkins and Stevens (1950). Except for the 5-kHz data, the

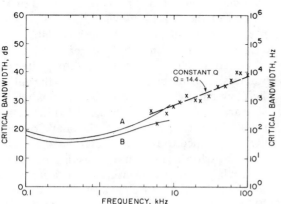

Figure 3.5. Critical ratios. (X) dolphin; (A) human monaural subjects (French and Steinberg, 1947); (B) human binaural subjects (Hawkins and Stevens 1950). (From Johnson 1968b.)

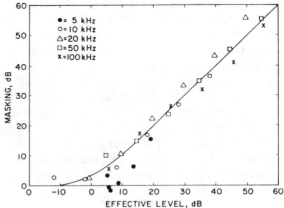

Figure 3.6. Plot of the shift in threshold resulting from the presence of masking noise versus the effective level of the masker. The solid curve gives results from experiments with human subjects. (From Johnson 1968b.)

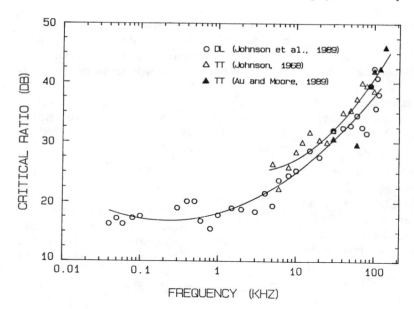

Figure 3.7. Critical ratios for two Atlantic bottlenose dolphins (TT) and a beluga (DL). (From Johnson 1968b; Johnson et al. 1989; Au and Moore 1990.)

dolphin results compare favorably with the human data, showing a linear shift in threshold as a function of the masker level.

Au and Moore (1990) also measured the critical ratios of an Atlantic bottlenose dolphin. The animal's masked threshold was determined using a staircase technique and a 1-dB stepsize. Time-outs were not used when the animal committed false alarm errors. Our results along with Johnson's results for *Tursiops* are shown in Figure 3.7. Also included in the figure are the critical ratios for a beluga (*Delphinapterus leucas*) measured by Johnson et al. (1989). They used a staircase technique with 5-dB stepsize and one noise level to obtain their results. The solid curves are second-order polynomial curve fits to the combined data of Johnson (1968b) and Au and Moore (1990) for the two *Tursiops* and for the *Delphinapterus leucas*. The data indicate that the critical ratios of the beluga are slightly lower than those of the bottlenose dolphin at low frequencies and approach those of the dolphin at high frequencies.

3.2.2 Critical Bandwidth

Fletcher (1940) also devised a more direct method to measure the width of the auditory filter. In this second technique, the bandwidth of the masking noise is varied and the masked thresh-

old determined for each bandwidth. For noise bandwidths that are narrower than the auditory filter at the test frequency, the masked threshold should increase proportionately as the width of the noise filter increases. When the width of the noise becomes equal to or greater than the auditory filter, the masked threshold should no longer increase as the noise filter increases since the filter cannot receive any more noise. Therefore, the masked threshold should be a constant with respect to the bandwidth of the noise. In this technique the masked threshold is plotted as a function of the noise bandwidth. Two straight lines are fitted to the masked threshold data, one to the varying portion and the other to the constant portion. The width of the auditory filter is the intersection of the two lines fitted to the data. The filter bandwidths determined by this technique are referred to as "critical bandwidths." As in the critical ratio technique, the shape of the filters is assumed to be rectangular.

Au and Moore (1990) used masking noise of different bandwidths to measure the critical bandwidth of a *Tursiops* at frequencies of 30, 60, and 120 kHz. The masked threshold was determined as a function of the noise bandwidth using the staircase procedure and a go/no-go response paradigm. The masked threshold as a function of noise bandwidth for the three test frequencies is displayed in Figure 3.8. Also included are the intersecting lines that were matched to the data in a minimum least square error manner. The bandwidth associated with each point of intersection is the critical bandwidth and represents an estimate of the animal's auditory filter bandwidth at the test frequencies. Our critical bandwidth (open triangles) and critical ratio (closed circles) data are plotted in Figure 3.9, along with Johnson's (1968b) critical ratio (open circles) data. The solid lines represent constant-Q curves that best fitted the dolphin data in a least mean square error manner. Except for three critical ratio outliers, the constant-Q curves fit the data relatively well. Taking the quotient of the two Q's, we find that the critical bandwidth is approximately 5.6 times or 7.5 dB greater than the critical ratio. Critical bandwidths for humans are typically 2.5 times or 4.0 dB greater than critical ratios (Scharf, 1970). The reason for the larger difference in the dolphin's case is not known. Perhaps Fletcher's assumption for humans that signal and noise power are equal at the detection threshold may not be valid for the dolphin. Larger critical ratios would result if the signal power is smaller than the noise power at threshold.

3.2.3 Masking by Pure Tone

Another technique that can be used to study the frequency analysis property of the dolphin auditory system involves masking of a pure tone signal by another pure tone. Johnson (1971) continued his auditory research on *Tursiops* by conducting a tone-on-tone masking experiment using the same experimental apparatus, procedure, and dolphin as in his original cw auditory sensitivity study. A 70-kHz tone was used as the masking tone with two masking levels, 40 and 80 dB above the animal's threshold. Two measures of the dolphin's sensitivity were made at the 80dB masking level. The results of the tone on tone masking experiment are shown in Figure 3.10, with the threshold shift from the unmasked condition plotted as a function of the frequency of the test tone. The curves indicate how many dB the intensity of the signal at a particular frequency must be raised above its absolute threshold to be detectable in the presence of the

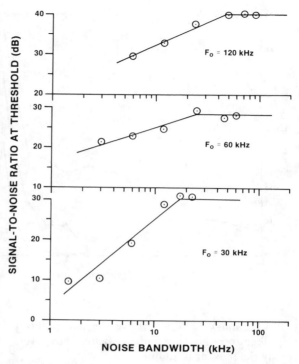

Figure 3.8. Masked hearing threshold as a function of noise bandwidth for three test frequencies. (From Au and Moore 1990.)

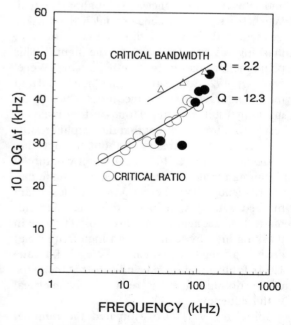

Figure 3.9. Critical bandwidth (triangles) and critical ratio (circles) for *Tursiops* (from Au and Moore, 1990; Johnson, 1968b).

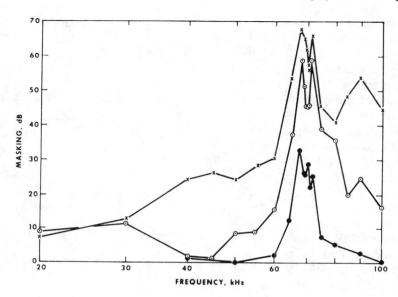

Figure 3.10. Threshold shift (masking) from threshold in quiet due to the presence of the masking tone for various test frequencies. Crosses: first run data; points in open circles: second run data at 80-dB masking level; closed circles: data taken at the 40-dB masking level. (From Johnson 1971.)

70-kHz masking tone. Above the curves, both masker and signal are audible. Below the curves, only the masker is audible.

Most of the masking occurred when the signal frequency approached the masker frequency; masking was at a maximum when the signal and masking frequencies were almost equal. When the signal frequency equaled the masker frequency, there was a dip in the amount of masking, as can be seen in the figure. This phenomenon also occurs with humans (Wegel and Lane 1924) and is the result of the signal and masker interacting to produce beats. The dolphin masking results resembled tone-on-tone masking results for humans except that the dolphin results were relatively symmetrical about the masking frequency compared to the asymmetric nature of the human results. Signals both lower and higher in frequency than the masking frequency are masked equally well in the dolphin. With humans, low frequency tones are more effective in masking high frequencies than high frequencies are in masking low frequencies (Wegel and Lane 1924). Johnson's data may indicate that the shape of the dolphin filters is relatively symmetrical with frequency.

Johnson (1971) also estimated the minimal intensity difference dolphins can perceived from the magnitude of the dips in the 70-kHz masking curves of Figure 3.10. Assuming that at 70 kHz

the animal's threshold is determined by its ability to resolve beats in the tone, the following relation was derived:

$$\Delta L = 20 \log \frac{1 + 10^{-M/20}}{1 - 10^{-M/20}} \qquad (3\text{-}5)$$

where ΔL is the variation in sound pressure level accompanying beats, and M is the difference in level of beating tones. Both ΔL and M are in dB. From equation (3-2), the values of ΔL for the three curves of Figure 3.9 are 1.0, 0.35, and 2.0 dB for the first and second run at the 80-dB masker level and for the 40-dB masker level, respectively.

3.2.4 Frequency Discrimination

The ability to distinguish one frequency from another is important in detecting and processing acoustic signals as well as sonar echoes. The frequency discrimination capability of the bottlenose dolphin was studied by Jacobs (1972), Herman and Arbeit (1972), and Thompson and Herman (1975). In all three studies, the dolphin was required to discriminate a constant frequency (CF) pure tone signal from a frequency modulated (FM) signal having the same center frequency as the CF signal. Jacobs (1972) and Thompson and Herman (1975) used a single projector; during each trial the dolphin was

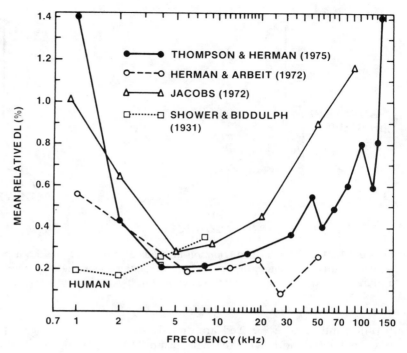

Figure 3.11. Acoustic frequency discrimination capability of *Tursiops truncatus* and human.

required to indicate whether a CF or FM signal was emitted by pressing one of two paddles. Herman and Arbeit (1972) used two spatially separated projectors and presented sounds sequentially during a trial. A CF signal was transmitted from one projector followed by a FM sound from the other projector, or vice versa. The dolphin was required to touch the paddle closest to the projector that transmitted the CF signal. In all three studies, the amount of frequency modulation was decreased in a staircase procedure until the animals could not discriminate the FM signal from the CF signal.

The difference limen (DL) is the difference between the upper and lower frequency of the FM signal at the animal's threshold. The relative difference limen is the Weber fraction $\Delta F/F$, where F is the center frequency of the FM signal. The results of the three frequency discrimination experiments with *Tursiops truncatus* are shown in Figure 3.11, with the relative difference limen in percent plotted as a function of frequency. The data of Herman and Arbeit and Thompson and Herman were obtained with the same animal. Their results, if averaged, would be a good representation of the dolphin's difference limens.

The data plotted in Figure 3.11 indicate that the dolphin's frequency discrimination capability is similar to that of humans, but shifted to higher frequencies by a factor of 5 to 10. The dolphin's best discrimination occurred between 2 and 55 kHz, with relative DLs ranging from 0.2% to 0.4%. Relative DLs remained within 1% for all frequencies between 1 and 140 kHz. According to Fay (1974), the frequency discrimination capability of the bottlenose dolphin is superior to that of any other vertebrate studied except man.

3.3 Directional Hearing

3.3.1 Receiving Beam Patterns

A directional hearing capability is an important property of a sonar system operating in either the passive or active mode. Its narrow transmission and reception beams allow the dolphin to localize objects in a three-dimensional volume, to separate objects spatially within a multi-object field, to resolve features of extended or elongated objects, and to minimize the amount of interference received. The amount of ambient noise from an isotropic noise field and the amount of rever-

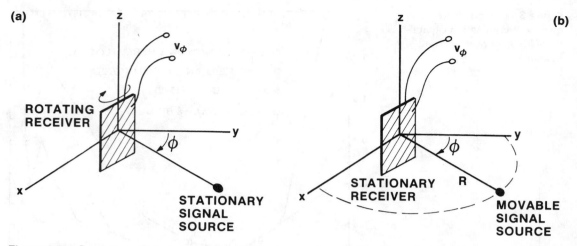

Figure 3.12. Geometry of two common methods of measuring the beam pattern of a hydrophone. (*A*) The acoustic source is fixed and the hydrophone is rotated about one of its axes. (*B*) The hydrophone is fixed and a source is moved in a circle about the hydrophone.

beration interference received are directly proportional to the width of the receiving beam. The effects of discrete or partially extended interfering noise or reverberant sources can be minimized by simply directing the beam away from the sources.

A common method of measuring the beam pattern of a hydrophone is to rotate the hydrophone about an axis and measure the voltage produced by a stationary sound source (Fig. 3.12A). If the hydrophone is too large or is integrated into some other structure, such as a ship, the beam pattern can be determined by fixing the position of the hydrophone and measuring the voltage produced by a sound source as it moves about the hydrophone along the arc of a circle (Fig. 3.12B). Unfortunately, the receiving beam pattern of a dolphin cannot be measured with either method unless an electrophysiological technique such as the evoked potential technique, using waterproof surface electrodes, is employed. Instead, the masked hearing threshold as a function of the azimuth about the animal's head needs to be measured. The relative masked hearing threshold as a function of azimuth is equivalent to the receiving beam pattern since the receiving beam pattern is the spatial pattern of hearing sensitivity.

The receiving beam pattern of a *Tursiops truncatus* was measured in both the vertical and horizontal planes by Au and Moore (1984). We measured the dolphin's masked hearing threshold as the position of either the noise or the signal sources varied in their angular position about the animal's head. The dolphin was required to voluntarily assume a stationary position on a bite plate constructed out of a polystyrene plastic material. The noise and signal transducers were positioned along an arc, with the center of the arc located approximately at the pan bone of the animal's lower jaw. A schematic drawing of the experimental apparatus is shown in Figure 3.13. In order to measure the dolphin's receiving beam in the vertical plane, the animal was trained to turn to its side before biting the specially designed vertical bite plate stationing device, as shown. A go/no-go response paradigm was used and the animal's masked threshold was determined with an up/down staircase psychophysical procedure. For measurements in the vertical plane, the position of the signal transducer was fixed directly in line with the bite plate and its acoustic output was held constant. Masked thresholds were measured for different angular positions of the noise transducer along the arc. The level of noise was varied in order to obtain the masked threshold. A threshold estimate was considered complete when at least 20 reversals (10 per session) at a test angle had been obtained over at least two consecutive sessions, and if the average reversal values of the two sessions were

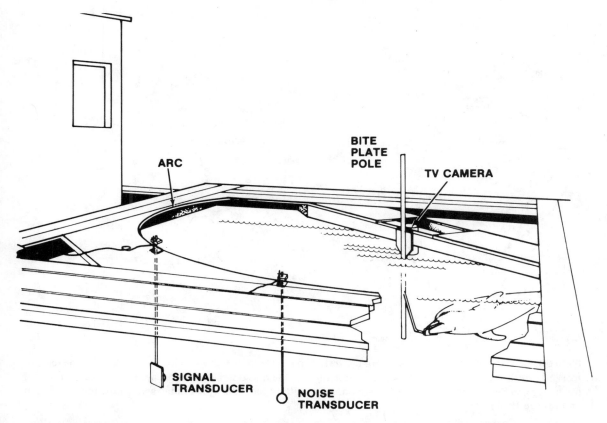

Figure 3.13. Experimental configuration for the receive beam pattern measurement of a bottlenose dolphin. The animal was trained to turn to its side and position on the vertical bite plate for the measurement in the vertical plane. For the measurement in the horizontal plane, a different bite plate assembly was used. (From Au and Moore 1984.)

within 3 dB of each other. After a threshold estimate was achieved, the noise transducer was moved to a new azimuth chosen from a set of randomly predesignated azimuths. As the azimuth about the dolphin's head increased, the hearing sensitivity of the dolphin tended to decrease, requiring higher levels of masking noise from a transducer located at a greater azimuth to mask the signal from a source located directly ahead of the animal.

The receiving beam patterns in the vertical plane were plotted for signal frequencies of 30, 60, and 120 kHz in Figure 3.14A. The data points of the beam patterns were fitted with a cubic spline function (Forsythe et al. 1977) to interpolate values between points. The beam patterns redrawn as smooth curves based on the cubic spline interpolation process are shown in Figure 3.14B. The radial axis of Figure 3.14 represents the difference in dB between the noise level needed to mask the test signal at any azimuth and the minimum noise level needed to mask the test signal at the azimuth corresponding to the major axis of the vertical beam. The shape of the beams indicates that the patterns were dependent on frequency, becoming narrower, or more directional, as the frequency increased. The beam of a planar hydrophone also becomes narrower as the frequency increases. The 3-dB beamwidths were approximately 30.4°, 22.7°, and 17.0° for frequencies of 30, 60, and 120 kHz, respectively. There was also an asymmetry between the portions of the beam above and below the dolphin's head. The shape of the beams dropped off more rapidly for increasing angles above the animal's head than for angles below the animal's head,

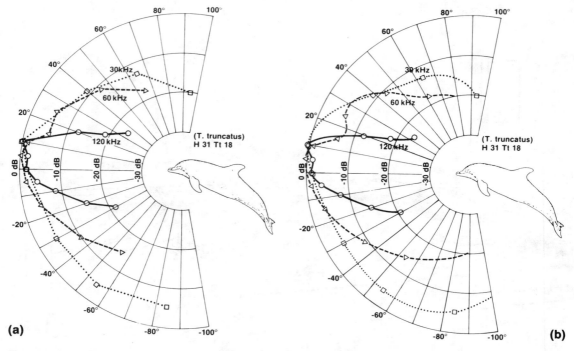

(a)   **(b)**

Figure 3.14. (*A*) Receiving beam patterns in the vertical plane for frequencies of 30, 60, and 120 kHz. The relative masked thresholds as a function of the elevation angle of the masking noise source are plotted for each signal frequency. (*B*) Smoothened beam patterns obtained by fitting the data with a cubic spline function. (From Au and Moore 1984.)

indicating a more rapid decrease in the animal's hearing sensitivity for angles above the head than for angles below the head. If the dolphin receives sounds through the lower jaw, the more rapid reduction in hearing sensitivity for angles above the head may have been caused by shadowing of the received sound by the upper portion of the head structure. The 60-kHz beam is slightly peculiar, since it shows almost the same masked threshold value for 15° and 25° elevation angles.

The major axis of the vertical beams in Figure 3.14 is elevated between 5° and 10° relative to the reference axis. The 30- and 120-kHz results place the major axis at 10° while the 60-kHz results place it at 5°. It will be shown in Chapter 5 that the major axis of the receiving beam in the vertical plane is elevated to approximately the same angle as the major axis of transmitted beam in the vertical plane.

In the horizontal beam pattern measurements, two noise sources were fixed at azimuth angles of

± 20°. The level of the noise sources was also fixed. The position of the signal transducer was varied from session to session. Masked thresholds were determined as a function of the azimuth of the signal transducer by varying the signal level of the signal transducer in a staircase fashion. As before, a threshold estimate was considered complete when at least 20 reversals at a test angle had been obtained over at least two consecutive sessions, and if the average reversal values of the sessions were within 3 dB of each other. After a threshold estimate was determined, the signal transducer was moved to a new azimuth chosen from a set of randomly predesignated azimuths. Two noise sources were used in order to discourage the dolphin from internally steering its beam in the horizontal plane; if the animal could have steered its beam, it would have received more noise from one of the two hydrophones and therefore not experienced any improvement in the signal-to-noise ratio. The

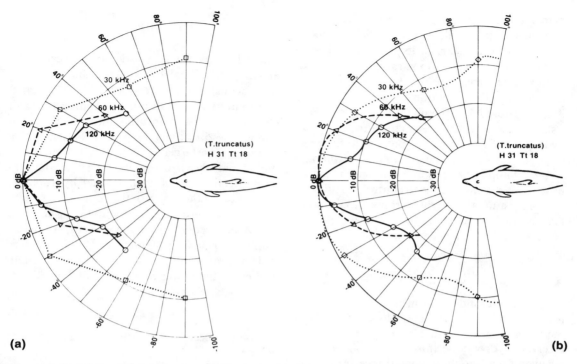

Figure 3.15. (*A*) Receiving beam patterns in the horizontal plane for frequencies of 30, 60, and 120 kHz. The relative masked thresholds as a function of the angle of the signal source are plotted for each signal frequency. (*B*) Smoothened beam patterns obtained by fitting the data with a cubic spline function. (From Au and Moore 1984.)

masking noise from the two sources was uncorrelated but equal in amplitude.

The receiving beam patterns in the horizontal plane for frequencies of 30, 60, and 120 kHz are shown in Figure 3.15A. The beam patterns redrawn as smooth curves based on a cubic spline interpolation process are shown in Figure 3.15B. The radial axis represents the difference in dB between the signal level at the masked threshold for the various azimuths and the signal level at the masked threshold for the 0° azimuth (along the major axis of the horizontal beam). The horizontal receiving beams were directed forward, with the major axis being parallel to the longitudinal axis of the dolphin. The beams were nearly symmetrical about the major axis; any asymmetry was within the margin of experimental error involved in estimating the relative thresholds. The horizontal beam patterns exhibited a similar frequency dependence as the vertical

beam patterns, becoming narrower, or more directional, as the frequency increased. The 3-dB beamwidths were 59.1°, 32.0°, and 13.7° for frequencies of 30, 60, and 120 kHz, respectively.

Zaytseva et al. (1975) measured the horizontal beam pattern of a dolphin by measuring the masked hearing threshold as a function of azimuth. Their beamwidth of 8.2° for a frequency of 80 kHz was much narrower than the 13.7° we found for a frequency of 120 kHz. The difference in beamwidth is even larger if the results of Au and Moore (1984) are linearly interpolated to 80 kHz. We calculated an interpolated beamwidth of 25.9° at 80 kHz, which was considerably greater than the 8.2° obtained by Zaytseva et al. The difference in beamwidths measured in these studies may be attributed to the use of only one noise source by Zaytseva et al. compared to the two noise sources used by Au and Moore. With a single masking noise source in the horizontal plane, there is a possibility that the animal performs a spatial filtering operation by internally steering the axis of its beam in order to maximize the signal-to-noise ratio. Another possibility arises from the fact that Zaytseva et al. did not use a fixed stationing device. Rather, the dolphin approached the signal hydrophone from a start line, always oriented in the direction of the hydrophone. The animal responded to the presence or absence of a signal by either swimming or not swimming to the hydrophone. In such a procedure, it is impossible to control the orientation of the animal's head with respect to the noise masker, so that the dolphin could have moved its head to minimize the effects of the noise.

3.3.2 Directivity Index

The directivity index is a measure of the sharpness of the beam or major lobe of either a receiving or transmitting beam pattern. For a receiving hydrophone, it is the ratio of the power received by an omnidirectional receiver to that received by a directional receiver in the same isotropic noise field, as described by the equation

$$DI = 10 \log\left(\frac{P_0}{P_D}\right) \tag{3-6}$$

where P_0 is the total acoustic power received by an omnidirectional hydrophone and P_D is the total acoustic power received by a directional hydrophone. Since a directional hydrophone will receive less noise than a nondirectional hydrophone, the directivity index will be a positive quantity. The larger the value of DI, the more directional the hydrophone. Assume that an omnidirectional and a directional hydrophone are located at the origin of the coordinate system shown in Figure 3.16. The amount of acoustic power received by the hydrophones in an isotropic acoustic field of intensity I_0 is $I_0 A$, where A is the area of a sphere about the origin. Therefore, the power received by the omnidirectional hydrophone will be

$$P_0 = 4\pi r^2 I_0 \tag{3-7}$$

The power received by the directional transducer will be

$$P_D = \int_0^{4\pi} \left(\frac{p(\theta,\phi)}{p_0}\right)^2 I_0 \, dS \tag{3-8}$$

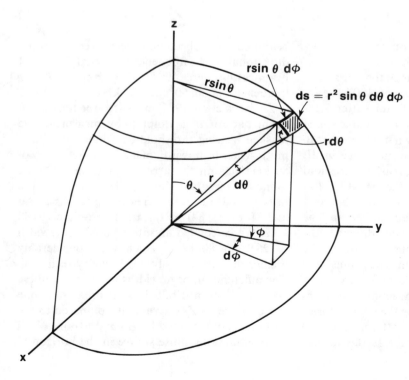

Figure 3.16. Coordinate system used to derive the directivity index of a transducer. The incremental area dS is located on the surface of the sphere.

Figure 3.17. Geometry for directivity index calculations (*A*) for dolphin data and (*B*) for a two-element rectangular array model.

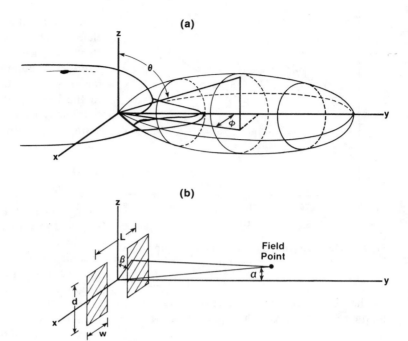

where $p(\theta, \phi)/p_0$ is the beam pattern of the directional transducer and dS is the incremental area of the sphere. The receiving intensity of the directional hydrophone is merely the beam pattern multiplied by the isotropic intensity I_0 in (3-8). From Figure 3.16, the incremental area on the surface of the sphere can be expressed as

$$dS = r^2 \sin \theta \, d\theta \, d\phi \tag{3-9}$$

Given (3-6) to (3-8), the directivity index can be expressed as

$$DI = 10 \log \frac{4\pi}{\displaystyle\int_0^{2\pi} \int_{-\pi/2}^{\pi/2} \left(\frac{p(\theta, \phi)}{p_0}\right)^2 \sin \theta \, d\theta \, d\phi} \tag{3-10}$$

Although the expression for the directivity index is relatively simple, using it to obtain numerical values can be a formidable task unless transducers of relatively simple shapes (cylinders, lines, and circular apertures) with symmetry about one axis is involved. Otherwise, the beam pattern needs to be measured as a function of both θ and ϕ. This is done by choosing various discrete values of θ and measuring the beam pattern as a function of ϕ, a tedious process.

Equation (3-10) can then be evaluated by numerically evaluating the double integral with a digital computer. Fortunately, directivity index measurements for simple planar transducers shaped as circles, rectangles, and ellipses have been performed and nomograms relating directivity index to beamwidth exist (see Bobber 1970).

The directivity indices associated with the dolphin's beam patterns in Figures 3.14b and 3.15b were estimated by Au and Moore (1984) using the geometry shown in Figure 3.17A. We assumed that the overall three-dimensional beam pattern was a product of the beam patterns in the horizontal and vertical planes, namely,

$$\left(\frac{p(\theta, \phi)}{p_0}\right)^2 = \left(\frac{p(\theta)}{p_0} \frac{p(\phi)}{p_0}\right)^2 \tag{3-11}$$

Expression (3-10) was then evaluated numerically with a digital computer and a two-dimensional Simpson's 1/3-rule algorithm (McCormick and Salvadori 1964). The results of the numerical evaluation using the smoothed beam patterns of Figures 3.14b and 3.15b are plotted as a function of frequency in Figure 3.18. DIs of 10.4, 15.3, and 20.6 dB were obtained for frequencies of 30, 60, and 120 kHz, respectively. A linear curve fitted

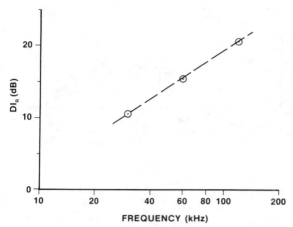

Figure 3.18. Receiving directivity index as function of frequency for a *Tursiops truncatus*.

to the computed DIs in a least square error manner is also shown. The equation of the curve is

$$\text{DI(dB)} = 16.9 \log f(\text{kHz}) - 14.5 \quad (3\text{-}12)$$

These results indicate that the dolphin's receiving directivity index increased with frequency in a manner similar to that of a linear transducer (Bobber 1970).

3.3.3 Equivalent Receiving Aperture

In this section, the directivity index for a two-element rectangular aperture array as depicted in Figure 3.17B will be applied to the dolphin. This procedure will allow us to estimate the size of a linear transducer array that has a directivity index equivalent to that of the dolphin. An

expression for DI can be derived for a two-element rectangular array by first using the Product Theorem for a transducer array. The Product Theorem states that the beam pattern for an array of directional transducers is equal to the beam pattern of an equivalent array of point sources multiplied by the beam pattern of a single directional transducer, as expressed by the equation

$$\frac{p(\alpha, \beta)}{p_0} = \left(\frac{p(\alpha, \beta)}{p_0}\right)_{array} \left(\frac{p(\alpha, \beta)}{p_0}\right)_{rect} \quad (3\text{-}13)$$

The geometry for the two–point source array is depicted in Fig. 3.19A and the geometry for the planar rectangular aperture in an infinite baffle is depicted in Figure 3.19B. The point sources are located along the x-axis, equidistant from the origin and separated from each other by a distance L. The derivation of the directivity index is detailed in the appendix to this chapter. The beam pattern of the two–point source array can be expressed as

$$\left(\frac{p(\alpha, \beta)}{p_0}\right)_{array} = \frac{\sin 2\psi_L}{2 \sin \psi_L} \quad (3\text{-}14)$$

(cf. (A-12)). The beam pattern for a rectangular aperture can be expressed as

$$\left[\frac{p(\alpha, \beta)}{p_0}\right]_{rect} = \frac{\sin \psi_w}{\psi_w} \frac{\sin \psi_d}{\psi_d} \quad (3\text{-}15)$$

(cf. (A-24)). We can insert (3-14) and (3-15) into (3-13) to obtain the total beam pattern and enter the resulting expression in (3-10) to obtain the directivity index for the two–rectangular aperture geometry (cf. A-25)

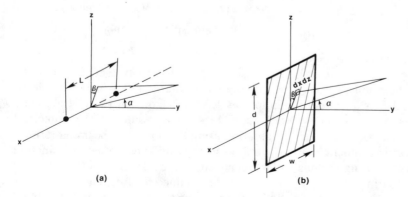

(a)   (b)

Figure 3.19. Geometry for the derivation of the receiver directivity index. (*A*) Two-element point source array; (*B*) rectangular aperture in an infinite baffle.

$$DI =$$

$$10 \log \frac{4\pi}{\int_0^{2\pi} \int_0^{\pi/2} \left(\frac{\sin 2\psi_L}{2\sin\psi_L} \frac{\sin\psi_w}{\psi_w} \frac{\sin\psi_d}{\psi_d} \right)^2 \sin\alpha \sin\beta \, d\alpha \, d\beta}$$

$$(3\text{-}16)$$

where
$$\psi_L = \frac{kL}{2} \sin\alpha \sin\beta$$

$$\psi_w = \frac{kw}{2} \sin\alpha \sin\beta$$

$$\psi_d = \frac{kd}{2} \sin\alpha \sin\beta$$

Measurements made at the thin oval area of the flared posterior end of the mandible of an adult *Tursiops truncatus* indicated that the distance between the pan bone areas of the right and left side of the mandible is approximately 12 cm. Using this value for L, (3-16) was evaluated for different values of d and w, keeping the ratio of d and w approximately the same as the ratio of DIs calculated separately in the vertical and horizontal planes. The solid line in Figure 3.20 represents the calculated DIs as a function of frequency for the two-element array model of Figure 3.17 with $L = 12$ cm, $d = 3.1$ cm, and $w = 2.6$ cm. The

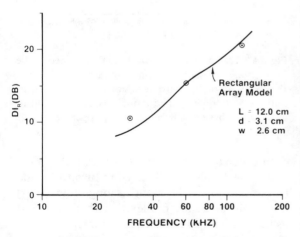

Figure 3.20. Receiving directivity index as a function of frequency calculated from the two-rectangular-element array model. The circles are DIs estimated from the horizontal and vertical beam patterns in conjunction with eq. (3-10). (From Au and Moore 1984.)

results indicate that each receptor of ultrasonic sound on the dolphin's head can be modeled by an effective aperture having an area of approximately 8.1 cm², with a spacing of 12 cm between apertures.

3.3.4 Sound Localization

The ability to localize or determine the position of a sound source is important in a sonar in order to resolve echoes from closely spaced targets, or from different portions of an extended target, and to determine the relative position of targets within the sonar beam. The capability to localize sounds has been studied extensively in humans and in many vertebrates (see Fay 1988). Lord Rayleigh (1907) proposed that humans localized sound in the horizontal plane by using interaural time differences for low frequency sounds and interaural intensity differences for high frequency sounds. Consider the geometry of Figure 3.21A, with a sound approaching a dolphin at an angle ϕ. If the frequency of the sound is high enough so the wavelength is smaller than the dimension of the dolphin's head, the sound reaching the more distant ear will be shadowed by the head and will have different intensity in each ear. Lower frequency sounds with longer wavelengths will merely diffract around the head with little loss, so the interaural intensity difference cue will be minimal and interaural time difference (or phase difference) cues must be used for localization. The distance between the pan bone areas on either side of the mandible is approximately 12 cm, the wavelength of an 11 kHz signal in water.

Renaud and Popper (1975) examined the sound localization capabilities of a *Tursiops truncatus* by measuring the *minimum audible angle* (MAA) in both the horizontal and vertical planes. The MAA is defined as the angle subtended at the subject by two sound sources—one being at a reference azimuth—at which the subject can just recognize the sound sources as discrete (Mills 1958). During a test trial, the dolphin was required to station on a bite plate facing two transducers positioned at equal angles from an imaginary line running through the center of the bite plate (Fig. 3.21B). An acoustic signal was then transmitted from one of the transducers and

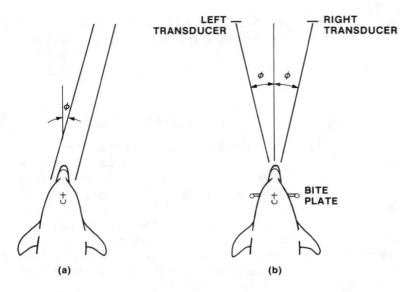

Figure 3.21. (*A*) Geometry for sound localization of a distant source. (*B*) Geometry for the minimum audible angle measurement in the horizontal plane.

the dolphin was required to swim and strike the paddle on the side of the emitting transducer. The angle between the transducers was varied in a modified staircase fashion. If the dolphin was correct for two consecutive trials, the transducers were moved an incremental distance closer together, decreasing the angle between them by 0.5°. After each incorrect trial, the transducers were moved an incremental distance apart, increasing the angle between them by 0.5°. This modified staircase procedure allowed threshold determination at the 70% level as discussed in Section 1.4.4. The threshold angle was determined by averaging a minimum of 16 reversals. A randomized schedule for sound presentation through the right or left transducer was used.

The localization threshold determined in the horizontal plane as a function of frequency for narrow band pure tone signals is shown in Figure 3.22. The MAA has a U-shaped pattern, with a large value of 3.6° at 6 kHz, decreasing to a minimum of 2.1° at 20 kHz and then slowly increasing again in an irregular fashion to 3.8° at 100 kHz. MAAs for humans vary between 1° and 3.7°, with the minimum at a frequency of approximately 700 Hz (Mills 1972). The region where the MAA decreases to a minimum in Figure 3.22 (about 20 kHz) may be close to the frequency at which the dolphin switches from using interaural time difference cues to using interaural intensity difference cues.

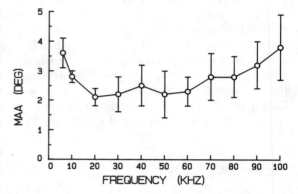

Figure 3.22. Localization threshold determined in the horizontal plane as a function of frequency. Shown are the mean ± one standard deviation for seven determinations per frequency. The animal faced directly ahead at 0° azimuth. (From Renauld and Popper 1978.)

In order to measure the MAA in the vertical plane, the dolphin was trained to turn to its side (rotate its body by 90° along its longitudinal axis) and to bite on a sheet of Plexiglas which was used as the stationing device. The experimental geometry was similar to Figure 3.21B, except that the dolphin was turned to its side. The results of the MAA measurements in the vertical plane as a function of frequency are shown in Figure 3.23. The MAA in the vertical plane varied from 2.3° at 20 kHz to 3.5° at 100 kHz. These results indicate that the animal could localize sound in

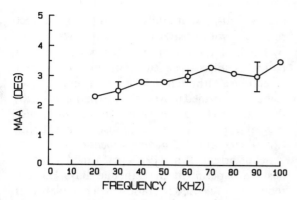

Figure 3.23. Localization threshold determined in the vertical plane as a function of frequency. Standard deviations are indicated for vertical data for 30, 60, and 90-kHz (seven sessions each). The dolphin's azimuth was 0°. (From Renauld and Popper 1978.)

the vertical plane nearly as well as in the horizontal plane. This was unexpected since binaural effects (interaural time or intensity differences) are not present in the vertical plane. The dolphin's ability to localize sounds in the vertical plane may be explained in part by the asymmetry in the vertical receiving beam patterns (Fig. 3.15). If the dolphin was able to rock its jaws slightly in the vertical plane, it would hear larger intensity variations for a sound approaching from a direction above the jaw than if the sound approached from below the jaw. Although Renaud and Popper (1975) used a vertical bite plate

station, the animal could conceivably still make the necessary movements. Another factor contributing to the relatively small MAA measured in the vertical plane may have been unforeseen and unwanted interference by the Plexiglas sheet used as vertical bite plate. Unlike the bite plate used by Au and Moore (1984), which was shaped to fit the dolphin's jaw (Fig. 3.14), Renaud and Popper (1975) used a Plexiglas sheet that extended at least 10 cm beyond the jaw on either side. The animal's geometry for their measurements in the vertical plane is shown schematically in Figure 3.24. It is possible that the plexiglass was not acoustically transparent and signals coming from above the animal's head might have been shadowed slightly by the bite plate. If shadowing occurred, then a signal from above the head would have been less intense than a signal approaching from below, resulting in an intensity difference cue. It would be prudent to continue investigation of the vertical localization capability of the dolphin.

Renaud and Popper (1975) also determined the MAA for a broadband transient signal or click signal, with a peak frequency of 64 Khz, and presented to the dolphin at a repetition rate of 333 Hz. The MAA was found to be aproximately 0.9° in the horizontal plane and 0.7° in the vertical plane. It is not surprising that a broadband signal should result in a lower MAA than a pure tone signal of the same frequency as the peak frequency of the click. The short onset time and

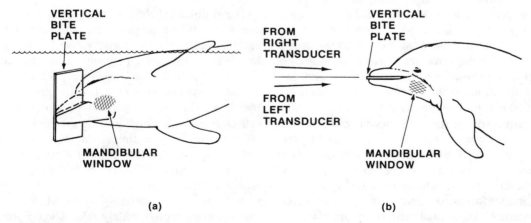

Figure 3.24. Schematic drawings of experimental geometries for the MAA measurements in the vertical plane by Renaud and Popper (1978). (*A*) oblique 3-D view; (*B*) side view.

the broad frequency spectrum of a click signal should provide additional cues for localization.

3.4 Response Bias and Sensitivity in Hearing

The various auditory experiments discussed in this chapter have been performed using techniques from classical psychophysics. We conclude with a section examining hearing sensitivity from a slightly different perspective. In Section 1.5 we discussed the fact that there is an alternative way to study the sensory capabilities of animals, that is, through Signal Detection Theory (SDT). We discussed the notion that a dolphin's performance in detecting signals may be influenced by nonsensory biasing factors such as motivation and expectation, and that SDT provides a method to separate sensitivity and response bias. Therefore, instead of a specific sensory threshold, the parameters d' and β are measured. Unfortunately, there has been very little research performed in SDT with dolphins. The purpose of this section is to emphasize that the auditory capabilities of dolphins can indeed be studied using SDT, and to examine the results of a hearing experiment using this technique.

Schusterman and his colleagues were the first to apply SDT to marine mammals involved with passive hearing tasks (Schusterman 1974). They experimented with different techniques to vary the response bias of their subjects by manipulating nonsensory variables, such as the signal presentation probability (Schusterman and Johnson 1975), the food reinforcement payoff matrix (Schusterman et al. 1975), and the probability of fish reinforcement for the two classes of correct responses—correct detections and correct rejections (Schusterman 1976). An animal can be manipulated to favor the "yes" or signal-present response (become more liberal) by having more signal-present than signal-absent trials, by providing correct detections with greater amounts of reward than correct rejections, and by having the probability of receiving a reward be greater for correct detections than for correct rejections. An animal can be manipulated to favor the "no" or signal-absent response (become more conservative) by having more signal-absent

than signal-present trials, by providing correct rejections with greater amounts of reward, and by having the probability of receiving a reward be higher for correct rejections than for correct detections. Although Schusterman and his colleagues performed most of their experiments with the California sea lion (*Zalophus californianus*), they did perform one response bias shaping experiment with a *Tursiops truncatus*.

Schusterman et al. (1975) performed a simple hearing experiment in Kaneohe Bay, Hawaii, requiring a *Tursiops* to station on a bite plate and respond in a go/no-go paradigm to an 8-kHz tone projected from a transducer 3.7 m away. The dolphin was trained to release the bite plate whenever it heard the stimulus and swim to a paddle. In stimulus-absent trials, the dolphin was required to remain at station until receiving a release tone. The sound pressure level of the signal was chosen to yield a d' value of approximately 1.0. The dolphin's response bias was manipulated by varying the payoff matrix for correct responses. The amount of reinforcement (number of fish) was varied in terms of the ratio of hits to correct rejections in the following way: 1:1, 1:4, and 4:1. The 1:1 ratio was considered as the baseline ratio, and was used in sessions immediately following testing with the other two payoff matrices. The results of the experiment are shown in the ROC plot (defined in Section 1.5) of Figure 3.25. The data points for the 1:4 and 4:1 payoff conditions represent 300 trials each. The point for the 1:1 condition represents 1,000 trials.

The dolphin's behavior was predictable and consistent. When the payoff matrix was 1:4, the animal was relatively conservative, showing a low false alarm rate along with a low hit rate. When the payoff matrix was 4:1, the dolphin became more liberal in reporting the presence of the signal, showing a relatively high false alarm rate coupled with a high hit rate. At the 1:1 payoff, the dolphin was almost unbiased, showing only a slight inclination toward a conservative response criterion. While the dolphin's response bias was being manipulated, its detection sensitivity remained relatively constant, with an average d' of approximately 1.32. The dolphin's performance results, including the probability of a correct response $P(C)$ (calculated with (1-33)),

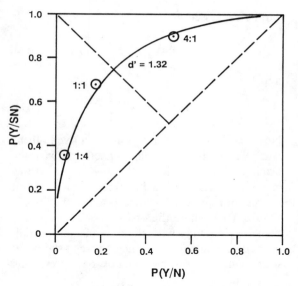

Figure 3.25. Effects of varying the payoff matrix for a dolphin's detection of an acoustic signal. (From Schusterman et al. 1975.)

Table 3.2. Results of dolphin hearing experiment conducted by Schusterman et al. (1975)

Payoff	$P(Y/n)$	$P(Y/sn)$	$P(C)$
1:4	0.04	0.36	0.66
1:1	0.18	0.68	0.75
1:4	0.52	0.90	0.69

are tabulated in Table 3.2. It is interesting to note how much $P(Y/sn)$ depended on the payoff ratio, which affected the animal's response bias; yet according to SDT, the dolphin's sensitivity was approximately constant for the three payoff conditions. $P(C)$ also showed some variation but not as much $P(Y/sn)$. There is a tendency for experimenters to train their animals to have a low false rate based on the notion that good stimulus control is thus being acquired and maintained. Unfortunately, extremely low false alarm rates sometimes lead to underestimates of an animal's sensitivity (Schusterman 1974).

An understanding of the important relationship just illustrated between nonsensory response bias and sensitivity measures is helpful in understanding the various sensory sensitivities dis-cussed in this chapter. We should be cautious and not naively accept sensitivity estimates without an awareness of the methodology and paradigm used in the behavioral experiment in question. Not least, this section is intended as a source of encouragement for the use of SDT methodology in estimating the sensitivity of the various sensory processes of dolphins.

3.5 Summary

Dolphins respond to pure tone signals in a similar manner as humans and other terrestrial mammals, the major difference being the wide frequency range of sensitivity, stretching over seven octaves. The audiogram of *Tursiops* has the same shape as a human audiogram but is shifted to higher frequencies by a factor of 10. The dolphin seems to integrate pure tone signals in a similar fashion as humans, having similar integration time constants for frequencies within the human auditory range. Spectral filtering in the dolphin's auditory system also seems to function similarly as in humans, with the dolphin critical ratio appearing like an extension of the human critical ratio to higher frequencies. The spectral filtering property of the dolphin ear can be modeled by a bank of contiguous constant-Q filters, as for humans. Other hearing characteristics that are similar for dolphins and humans include frequency discrimination and sound localization capabilities: curves of the dolphin frequency discrimination capability as a function of frequency have a similar shape and show similar sensitivity as the one for humans, but are shifted to higher frequencies by a factor of about 5 to 10; and finally, dolphins can localize sound in three-dimensional space about as well as humans.

The functional similarities between the auditory systems of dolphins and humans, as well as other terrestrial mammals, provide indirect evidence for the sensitive and sophisticated nature of the dolphin auditory system. The human auditory system, including its processing centers, is an extremely sensitive, sophisticated, and delicate instrument. Humans can detect, recognize, and process extremely weak sounds in noisy and reverberant environments much better than any instrument yet developed. The superiority in performance of the human auditory system over any man-made instruments is probably paralleled by the dolphin's superiority over any man-made passive underwater acoustic detection and recognition system.

Appendix: Derivation of the Receiving Directivity Index

The acoustic pressure for the two–point source array at the observation point in Figure 3.19A is the sum of the pressure from both point sources,

$$p(\alpha, \beta) = A e^{j\omega t}\left(\frac{e^{-jkr_1}}{r_1} + \frac{e^{-jkr_2}}{r_2}\right) \quad \text{(A-1)}$$

where r_1 and r_2 are the distances from the observation point to each point source, and k is the wave number

$$k = \frac{2\pi}{\lambda} \quad \text{(A-2)}$$

The wavelength λ is equal to c/f, where c is the velocity and f is the frequency of the acoustic signal. We will restrict our analysis to the far field, assuming that $r \gg L$, so that the lines from the two point sources and from the origin to the observation point are essentially parallel. Our analysis will be simplified if we consider the case in which the observation point is in the x–z plane of Figure 3.19A so that we can use the geometry shown in Figure A.1A. In this case, the distance from the point sources to the observation point

can be expressed as

$$r_1 = r + \Delta l = r + \frac{L}{2}\sin\beta \quad \text{(A-3)}$$

and

$$r_2 = r - \Delta l = r - \frac{L}{2}\sin\beta \quad \text{(A-4)}$$

Now, let the observation point be moved off the x–z plane so that the line connecting it with the origin makes an angle α with the y-axis, as shown in Figure A.1B. The dashed lines from the point sources in Figure A.1B correspond to the lines in Figure A.1A where the observer was in the x–z plane. By analogy with the case of the observer in the x–z plane, the distance between the point sources and the observer can now be expressed as

$$r_1 = r + \Delta r = r + \frac{L}{2}\sin\alpha\sin\beta \quad \text{(A-5)}$$

and

$$r_2 = r - \Delta r = r - \frac{L}{2}\sin\alpha\sin\beta \quad \text{(A-6)}$$

Since $r \gg L$, we can replace r_1 and r_2 in the denominator of (A-1) by r. Let

$$\psi_L = \frac{kL}{2}\sin\alpha\sin\beta \quad \text{(A-7)}$$

Insert (A-5) and (A-6) into (A-1) and use the identity of (A-7) to derive

$$p(\alpha, \beta) = \frac{2A}{r}e^{j(\omega t - kr)}\cos\psi_L \quad \text{(A-8)}$$

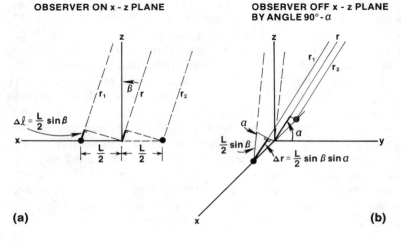

OBSERVER ON x - z PLANE

OBSERVER OFF x - z PLANE BY ANGLE 90° - α

$\Delta l = \frac{L}{2}\sin\beta$

$\frac{L}{2}\sin\beta$

$\Delta r = \frac{L}{2}\sin\beta\sin\alpha$

(a)

(b)

Figure A.1. (*A*) Geometry of the two–point source array with the observer in the x–z plane; parallel lines connect the point sources with the observer. (*B*) Geometry with the observer off the x–z plane so that the lines connecting the point sources with the observer are at an angle α with respect to the y-axis.

where the following trigonometric identity was used

$$\cos \psi_L = \frac{e^{j\psi_L} + e^{-j\psi_L}}{2} \qquad \text{(A-9)}$$

It is clear that the pressure is at a maximum when the observer is in the y–z plane since the distances r_1 and r_2 will be equal and the pressures from the two point sources add in phase. In this plane $\beta = 0$, so that from (A-7) and (A-8) the maximum pressure at a distance r is

$$p_0 = \frac{2A}{r} e^{j(\omega t - kr)} \qquad \text{(A-10)}$$

Therefore, the beam pattern for the two–point source array can be expressed as

$$\frac{p(\alpha, \beta)}{p_0} = \cos \psi_L \qquad \text{(A-11)}$$

Using the trigonometric identity $\sin 2\psi = 2 \sin \psi \cos \psi$, we can derive an alternative expression for the beam pattern:

$$\left(\frac{p(\alpha, \beta)}{p_0} \right)_{array} = \frac{\sin 2\psi_L}{2 \sin \psi_L} \qquad \text{(A-12)}$$

Let us now examine the rectangular aperture of Figure 3.19B. The pressure in the far field can be derived by dividing the aperture into infinitesimal elements, with each element of area $dxdz$ being an omnidirectional source of strength $A \cdot dxdz$, where A is the pressure per unit area on the face of the aperture. The pressure can be expressed as

$$p(\alpha, \beta) = A \iint \frac{e^{(\omega t - kr')}}{r'} dxdy \qquad \text{(A-13)}$$

We will again restrict our analysis to the far field, assuming that $r \gg d$ and w, so that the lines in Figure A.2 from the infinitesimal area $dxdy$ and the origin to the observation point are essentially parallel. With the observer on the x–z plane, the distance from the infinitesimal area $dxdy$ in Figure A.2 to the observer can be expressed as

$$r' = r - \Delta l = r - z \cos \beta + x \sin \beta \qquad \text{(A-14)}$$

Now, if the observer is moved off the x–z plane so that a line between it and the origin makes an angle α with the y-axis as shown in Figure A.2B, the distance between the observer and the infinitesimal area can be expressed as

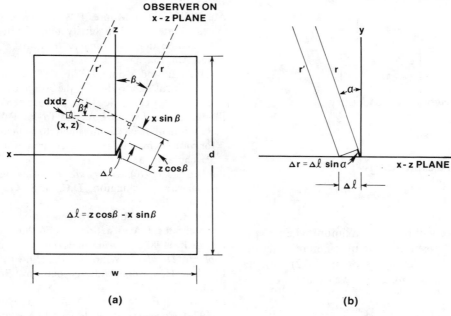

Figure A.2. Geometry of the rectangular aperture (A) with the observer in the x–z plane and (B) with the observer off the x–z plane and the line from the origin to the observer making an angle α with the y-axis.

$$r' = r - \Delta l \sin \alpha$$

$$= r - z \cos \beta \sin \alpha + x \sin \beta \sin \alpha \quad \text{(A-15)}$$

Substituting r' in (A-15) into (A-13) results in

$$p(\alpha, \beta) = \frac{A}{r} e^{j(\omega t - kr)} \int_{-w/2}^{w/2} e^{-jx \sin \beta \sin \alpha} \, dx$$

$$\cdot \int_{-d/2}^{d/2} e^{jz \cos \beta \sin \alpha} \, dz \quad \text{(A-16)}$$

where r' in the denominator of (A-13) was replaced by r. Performing the integration, we obtain

$$p(\alpha, \beta) = \frac{A}{r} e^{j(\omega t - kr)} \frac{e^{-jx \sin \beta \sin \alpha}}{-j \sin \beta \sin \alpha} \Big|_{-w/2}^{w/2}$$

$$\cdot \frac{e^{jz \cos \beta \sin \alpha}}{j \cos \beta \sin \alpha} \Big|_{-d/2}^{d/2} \quad \text{(A-17)}$$

Letting

$$\psi_w = \frac{kw}{2} \sin \alpha \sin \beta \quad \text{(A-18)}$$

and

$$\psi_d = \frac{kd}{2} \sin \alpha \cos \beta \quad \text{(A-19)}$$

(A-17) can be expressed as

$$p(\alpha, \beta) = \frac{Adw}{r} e^{j(\omega t - kr)} \left[\frac{e^{j\psi_w} - e^{-j\psi_w}}{2j\psi_w} \right]$$

$$\cdot \left[\frac{e^{j\psi_d} - e^{-j\psi_d}}{2j\psi_d} \right] \quad \text{(A-20)}$$

Using the trigonometric identity

$$\sin \psi = \frac{e^{j\psi} - e^{-j\psi}}{2j} \quad \text{(A-21)}$$

(A-20) can be expressed as

$$p(\alpha, \beta) = \frac{Adw}{r} e^{j(\omega t - kr)} \frac{\sin \psi_w}{\psi_w} \frac{\sin \psi_d}{\psi_d} \quad \text{(A-22)}$$

The pressure will be at a maximum when $\alpha = 0$. Applying L'Hospital's rule by differentiating the numerator and denominator of (A-22) with respect to ψ and letting $\alpha = 0$, we obtain

$$p_0 = \frac{Adw}{r} e^{j(\omega t - kr)} \quad \text{(A-23)}$$

Dividing (A-23) into (A-22), the beam pattern for

a rectangular aperture can be expressed as

$$\left[\frac{p(\alpha, \beta)}{p_0} \right]_{rect} = \frac{\sin \psi_w}{\psi_w} \frac{\sin \psi_d}{\psi_d} \quad \text{(A-24)}$$

Finally, the beam pattern for an array of two rectangular apertures can be derived by inserting (A-12) and (A-24) into (3-13) and then inserting the resultant equation into (3-10) to get

$$\text{DI} =$$

$$10 \log \frac{4\pi}{\int_0^{2\pi} \int_0^{\pi/2} \left(\frac{\sin 2\psi_L}{2 \sin \psi_L} \frac{\sin \psi_w}{\psi_w} \frac{\sin \psi_d}{\psi_d} \right)^2 \sin \alpha \sin \beta \, d\alpha \, d\beta}$$

$$\text{(A-25)}$$

where

$$\psi_L = \frac{kL}{2} \sin \alpha \sin \beta$$

$$\psi_w = \frac{kw}{2} \sin \alpha \sin \beta$$

$$\psi_d = \frac{kd}{2} \sin \alpha \cos \beta$$

References

Anderson, S. (1970). Auditory sensitivity of the harbor porpoise, *Phocoena phocoena*. Invest. Cetacea 2: 255–259.

Au, W.W.L., and Moore, P.W.B. (1984). Receiving beam patterns and directivity indices of the Atlantic bottlenose dolphin *Tursiops truncatus*. J. Acoust. Soc. Am. 75: 255–262.

Au, W.W.L., and Moore, P.W.B. (1990). Critical ratio and critical bandwidth for the Atlantic bottlenose dolphin. J. Acoust. Soc. Am. 88: 1635–1638.

Blodgett, H.C., Jeffress, L.A., and Taylor, R.W. (1958). Relation of masked threshold to signal-duration for various interaural phase-combinations. Am. J. Psychol. 71: 283–290.

Bobber, R.J. (1970). *Underwater Electroacoustic Measurements*. Washington, D.C.: U.S. Government Printing Office.

Fay, R.R. (1974). Auditory frequency discrimination in vertebrates. J. Acoust. Soc. Am. 56: 206–209.

Fay, R.R. (1988). *Hearing in Vertebrates: A Psychophysics Databook*. Winnetka, Ill.: Hill-Fay Assoc.

Fletcher, H. (1940). Auditory patterns. Rev. Mod. Phys. 12: 47–65.

Forsythe, F.E., Malcolm, M.A., and Moler, C.B. (1977). *Computer Methods for Mathematical Computations*. Englewood Cliffs, N.J.: Prentice-Hall.

French, N.R., and Steinberg, J.C. (1947). Factors governing the intelligibility of speech sound. J. Acoust. Soc. Am. 19: 90–119.

Hall, J.D., and Johnson, C.S. (1971). Auditory thresholds of a killer whale. J. Acoust. Soc. Am. 51: 515–517.

Hamilton, P.M. (1957). Noise masked thresholds as a function of tonal duration and masking band width. J. Acoust. Soc. Am. 29: 506–511.

Hawkins, J.E., and Stevens, S.S. (1950). The masking of pure tones and speech by white noise. J. Acoust. Soc. Am. 22: 6–13.

Herman, L.M., and Arbeit, W.R. (1972). Frequency difference limens in the bottlenose dolphin: 1–70 kHz. J. Aud. Res. 12: 109–120.

Hughes, J.W. (1946). The threshold of audition for short periods of stimulation. Proc. Roy. Soc. (London) B133: 486–490.

Jacobs, D.W. (1972). Auditory frequency discrimination in the Atlantic bottlenose dolphin, *Tursiops truncatus* Montagu: a preliminary report. J. Acoust. Soc. Am. 53: 696–698.

Jacobs, D.W., and Hall, J.D. (1972). Auditory thresholds of a fresh water dolphin, *Inia geoffrensis* Blainville. J. Acoust. Soc. Am. 51: 530–533.

Johnson, S.C. (1967). Sound detection thresholds in marine mammals. In: W. Tavolga, ed., *Marine BioAcoustics*. New York: Pergamon Press, pp. 247–260.

Johnson, S.C. (1968a). Relation between absolute threshold and duration of tone pulse in the bottlenosed porpoise. J. Acoust. Soc. Am. 43: 757–763.

Johnson, S.C. (1968b). Masked tonal thresholds in the bottlenosed porpoise. J. Acoust. Soc. Am. 44: 965–967.

Johnson, S.C. (1971). Auditory masking of one pure tone by another in the bottlenose porpoise. J. Acoust. Soc. Am. 49: 1317–1318.

Johnson, S.C. (1986). Dolphin audition and echolocation capacities. In: *Dolphin Cognition and Behavior: a Comparative Approach*, R.J. Schusterman, J.A. Thomas and F.G. Wood, eds., Hillsdale, N.Y.: Lawrence Erlbaum Associates, pp. 115–136.

Johnson, S.C., McManus, M.W., and Skaar, D. (1989). Masked tonal thresholds in the belukha whale. J. Acoust. Soc. Am. 85: 2651–2654.

Kellogg, W.N., Kohler, R., and Morris, H.N. (1953). Porpoise sounds as sonar signals. Science 117: 239–243.

Ljungblad, D.K., Scoggins, P.D., and Gilmartin, W.G. (1982). Auditory thresholds of a captive eastern Pacific bottle-nosed dolphin, *Tursiops* spp. J. Acoust. Soc. Am. 72: 1726–1729.

McCormick, J.M., and Salvadori, M.G. (1964). *Numerical Methods in FORTRAN*. Englewood Cliffs, N.J.: Prentice-Hall.

Mills, A.W. (1958). On the minimum audible angle. J. Acoust. Soc. Am. 30: 237–246.

Mills, A.W. (1972). Auditory Localization. In: J.V. Tobias, ed., *Foundations of Modern Auditory Theory*, Vol 2. New York: Academic Press, pp. 303–348.

Patterson, R.D., and Moore, B.C.J. (1986). Auditory filters and excitation patterns as representations of frequency resolution. In: B.C.J. Moore, ed., *Frequency Selectivity in Hearing*. New York: Academic Press, pp. 123–177.

Plomp, R., and Bouman, M.A. (1959). Relation between hearing threshold and duration for tone pulses. J. Acoust. Soc. Am. 31: 749–758.

Rayleigh, Lord. (1907). Our perception of sound direction, Philosophical Magazine 13, 214–232.

Renaud, D.L., and Popper, A.N. (1975). Sound localization by the bottlenose porpoise *Tursiops truncatus*. J. Exp. Biol. 63: 569–585.

Scharf, B. (1970). Critical bands. In: *Foundation of Modern Auditory Theory*. Vol 1, J.V. Tobias, ed., N.Y.: Academic Press, pp. 159–202.

Schevill, W.E., and Lawrence, B. (1953). Auditory response of a bottle-nosed porpoise, *Tursiops truncatus*, to frequencies above 100 kc. J. Exper. Zool. 124: 147–165.

Schusterman, R.J. (1974). Low false-alarm rates in signal detection by marine mammals. J. Acoust. Soc. Am. 55: 845–848.

Schusterman, R.J. (1976). California sea lion underwater auditory detection and variation of reinforcement schedules. J. Acoust. Soc. Am. 59: 997–1000.

Schusterman, R.J., and Johnson, B.W. (1975). Signal probability and response bias in California sea lions. Psychol. Rec. 25: 39–45.

Schusterman, R.J., Barrett, R., and Moore, P.W.B. (1975). Detection of underwater signals by a California sea lion and a bottlenose porpoise: variation in the payoff matrix. J. Acoust. Soc. Am. 57: 1526–1532.

Shower, E.G., and Biddulph, R. (1931). Differential pitch sensitivity of the ear. J. Acoust. Soc. Am. 3: 275–287.

Sivian, L.J., and White, S.D. (1933). On minimum audible sound fields. J. Acoust. Soc. Am. 4: 288–321.

Thomas, J., Chun, N., Au, W., and Pugh, K. (1988). Underwater audiogram of a false killer whale (*Pseudorca leucas*). J. Acoust. Soc. Am. 84: 936–940.

Thompson, R.K.R., and Herman, L.M. (1975). Underwater frequency discrimination in the bottlenose dolphin (1–140 kHz) and the human (1–8 kHz). J. Acoust. Soc. Am. 57: 943–948.

Wegel, R.L., and Lane, C.E. (1924). The auditory masking of one pure tone by another and its probable relation to the dynamics of the inner ear. Phys. Rev. 23: 266–285.

White, M.J. Jr., Norris, J., Ljungblad, D., Baron, K., and di Sciara, G. (1978). Auditory thresholds of two beluga whales (*Delphinapterus leucas*). HSWRI Techn. Rep. No. 78–109, Hubbs Marine Research Institute, San Diego, Cal.

Zaytseva, K.A., Akopian, A.I., and Morozov, V.P. (1975). Noise resistance of the dolphin auditory analyzer as a function of noise direction. BioFizika 20: 519–521.

Zwislocki, J. (1960). Theory of temporal auditory summation. J. Acoust. Soc. Am. 32: 1046–1060.

4

Characteristics of the Receiving System for Complex Signals

The discussion of the dolphin's receiving system so far has dealt primarily with relatively simple sinusoidal signals. In this section, we will discuss characteristics of the dolphin's receiving system for complex sounds. Any broadband, nonsinusoidal sound, such as a short click, a combination of clicks, or broadband noise, is defined as a complex sound. Since we will not discuss the characteristics of dolphin sonar pulses until Chapter 5, it is sufficient to state at this time that dolphins typically echolocate with short duration broadband transient-like pulses or clicks. In fact, the experiments that will be discussed in this section were performed to answer questions concerning various facets of the dolphin's echolocation capabilities; however, since they were hearing or passive sonar experiments and measured the characteristics of the animals' receiving system, it is appropriate to present them here.

Vel'min and Dubrovskiy (1975, 1976) have coined the phrase "active hearing" to emphasize the possibility that dolphins may process long duration, tone-like signals differently than broadband short duration transient signals associated with the animals' active sonar system. They believe that dolphins may have two functionally independent auditory subsystems, one for long duration non-echolocation signals such as emotional and communication sounds (whistles) and the other for short duration click signals used in

active echolocation. Bel'kovich and Dubrovskiy (1976) offer a good summary of some of the research performed in the former Soviet Union. For our purposes, no differentiation will be made between "passive" and "active" hearing because the nomenclature is misleading. In the sonar field, an active sonar is one that transmits signals and receive echoes. A passive sonar is one that only receives, or "listens" to, sounds created by external sources. By contrast, Vel'min and Dubrovskiy (1975, 1976) associate active hearing with the reception of broadband pulses or clicks having similar properties as typical biosonar pulses, regardless of whether the dolphin is echolocating or not, and passive hearing with the reception of long duration, tone-like signals. I am of the opinion that this distinction is artificial at best since "hearing is hearing," even though the animal's auditory system may treat short, broadband sounds differently from long, tonal sounds. It is possible, however, that in the process of emitting sonar signals, a dolphin's auditory system may become more sensitive to the reception, recognition, and classification of broadband echoes or other sounds. It is also possible that a dolphin may be able to focus its attention on the reception of echoes in a manner that enhances its auditory processing capabilities. Thus the sensitivity of the dolphin's auditory system may vary slightly depending on whether or not the animal is using its active sonar.

4.1 Perception of Click Signals

4.1.1 Temporal Summation

Vel'min et al. (1975a) studied the temporal summation processes of dolphins by measuring their detection sensitivity to click signals as a function of the repetition rate. The waveform and frequency spectrum of the click signal used is shown in Figure 4.1. The signal had a very short duration of about 40 μs with a peak frequency close to 92 kHz. The signal was produced by means of impact excitation of a spherical piezoelectric crystal transducer with a unipolar pulse lasting about 7 μs. The experimental geometry consisted of a tank with a transducer on either side of a dividing net (Fig. 4.2); the length of the net varied from 6.7 to 7.4 m. On any particular trial, only one transducer would transmit a signal and the dolphin was required to indicate which one by swimming to the side of the net where the transmitting transducer was located. The dolphin was rewarded with fish for making a correct response and not rewarded when it made an incorrect response. The choice of which transducer would be transmitting was randomized and the amplitude of the signal for a particular pulse repetition rate was progressively reduced until a threshold, defined as the 75% level of correct responses, was achieved.

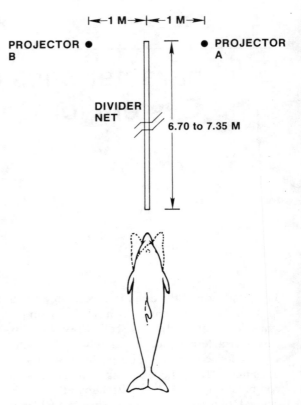

Figure 4.2. Experimental configuration for temporal summation and other studies conducted by Vel'min and his colleagues.

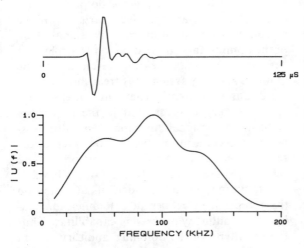

Figure 4.1. Click signal simulating an echolocation signal used by Vel'min and et al. (1975a). (From Bel'kovich and Dubrovskiy 1976.)

The results from experiments measuring the change in the perception of clicks by bottlenose dolphins as a function of the repetition rate are shown in Figure 4.3, from Dubrovskiy (1990). Dubrovskiy modified the original figure presented by Vel'min and Dubrovskiy (1975) by including the results of Zanin et al. (1977). He also considered the results from these two dolphin experiments in a composite manner and curve-fitted the data from the three dolphins involved with the solid line shown in Figure 4.3. From Dubrovskiy's point of view, any differences in the performance of the three dolphins could be attributed to normal variances typically observed in behavioral experiments. In contrast, Vel'min et al. originally emphasized the difference in the amount of threshold shift experienced by the dolphins they studied. The absolute threshold sensitivity for dolphins 1 and 2 at 1 Hz was approximately 77 dB and 82 dB re 1 μPa,

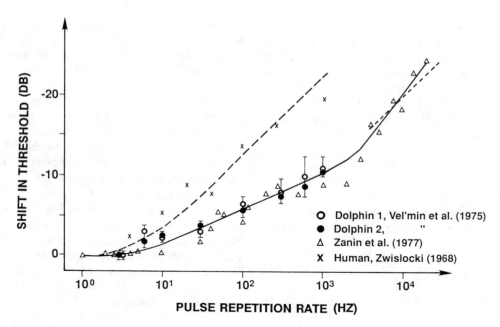

Figure 4.3. Shift in the relative detection threshold for click signals as a function of the signal repetition rate. (From Dubrovskiy 1990.)

respectively, probably calculated on a peak-to-peak basis. In the range of a repetition rate of 1 to 6 Hz, the shift in threshold was 12.6 dB for one dolphin and 8 dB for the other. Vel'min et al. (1975a) described the summation process of dolphin 1 as reaching the energetic level at which the energies of successive pulses are built up for a time that equals the summation time constant. For dolphin 2, the summation was close to coherent, implying that the individual pulses were first summed coherently and then squared and integrated in order to determine their energy. Dubrovskiy's interpretation of the dolphin data is more acceptable since there is no evidence that a mammal can process auditory information coherently as suggested by Vel'min et al. (1975a).

Zwislocki (1963) found that for click summation occurring with a time constant τ, the signal intensity I_v perceived by humans will increase according to

$$I_v = I_1/[1 - e^{-\frac{1}{v\tau}}] \qquad (4\text{-}1)$$

where I_1 is the intensity at a pulse repetition rate of 1 Hz. Equation (4-1) can be used to determine how much the intensity of a single click needs to be reduced if we want I_v to remain at the same

value as the repetition rate is increased. Invert (4-1) so that

$$I_1 = I_v[1 - e^{-\frac{1}{v\tau}}] \qquad (4\text{-}2)$$

If the repetition rate is high so that $v\tau \gg 1$, the exponential expanded in a power series will be

$$e^{-\frac{1}{v\tau}} \approx 1 - \frac{1}{v\tau} \qquad (4\text{-}3)$$

so that $I_1 \approx I_v/(v\tau)$. Therefore, the intensity of the single click must be increased by 3 dB per doubling of the repetition rate v. The dashed line through the human data in Figure 4.3 is a plot of (4-2) for $\tau = 200$ ms. As the pulse repetition rate increases from 1 to 100 Hz in the figure, the shift in the dolphin auditory threshold due to auditory temporal summation decreases (sensitivity increases) on the average by 1.2 dB per doubling of the repetition rate. This amount of summation is considerably less than energy summation of 3 dB per doubling of pulse repetition rate observed in humans (Zwislocki, 1963). As the pulse repetition rate increases beyond 300–400 Hz, the slope of the summation effect changes abruptly and the summation effect becomes close to that of energy summation. The short dashed line on the right side represents equation (4-2) for $\tau = 250$ μs. We shall see later in section 7.1 that dolphins typically echolocate with click repetition rates of 5 pulses per second (pps) for a target range of 120 m, 26 pps for a range of 10 m, and 50 pps for a range of 1 m. For repetition rates typically used to detect targets beyond a range of about 10 m, temporal summation will not significantly affect the sensitivity of active hearing and its resistance to noise in an echolocating dolphin (Dubrovskiy 1990). This implies that the detection threshold (signal-to-noise ratio at the detection threshold level) for an echolocating dolphin will not vary significantly (by at most 2 dB) for different target ranges beyond 10 m.

4.1.2 Critical Interval

Vel'min and Dubrovskiy (1975, 1976, 1978) performed a series of experiments using broadband pulse signals that simulated sonar signals to study how dolphins process sonar echoes. Their first experiment consisted of a forward masking study in which a higher amplitude click was used to mask a lower amplitude click having the same waveform. The time domain waveform of the stimuli is shown in Figure 4.4. The peak frequency of the waveform was not given and cannot be ascertained from the waveform since time scales were not included. The experimental geometry was similar to the one in Figure 4.2. One of the transducers projected a double click consisting of the masker followed by the signal click, and the other transducer projected just the masking click. The masking clicks were projected

in synchrony and a stimulus repetition rate of 30 Hz was used. The dolphin was required to swim to the side of the 7-m long net where the transducer projecting the masker plus signal was located. A masker level of 121 dB re 1 μPa, measured at the end of the net nearest the dolphin, was used, and for a specific delay time τ, the amplitude of the signal was progressively decreased until a threshold was reached. A randomized schedule was used to choose the transducer projecting the masker plus signal.

The results of the forward masking experiment for two dolphins are shown in Figure 4.5, with the shift in threshold plotted as a function of the delay time between the masker and the signal. The threshold obtained for a repetition rate of 30 Hz in Figure 4.3 was used as the 0-dB reference level here. Apparently Vel'min and his coworkers used the same bottlenose dolphins in both experiments, and the threshold was probably defined as the 75% correct response level. For $\tau \geq 500$ μs, the masker did not affect the detection of the signal. But as the delay decreased from 500 μs to 250 μs, the threshold shifted by 10 dB for dolphin 1 and by 12 dB for dolphin 2. As the delay decreased to 100 μs, the masking threshold rose to 23–26 dB. Upon further decrease in the

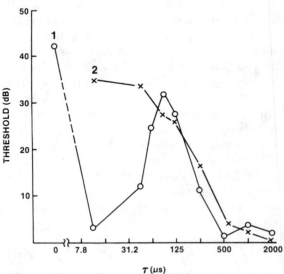

Figure 4.5. Shift in threshold as a function of the delay time between the masker and the signal. (From Vel'min and Dubrovskiy 1975.)

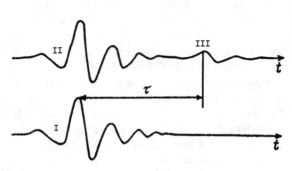

Figure 4.4. Time domain waveform of the stimuli used in the forward masking experiment of Vel'min and Dubrovskiy (1975). Click III is the signal; clicks II is the masker. (From Bel'kovich and Dubrovskiy 1976.)

delay time, the threshold of dolphin 1 first shifted back toward the unmasked level, reaching it at $\tau = 10$ μs, and then shifted up toward higher levels. Apparently dolphin 1 was able to separate the masker from the signal at delays between 10 and 100 μs. By contrast, the threshold for dolphin 2 continued to rise as the delay decreased from 500 μs to 10 μs. Vel'min and Dubrovskiy (1975, 1976) did not elaborate on the dip in the curve of dolphin 1 except for stating that dolphin 1 solved the problem of differentiating between the merged pair of pulses for delays between 10 and 125 μs. The performance of dolphin 1 in this delay region will be discussed further in Section 4.2, where we will offer a suggestion on how the dolphin solved the problem.

A significant shift in the dolphins' performance shown in Figure 4.5 occurred between delay times of 250 and 500 μs. Over this time interval the threshold shifted by approximately 10 dB. For delays greater than 500 μs the masker did not affect the perception of the signal. As the delay time decreased below 500 μs, however, the marker began to affect the detection of the signal. Vel'min and Dubrovskiy (1975) called the interval between 250 and 500 μs the "*critical interval*," which they defined as a measure of the temporal resolving power for click signals, or a measure of the interval in which echo components (multiple clicks) merge into a single auditory pattern (Vel'min and Dubrovskiy 1976). Moore et al. (1984) performed a backward masking sonar target detection experiment using noise bursts that arrived at the dolphin at various intervals after the target echo. They found that the 70% correct response performance occurred when the masking noise followed the echo by 260 μs. Au et al. (1988) measured the integration time of a dolphin's detector while performing a target detection task. They obtained an integration time of 264 μs. The backward masking interval of Moore et al. (1984), the integration time of Au et al. (1988), and the critical interval of Vel'min and Dubrovskiy (1975) may all be related to the same auditory phenomenon, the integration time of the dolphin's receiver (Au et al. 1988). The experiments of Moore et al.(1984) and Au et al. (1988) will be discussed in detail in Chapter 8, where the active sonar capabilities of dolphins are considered.

Table 4.1. Mean and standard deviation of the absolute time difference threshold ($\Delta\tau$) and relative time difference threshold in percent ($\Delta\tau/\tau \times 100\%$) for different standard time intervals

	Mean interval (τ)		
	50 μs	100 μs	200 μs
$\Delta\tau$ (μs)	4.8 ± 0.8	9.3 ± 1.7	11.5 ± 2.4
$\Delta t/\tau \times 100\%$	9.6 ± 1.6	9.4 ± 1.7	5.7 ± 1.2

In a second experiment, Vel'min and Dubrovskiy (1975, 1976) measured the time resolution capabilities of a dolphin. Pairs of clicks with different interclick intervals were emitted at a repetition rate of 30 Hz from both transducers on either side of the net (Fig. 4.3). Dolphin 2 was required to select the transducer that emitted click pairs with the greater interval. Standard intervals were 50, 100, 200, and 500 μs. The time interval difference threshold was defined as the 75% correct response level. The results of the time resolution experiment with dolphin 2 are given in Table 4.1. The time interval difference thresholds were 4.8, 9.3, and 11.5 μs for standard intervals of 50, 100, and 200 μs, respectively. This translates to a relative time interval difference threshold of 9.6, 9.4, and 5.7%, respectively. For a standard interval of 500 μs, the dolphin could not be trained to discriminate any interval differences. Vel'min and Dubrovskiy (1975) concluded that a certain critical interval existed between 200 and 300 μs.

Vel'min and Dubrovskiy (1976) continued with another time interval experiment in which a double click signal with a standard interval of 100 μs was compared with another double click signal having a test interval that varied in increments from 150 to 400 μs. Using the same procedure as in the previous experiment, with the double clicks emitted at 30 Hz, they obtained the results shown in Figure 4.6, where percent correct response is plotted as a function of the delay time between clicks. The 90% confidence levels are shown as solid vertical lines. From this figure, using the 50% correct response level as their threshold, Vel'min and Dubrovskiy (1976) estimated a critical interval of 260 ± 25 μs. It will be shown in Chapter 6 that their critical interval is probably the integration time constant of the

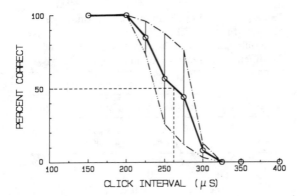

Figure 4.6. Results of time interval discrimination experiment in which a dolphin was required to distinguish the transducer that emitted double clicks with the standard 100-μs interclick interval from the transducer that emitted double clicks with different test intervals. (From Vel'min and Dubrovskiy, 1976.)

dolphin auditory system for broadband click signals. It may seem strange to have a dolphin perform at a 0% correct response level as in Figure 4.6. However, this is possible if the test interval is chosen randomly for each trial in a session, or if a small number of data points are collected for each test interval and the dolphin happens to be "locked" into a particular response. Otherwise, if an animal does not receive any reward for many consecutive trials it will probably begin prospecting or guessing and certainly improve on its performance.

4.1.3 Intensity Difference Discrimination

The ability of bottlenose dolphins to discriminate intensity differences in click signals was tested by Vel'min et al. (1975a), using once again the experimental configuration shown in Figure 4.2. Both transducers emitted a click signal at a repetition rate of 30 Hz, with one signal being of higher intensity than the other. The dolphins were required to swim toward the transducer that emitted the higher intensity signal, randomized from trial to trial. Several levels of intensity above the dolphins' auditory threshold shown in Figure 4.3 were used. The intensity discrimination threshold for two bottlenose dolphins, defined as the 75% correct response level, is given

Table 4.2. Intensity discrimination threshold in dB for click signals of different intensity (from Vel'min et al. 1975b)

Dolphin No.	Signal Level above Detection Threshold			
	5 dB	20 dB	36 dB	45 dB
1	2.5	2.1	2.1	1.3
2	1.7	0.8	0.7	

in Table 4.2 for different values of the high intensity stimulus. The dolphins were generally able to make finer intensity difference distinctions as the intensity increased. For signals 36 to 45 dB above the detection threshold, the average intensity difference threshold was about 1 dB.

4.1.4 Spatial Selectivity

Vel'min and Dubrovskiy (1977) examined the spatial selectivity of a bottlenose dolphin in detecting a click signal in the presence of two maskers also consisting of clicks. The configuration of their experiment is depicted in Figure 4.7, showing a dividing net with two transducers designated S and N on both sides of the net. The N transducers were used to transmit a masker each from a fixed location, 0.3 m from the net. Click signals were transmitted from one of the S transducers, chosen on a random basis from trial to trial. The S transducers were positioned symmetrically about the net, but their distance from the net was varied from session to session. The masker clicks and the signal clicks were transmitted simultaneously at a repetition rate of 30 Hz, each click having a peak frequency of approximately 80 kHz. The dolphin was required to swim to the side of the net corresponding to the transmitting S transducer. For each azimuth separation, the relative intensity ratio between signal and masker was varied and the dolphin's performance was recorded. The threshold was defined as the 75% correct performance level.

The dolphin's detection threshold as a function of the azimuth between the S and N transducers for the experiment of Vel'min and Dubrovskiy (1977) is shown as curve 1 in Figure 4.8. The reference threshold was taken when the azimuth separation between the S and N transducers was 0°. When the S and N transducers were separated

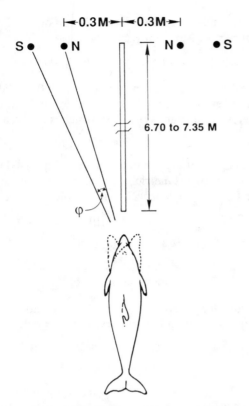

Figure 4.7. Configuration of the spatial selectivity experiment of Vel'min and Dubrovskiy (1977).

by 3.2°, the masking level decreased by 8 to 10 dB. When the separation azimuth was 15°, the masking level decreased by 30 dB. The spatial selectivity for pure tone signals was not nearly as good as for the click signal. The width of the curves 8 dB below the maximum was 6° for the click signal, 23° for the 120-kHz, and 40° for the 80-kHz tone pulses. Dubrovskiy (1990) interpreted the results of Figure 4.8 as further confirmation of the duplex hearing theory of Vel'min and Dubrovskiy (1975, 1976), in which short duration broadband click signals resembling echolocation emissions are processed differently from narrow-band tonelike signals. The narrow width of curve 1 may not be entirely due to the increased spatial sensitivity of a dolphin to click signals. It is possible that the dolphin treated the problem as a time-resolution task. The masker from transducer N and the click signal from transducer S were projected simultaneously so that there was a propagational-path difference between the animal and the transducers. Therefore, the task conveyed to the dolphin may have been to discriminate a double click from a single click. If the same was true for the 80- and 120-kHz tonal pulses used to obtain curves 2 and 3, the superior time resolution of a short duration

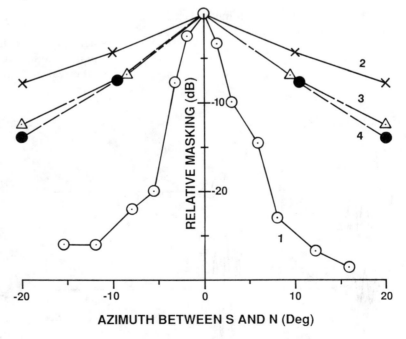

Figure 4.8. Masking of a click signal by an azimuthally spaced masker click versus angular separation of the signal and masker transducers; 1, click data from Vel'min and Dubrovskiy (1977); 2, 80-kHz pulse tone data from Zaytseva et al. (1975); 3, 120-kHz pulse tone data from Zaytseva et al. (1975); 4, directivity pattern for 80-kHz tone pulse from Akopian et al. (1977). (After Dubrovskiy 1990.)

click would have made the discrimination task easier with click signals than with pure tone signals. We can certainly state that if there was some kind of time difference cue involved in the experiments represented in Figure 4.8, the dolphin would have had an advantage with click signals over tonal signals.

Phase Difference or Temporal Order Discrimination?

Several techniques have been used to study the sensitivity of the human ear to phase spectra, involving stimuli with identical energy spectra but different phase spectra (Ronken 1970; Patterson and Green 1970). In studying monoaural detection of a phase difference between clicks in humans, Ronken (1970) made use of the fact that a waveform reversed in time has the same energy spectrum, but a different phase spectrum, as the unreversed waveform. One stimulus he used was a high amplitude click followed by a low amplitude click, and the other stimulus consisted of a low amplitude click followed by a high amplitude click. Consider the two stimuli denoted $s_1(t)$ and $s_2(t)$ in Figure 4.9. The stimuli can be expressed as

$$s_1(t) = s(t) + as(t - \tau) \qquad (4\text{-}4)$$

$$s_2(t) = as(t) + s(t - \tau) \qquad (4\text{-}5)$$

where $s(t)$ represents each individual click, a is an attenuation constant, and

$$s(t - \tau) = \begin{cases} 0 & \text{for } t < \tau \\ s(t) & \text{for } t \geq \tau \end{cases} \qquad (4\text{-}6)$$

is the time-shifted or delayed version of s(t). The Fourier transform of $s_1(t)$ in (4-4) is

$$\Im[s_1(t)] = S_1(f) = S(f) + a\,e^{-j2\pi f\tau}S(f) \qquad (4\text{-}7)$$

The Fourier transform of the time-shifted click was determined from the shift property in Fourier analysis (see Bracewell 1968), which is stated as

$$\Im[s(t - \tau)] = e^{-j2\pi f\tau}S(f) \qquad (4\text{-}8)$$

The exponential term in (4-8) can be expressed in terms of trigonometric functions (cf. (1-21)), and the Fourier transform $S(f)$ is complex so that

$$S_1(f) = [S_R(f) + j\,S_I(f)][1 + a\cos(2\pi f\tau)$$
$$- j\,a\sin(2\pi f\tau)] \qquad (4\text{-}9)$$

where $S_R(f) = $ real part of $S(f)$
          $S_I(f) = $ imaginary part of $S(f)$

Performing the complex multiplication and separating the real and imaginary parts, we obtain

$$S_1(f) = S_R(f)[1 + a\cos(2\pi f\tau)]$$
$$+ S_I(f)a\sin(2\pi f\tau)$$
$$+ j\{S_I(f)[1 + a\cos(2\pi f\tau)]$$
$$- S_R(f)a\sin(2\pi f\tau)\} \qquad (4\text{-}10)$$

Any complex function can be expressed in terms of its magnitude and phase, so

$$S_1(f) = |S_1(f)|e^{j\phi(f)} \qquad (4\text{-}10)$$

where $|S_1(f)|$ is the magnitude of $S_1(f)$ and is equal to the square root of the sum of the squares of the real and imaginary parts (cf. (1-22)), and

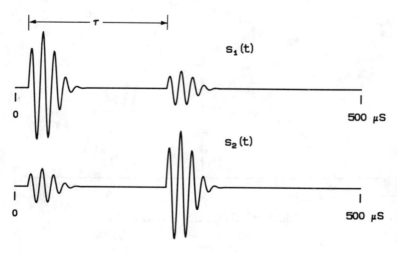

Figure 4.9. Example of acoustic stimuli with equal energy spectra but different phase spectra.

ϕ is the phase spectrum that is equal to the arctangent of the real part divided by the imaginary part. From (4-10), the magnitude and phase of $S_1(f)$ can be expressed as

$$|S_1(f)|$$
$$= \sqrt{[S_R(f)^2 + S_I(f)^2][1 + a^2 + 2a\cos(2\pi f\tau)]}$$
$$(4\text{-}12)$$

and

$$\phi_1(f) =$$
$$\tan^{-1}\frac{S_I(f)[1 + a\cos(2\pi f\tau)] - S_R(f)a\sin(2\pi f\tau)}{S_R(f)[1 + a\cos(2\pi f\tau)] + S_I(f)a\sin(2\pi f\tau)}$$
$$(4\text{-}13)$$

Now consider signal $s_2(t)$ in (4-5) and determine its Fourier transform:

$$S_2(f) = a\,S(f) + e^{-j2\pi f\tau}\,S(f) \qquad (4\text{-}14)$$

Performing the same algebraic operation as with $S_1(f)$ in (4-7), we obtain

$$|S_2(f)|$$
$$= \sqrt{[S_R(f)^2 + S_I(f)^2][1 + a^2 + 2a\cos(2\pi f\tau)]}$$
$$(4\text{-}15)$$

and

$$\phi_2(f) =$$
$$\tan^{-1}\frac{S_I(f)[a + \cos(2\pi f\tau)] - S_R(f)\sin(2\pi f\tau)}{S_R(f)[a + \cos(2\pi f\tau)] + S_I(f)\sin(2\pi f\tau)}$$
$$(4\text{-}16)$$

Note that the expressions for magnitude given in (4-12) and (4-15) are identical but that the phase terms in (4-13) and (4-16) are different.

Ronken (1970) required his human subjects to discriminate one stimulus from the other in a two-interval forced-choice experiment. He found that when the difference between the higher and lower amplitude clicks was greater than 2 to 3 dB, the subjects could easily discriminate between the two stimuli. He therefore concluded that humans can discriminate between very short wideband signals that differ only in phase.

Dubrovskiy et al. (1978) conducted an experiment to determine whether a bottlenose dolphin could discriminate the two signals $s_1(t)$ and $s_2(t)$ described by (4-4) and (4-5). For a particular delay time τ, the relative difference between the amplitude of the first and second clicks was varied and the animal's discrimination performance measured. The threshold was defined as the amplitude difference in dB at which the dolphin's performance was 75% correct. The threshold amplitude as a function of the delay time between signals $s_1(t)$ and $s_2(t)$ is shown in Figure 4.10. When $\tau < \tau_{\text{cr}}$, where τ_{cr} is the critical interval of Vel'min and Dubrovskiy, the difference in amplitude between the first and second click had to be at least 14.5 dB before the dolphin could discriminate the two sets of signals. As τ increased to 250 μs, the threshold decreased to 8 dB; it continued to decrease to 1.6 dB for $\tau = 500$ μs. Dubrovskiy (1990) concluded that for $\tau < \tau_{\text{cr}}$, the dolphin could not discriminate the

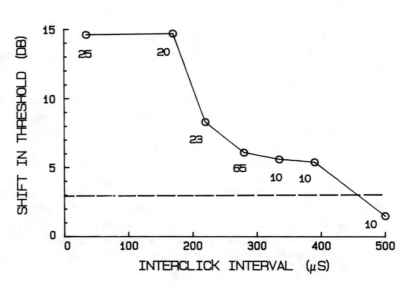

Figure 4.10. Threshold amplitude difference between the first and second clicks of the $S_1(t)$ and $S_2(t)$ stimuli for the dolphin to perform the discrimination task at the 75% correct response level. The number near each data point is the number of trials used to estimate the threshold. The dashed line denotes the DL for intensity. (From Dubrovskiy 1990.)

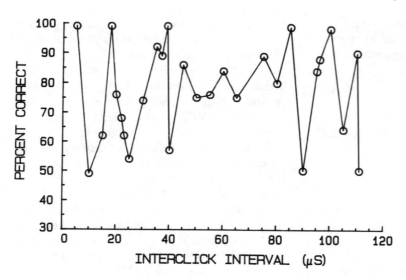

Figure 4.11. Discrimination of the $s_1(t)$ and $s_2(t)$ stimuli for a 10-dB difference in amplitude of the first and second clicks. An average of 50 trials was taken for each point. (From Dubrovskiy 1990.)

stimuli through their time profile or through differences in the phase spectra. For signals with $\tau > 250$ μs, the dolphin could readily discriminate the stimuli, but Dubrovskiy (1990) did not suggest the use of any particular cue.

The experiment for discriminating the $s_1(t)$ and $s_2(t)$ signals of (4-4) and (4-5) was continued with the amplitude difference between the first and second click fixed at 10 dB and τ varied. The dolphin's performance with the correct response percentage plotted as a function of τ is shown in Figure 4.11. The results exhibited a τ-dependent oscillation of 15-μs period. Dubrovskiy (1990) concluded that the dolphin was able to recognize a difference in the time profile features or the phase spectra even for a time separation less than τ_{cr}.

A similar experiment with an Atlantic bottlenose dolphin was performed by Johnson et al. (1988). Their underwater acoustic stimuli had a peak frequency of 60 kHz, with a 10-dB difference between large and small clicks, and were similar to the signals displayed in Figure 4.9. A two-interval forced-choice "same/different" procedure was used with a go/no-go response paradigm. At the beginning of a trial the dolphin stationed on a bite plate; at the appropriate moment the signals would be presented to the animal. The stimulus had a duration of 3.5 s which was divided into three intervals. During the first 1.5-s interval, the stimulus was always a $s_1(t)$ (large click followed by small click). The second interval

consisted of a 0.5-s silent period followed by a third 1.5-s interval containing a signal. In the third interval, either $s_1(t)$, the same signal as before, or $s_2(t)$ (small click followed by large click) was projected on a randomized schedule, each being used 50% of the time. If the signal in the third interval was the same—$s_1(t)$—as in the first interval, the dolphin was required to remain at station until the interval was over (no-go response). If the signal in the third interval was different—$s_2(t)$—the dolphin was required to leave the station before it was over and hit a response paddle (go response). A fish reward was presented after each correct response. The signals were presented to the dolphin at a repetition rate of 50 Hz.

The dolphin's ability to discriminate the signals was tested for two different separation times between the double clicks, 200 and 50 μs. The average correct response for the last 10 sessions of 30 trials each was $79\% \pm 8\%$ for the 200 μs separation time, and $78\% \pm 6\%$ for the 50 μs separation time. The dolphin performed just slightly above the discrimination threshold of 75% correct.

Johnson et al. (1988) interpreted their experiment in the same manner as Patterson and Green (1970), who suggested that the differences in temporal order, and not differences in the phase of the stimuli, allowed the subjects to perform the discrimination. The energy spectrum of a double click signal and its time-reversed version will

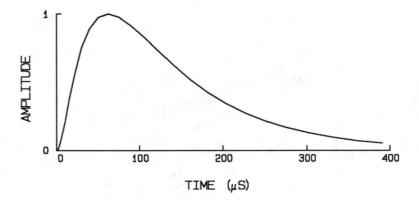

Figure 4.12. The chi-square analysis window used by Johnson et al. (1988).

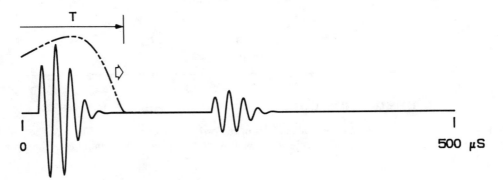

Figure 4.13. Pictorial description of a signal sliding past the chi-square analysis window.

only be perceived to be the same if the ear receives both clicks before performing any processing or analysis on the stimulus. In other words, in order for the energy spectrum of a double click signal to be the same as its time-reversed version, the signal must be completely received before it is processed. Humans (and probably dolphins) do not operate in such a manner, though. Rather, signals are processed or analyzed as soon as they are received. A short-time spectral analysis procedure (Schroeder and Atal 1962) would thus be more representative of how a mammalian auditory system processes signals (Patterson and Green 1970). In a short-time spectral analysis procedure, the received signal is shifted through an analysis window and processed at different time increments. Therefore the spectrum of the signal is continuously changing as the signal slides past the analysis window. Johnson et al. (1988) used a chi-square window that is depicted in Figure 4.12. The procedure of sliding the signal

past the analysis window is shown in Figure 4.13. Note that the window here is a time-reversed version of that in Figure 4.12. This representation is appropriate since the beginning part of the signal (part closest to 0 μs) should enter the beginning part of the window first, followed by the rest of the signal, as it does in Figure 4.13. The energy spectra for different values of T are shown for both signals in Figure 4.14. It is obvious that the energy spectra are considerably different for the two signals if a short-time spectrum analysis is performed. Johnson et al. obtained similar results with energy spectra calculated at 2-μs intervals; they plotted the results in a spectrogram format (frequency versus time), with the relative amplitude represented by different colors on a color computer monitor. The short-time spectral analysis of signal $s_1(t)$ indicates a high degree of rippling as the second click enters the chi-square window ($\tau \geq 300$ μs) and very little rippling for the $s_2(t)$ signal. Johnson et al. (1988) suggested

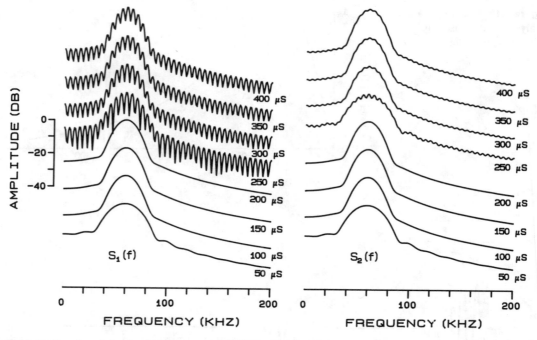

Figure 4.14. Results of the short-time spectral analysis of the two signals displayed in Fig. 4.7. The parameter in μs represents the varying position of the chi-square window with respect to the beginning of the signal.

that the dolphin probably used time separation pitch (a phenomenon that will be discussed in the next section), generated by the ripples in the frequency spectrum, to discriminate between the two signals. This interpretation of the discrimination results seems more plausible than the use of phase difference cues and is also applicable to the experiment of Dubrovskiy et al. (1978).

4.2 Perception of Time Separation Pitch

The reflection of sounds by submerged targets is a complex process which will be discussed in Chapter 7. For the present it is sufficient to know that the use of short clicks as sonar signals will often produce echoes with many highlights or echo components, each resembling the incident sonar signal. In Au and Hammer (1980) and Hammer and Au (1980) we suggested that dolphins may use *time separation pitch* (TSP) cues to discriminate targets. Thurlow and Small (1955) were the first to demonstrate that human subjects

could perceive a time separation pitch (TSP) of $1/T$ Hz when presented with two highly correlated broadband pulses separated by time T. McCellan and Small (1965) demonstrated that correlated noise bursts generated TSP in humans. We now know that any broadband auditory stimulus, including noise, added to a correlated delayed version of itself (delayed by time T) will produce a TSP of $1/T$ in human observers. The perceived pitch is considered a virtual pitch since no tonal signal is usually present at that frequency. The phenomenon of TSP has been referred to by many different names, such as periodicity pitch, infrapitch echo, and repetition pitch.

The stimulus used to generate TSP (a broadband signal summed with its delayed replica) has a rippled frequency spectrum which varies sinusoidally with frequency; the spacing between ripples is equal to $1/T$ (Johnson and Titlebaum 1976; Yost el al. 1978). In Section 4.1 we derived the frequency spectrum of a broadband signal summed with its delayed replica. From (4-11) the magnitude of the frequency spectrum is

Figure 4.15. Example of a click signal plotted in the time and frequency domains (solid line). The click delayed and added to itself in the time domain is also shown, along with its Fourier transform (dashed line).

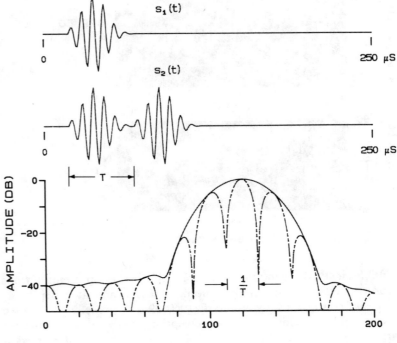

$$|\Im[s(t) + as(t - T)]|$$
$$= \sqrt{1 + a^2 + 2a\cos(2\pi fT)}\,|S(f)| \quad (4\text{-}17)$$

Consider as an example the click signal and its frequency spectrum shown in Figure 4.15. The click added to a delayed version of itself will have the frequency spectrum depicted by the dashed line. The ripples in the frequency spectrum are caused by the $\cos(2\pi fT)$ term in (4-17). From the equation we can see that the maxima, or peaks, in the spectrum occur when $\cos(2\pi fT) = 1$, or

$$f = (n - 1)\frac{1}{T} \qquad n = 1, 2, 3, \ldots \quad (4\text{-}18)$$

Minima, or troughs, in the frequency spectrum occur when $\cos(2\pi fT) = -1$, or

$$f = \frac{(2n - 1)}{2T} \qquad n = 1, 2, 3, \ldots \quad (4\text{-}19)$$

The strength of the perceived TSP depends on the peak-to-trough ratio (Yost and Hill 1978), which from (4-17) can be expressed as

$$k = \frac{1 + a}{1 - a} \quad (4\text{-}20)$$

It has a maximum value (infinity) for $a = 1$ and a minimum value (1) for $a = 0$ (no ripples). Often the amplitude of the delayed signal is expressed in dB relative to the first or original signal so that $a = 10^{-A/20}$, where A is the amount of attenuation in dB.

The delayed signal can also be subtracted from its original version; the total signal will still be a TSP stimulus with ripples in its frequency spectrum. In this case, an analogous expression to (4-17) is

$$|\Im[s(t) - as(t - T)]|$$
$$= \sqrt{1 + a^2 - 2a\cos(2\pi fT)}\,|S(f)| \quad (4\text{-}21)$$

The peaks will occur at $f = n/(2T)$ and the troughs at $f = (n - 1)/T$. The spectrum of equation (4-17) is referred to as a "cos+" signal and the spectrum of equation (4-21) is referred to as a "cos−" signal. The signal $s(t)$ need not be a click or pulse signal but can be any broadband

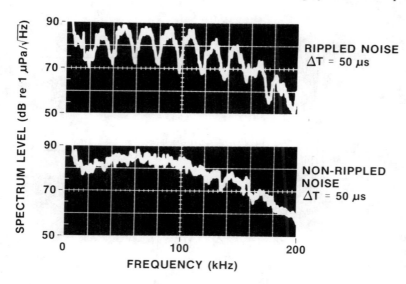

Figure 4.16. Examples of rippled and nonrippled noise stimuli measured with a spectrum analyzer and an H-52 hydrophone located at the hoop station. A delay of 50 μs was used to create the rippled spectrum; the spacing between ripples should be 1/50 μs, or 20 kHz. (From Au and Pawloski 1989.)

signal, for instance a continuous noise signal. The notion of echolocating animals utilizing TSP cues was first suggested by Normark (1960, 1961) as a possible mechanism for the perception of target range by an echolocating animal. Johnson and Titlebaum (1976) further expanded on this notion by considering the rippled energy spectrum associated with the total signal consisting of the transmitted signal and a specularly reflected echo. However, the idea of TSP as a distance-indicating mechanism for dolphins does not seem to be a viable notion when certain important factors are considered. The TSP associated with the summation of a transmitted signal and a target echo would be very low for realistic target ranges. For instance, a target at a range of 10 m would result in a TSP of approximately 76 Hz. It is doubtful that a dolphin can perceive low-level sounds (even virtual sounds) below a frequency of 100 Hz.

In human TSP experiments, subjects are required to match the sound from a tone generator to a perceived TSP. The concept of matching a tone to a perceived TSP by mechanically manipulating a signal generator seemed extremely difficult to convey to a dolphin and the procedure also seemed difficult to train. Therefore, Au and Pawloski (1989) used the simpler task of having a *Tursiops truncatus* discriminate between rippled and nonrippled noise in order to investigate the ability of dolphins to perceive TSP. The animal was required to station in a

hoop, facing a transducer 2.3 m from the hoop and at a depth of 1 m. A go/no-go response procedure was used in which the dolphin was required to leave the hoop and strike a response paddle if rippled noise was projected (go response) or remain in the hoop if nonrippled noise was projected (no-go response). Examples of rippled and nonrippled noise stimuli are shown in Figure 4.16 for a delay of 50 μs.

Three separate experiments were performed. In the first experiment, the dolphin's ability to discriminate rippled and nonrippled noise as a function of the delay time T of the delayed component of the noise signal was examined. Both $\cos +$ and a $\cos -$ stimuli were used, and the rippled and nonrippled noise were presented in a randomized schedule with an equal number of trials for each stimulus in each session. The dolphin's performance as a function of the delay time T is depicted in Figure 4.17 for the $\cos +$ and $\cos -$ signal. The dolphin performed above 90% correct for delay times between 50 and 400 μs, with an upper 75% correct response threshold at 500 μs and a lower 75% threshold at 15 μs for the $\cos -$ stimulus. The dolphin's perception of the $\cos +$ stimulus did not cover as broad a range of delay times as that of the $\cos -$ stimulus. The upper delay time threshold was approximately 190 μs and the lower delay time threshold was 13 μs. These delay times correspond to rippled intervals of 2 and 67 kHz for the $\cos -$ stimulus and 5 and 77 kHz for the $\cos +$ stimulus. Humans

Figure 4.17. Dolphin's performance in discriminating between rippled and nonrippled noise as a function of the delay time of the delayed component used to generate the rippled noise spectrum. (From Au and Pawloski 1989.)

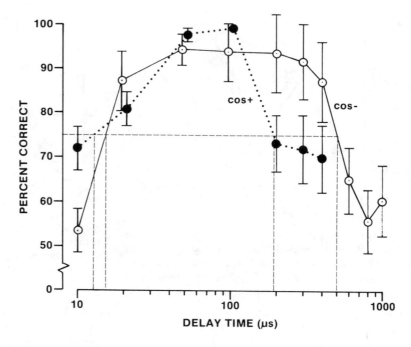

cannot perceive TSP above 2 kHz and below 50 Hz (Yost and Hill 1978). The ratio of upper and lower frequency threshold for the perception of TSP in humans is approximately 40, which compares well with the ratio of 38.5 (77 to 2 kHz) for the dolphin. That the lowest detectable ripple frequency is 2 kHz (500 μs delay) does not support (in the case of dolphins) the theory of Normark (1960, 1961) that TSP could be used to perceive target range. A threshold of 500 μs would correspond to a maximum target range of approximately 0.4 m. We will see in Chapter 6 that dolphins can actually detect targets at ranges larger than 100 m.

The cos + results in Figure 4.17 may be used to explain the results of Vel'min and Dubrovskiy (1975) shown in Figure 4.5. *Tursiops* can discriminate a rippled stimulus with ripple spacing between 5.3 and 77 kHz, corresponding to delay times between 190 μs and 13 μs (Fig. 4.16). The dip in the performance curve of Figure 4.5 occurred between 10 and 100 μs for dolphin 1, which roughly corresponds to the delay times of 13 and 190 μs for the perception of a rippled spectrum measured by Au and Pawloski (1989). This sudden improvement in performance suggest that dolphin 1 may have keyed in on the

presence of the rippled spectrum or TSP for the double click stimulus whereas dolphin 2 could not perceive either a rippled spectrum or TSP. The short-time spectral analysis shown in Figure 4.14 certainly indicates that the spectrum for the double click stimulus used by Vel'min and Dubrovskiy (1976; cf. Fig. 4.4) was rippled.

In a second experiment, Au and Pawloski (1989) examined the sensitivity of the dolphin to rippled noise by progressively attenuating the level of the delayed component of the noise until the dolphin could no longer discriminate between rippled and nonrippled stimuli. The delayed component of the noise was attenuated in a staircase manner with a 1-dB stepsize. Ten reversals were obtained in each session, and a threshold was defined as the average attenuation at the reversal points for two consecutive sessions in which the averaged reversal of one session did not differ by more than 3 dB from the other. Thresholds were obtained for five delay times between 20 and 300 μs, and the order in which the delay times were tested was randomized, as was the schedule of rippled and nonrippled noise for each trial. The results of the experiment to test the strength of the perceived rippled pitch are shown in Figure 4.18. The pitch associated with

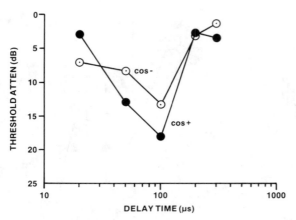

Figure 4.18. Results of the experiment to test the strength of the rippled pitch produced by different delay times. The threshold attenuation needed to extinguish the perception of rippled noise is plotted as a function of the delay time. (From Au and Pawloski 1989.)

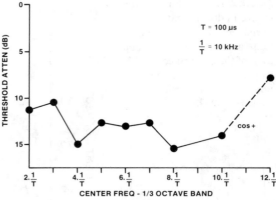

Figure 4.19. Results of the experiment to determine whether dolphins have a dominant frequency range for enhanced perception of a rippled spectrum. (From Au and Pawloski 1989.)

a delay of 100 μs (10 kHz) was the strongest, requiring an attenuation of 14 and 18 dB for the cos$-$ and cos$+$ stimuli, respectively, before reaching threshold. The shapes of the curves are similar to those obtained with humans except for a difference in the range of delay time (Yost and Hill 1978). Bilsen and Ritsma (1970) found that the human maximum sensitivity occurred at 200 Hz, whereas Yost and Hill (1978) obtained maximum sensitivity at 500 Hz. Therefore, the dolphin's maximum threshold occurred at a frequency 20 to 50 times higher than that of humans. The dolphin's maximum threshold of 18 dB for the cos$+$ and 14 dB for the cos$-$ stimuli compared well with the human maximum threshold of 22 and 14 dB for cos$+$ and cos$-$ stimuli, respectively.

A third experiment determined whether or not there is a dominant spectral region for the noise stimulus that enhances the perception of a rippled spectrum. Bilsen and Ritsma (1967/1968) found that TSP was best perceived by humans if the center frequency of a broadband stimulus was approximately $4/T$, where $1/T$ is the TSP. Yost and Hill (1978) obtained similar results with a broadband noise stimulus. We used a 100-μs time delay to produce the rippled noise, and the stimuli were bandpass filtered in 1/3-octave

bands. The same up/down staircase technique as in the second experiment was used to obtain a threshold for each center frequency.

The dolphin's discrimination threshold for the cos$+$ rippled noise is plotted as a function of the center frequency of the 1/3-octave bandwidth in Figure 4.19. The results indicate that there was no dominant frequency at which the perception of a rippled spectrum was enhanced. The dolphin's threshold for center frequencies between $4/T$ and $10/T$ differed by less than 2.5 dB. The absence of a dominant frequency for perceiving TSP may be advantageous to a dolphin by allowing it to use echolocation clicks with widely different peak frequencies without degrading its discrimination capabilities.

4.3 Summary

The auditory system of dolphins seem to have certain adaptive specializations geared toward the reception and processing of broadband transient signals used in echolocation. Awareness of this specialization has come from experiments conducted by Vel'min and his colleagues in Russia. There is still much to be learnt, however, concerning the performance characteristics, anatomy, and physiology associated with this auditory specialization. There are also other processes connected with the reception and processing of broadband sonar echoes that are not yet well understood: for instance, the issue concerning the detection of relative phase differences in echoes has not been resolved satisfactorily, and the question whether dolphins are able to perceive TSP still needs to be answered. This last point is an important issue since the ability to perceive TSP may be one of the major ways in which dolphins discriminate targets while using their active sonar.

References

Akopian, A.J., Zaytseva, K.A., Morozov, V.P., and Titov, A.A. (1977). Spatial directivity of the dolphin's auditory system in perception of varied frequencies signal in noise. Proc. of the Ninth All-Union Acoust. Conf., Moscow.

Au, W.W.L., and Hammer, C.E., Jr. (1980). Target recognition via echolocation by *Tursiops truncatus*. In: G. Busnel and J.F. Fish, eds., *Animal Sonar Systems*. New York: Plenum Press, pp. 855–858.

Au, W.W.L., and Pawloski, J.L. (1989). Detection of ripple noise by an Atlantic bottlenose dolphin. J. Acoust. Soc. Am. 86: 591–596.

Au, W.W.L., Moore, P.W.B., and Pawloski, D.A. (1988). Detection of complex echoes in noise by an echolocating dolphin. J. Acoust. Soc. Am. 83: 662–668.

Bel'kovich, V.M., and Dubrovskiy, N.A. (1976). *Sensory Basis of Cetacean Orientation*. Leningrad: Nauka.

Bilsen, F.A., and Ritsma, R.J. (1967/1968). Repetition pitch mediated by temporal fine structure at dominant spectral regions. Acustica 19: 114–115.

Bilsen, F.A., and Ritsma, R.J. (1970). Some parameters influencing the perception of pitch. J. Acoust. Soc. Am. 47: 469–476.

Bracewell, R.M. (1978). *The Fourier Transform and its Application*, New York: McGraw Hill.

Dubrovskiy, N.A. (1990). On the two auditory subsystems of dolphins. In: J.A. Thomas and R. Kastelein, eds., *Sensory Ability of Cetacean: Laboratory and Field Evidence*, edited by New York: Plenum Press, pp. 233–254.

Dubrovskiy, N.A., Krasnov, P.S., and Titov, A.A. (1978). Auditory discrimination of acoustic stimuli with different phase structures in a bottlenose dolphin. In: *Marine Mammals*, 114–115.

Hammer, C.E., Jr., and Au, W.W.L. (1980). Porpoise echo-recognition: an analysis of controlling target characteristics. J. Acoust. Soc. Am. 68: 1285–1293.

Johnson, R.A. (1980). Energy spectrum analysis in echolocation. In: R.G. Busnel and J.F. Fish, eds., *Animal Sonar Systems*. New York: Plenum Press, pp. 673–693.

Johnson, R.A., and Titlebaum, E.L. (1976). Energy spectrum analysis: a model of echolocation processing. J. Acoust. Soc. Am. 60: 484–491.

Johnson, R.A., Moore, P.W.B., Stoermer, M.W., Pawloski, J.L., and Anderson, L.C. (1988). Temporal order discrimination within the dolphin critical interval. In: P.E. Nachtigall and P.W.B. Moore, eds., *Animal Sonar: Processes and Performance*. New York: Plenum Press, pp. 317–321.

McCellan, M.E., and Small, A.M. (1965). Time separation pitch associated with correlated noise bursts. J. Acoust. Soc. Am. 38: 142–143.

Moore, P.W.B., Hall, R.W., Friedl, W.A., and Nachtigall, P.E. (1984). The critical interval in dolphin echolocation: what is it? J. Acoust. Soc. Am. 76: 314–317.

Normark, J. (1960). Perception of distance in animal echolocation. Nature 183: 1009–1010.

Normark, J. (1961). Perception of distance in animal echolocation. Nature 190: 363–364.

Patterson, J.H., and Green, D.M. (1970). Detection of transient signals having identical energy spectra. J. Acoust. Soc. Am. 48: 894–90 .

Ronken, D.A. (1970). Monaural detection of a phase difference between clicks. J. Acoust. Soc. Am. 46: 1091–109.

Schroeder, M.R., and Atal, B.S. (1962). Generalized short-time spectra and autocorrelation functions. J. Acoust. Soc. Am. 34: 1679–168.

Small, A.M., and McClellan, M.E. (1963). Pitch associated with time delay between two pulse trains. J. Acoust. Soc. Am. 35: 1246–1255.

Thurlow, W.R., and Small, A.M., Jr. (1955). Pitch perception for certain periodic auditory stimuli. J. Acoust. Soc. Am. 27: 132–137.

Vel'min, V.A., and Dubrovskiy, N.A. (1975). On the analysis of pulsed sounds by dolphins. Dokl. Adak. Nauk. SSSR 225: 470–473.

Vel'min, V.A., and Dubrovskiy, N.A. (1976). The critical interval of active hearing in dolphins. Sov. Phys. Acoust. 2: 351–352.

Vel'min, V.A., and Dubrovskiy, N.A. (1977). Spatial selectivity of active hearing in a bottlenose dolphin. Proc. of the Ninth All–Union Acoust. Conf., Moscow, 5–8.

Vel'min, V.A., and Dubrovskiy, N.A. (1978). Auditory perception by bottlenosed dolphins of pulsed signals. In: *Marine Mammals: Results and Methods of Study, V. Ye.*

Vel'min, V.A., Titov, A.A., and Yurkevich, L.I. (1975a). Time summation of pulses in the bottlenose dolphin.

In: Morskiye mlekopitayushciye. Mater. 6-go Vses. soveshch. po izuch. morsk. mlekopitayushchikh, Part 1. Kiev: Naukova Dumka, pp. 78–80.

Vel'min, V.A., Titov, A.A., and Yurkevich, L.I. (1975b). Differential intensity thresholds for short pulsed signals in bottlenose dolphins. In: Morskiye mlekopitayushciye. Mater. 6-go Vses. soveshch. po izuch. morsk. mlekopitayushchikh, Part 1. Kiev: Naukova Dumka, pp. 72–74.

Yost, W.A., and Hill, R. (1978). Strength of pitches associated with rippled noise. J. Acoust. Soc. Am. 64: 485–492.

Yost, W.A., Hill, R., and Perez-Falcon, T. (1978). Pitch and pitch discrimination of broadband signals with ripped power spectra. J. Acoust. Soc. 63: 1166–1173.

Zanin, A.V., Zaslavskiy, G.L., and Titov, A.A. (1977). Temporal summation of pulses in the auditory system of the bottlenose dolphin. 9th All-Union Acoust. Conf., Moscow, pp. 21–23.

Zaytseva, K.A., Akopian, A.I., and Morozov, V.P. (1975). Noise resistance of the auditory analyzer in dolphins as a function of noise presentation angle. Biofizika 20: 519–522.

Zwislocki, J. (1963). Analysis of some auditory characteristics. In: *Handbook of Mathematical Psychology*. New York: John Wiley & Sons, Chapter 17.

5

The Sonar Signal Transmission System

We will now begin to examine another major subsystem of the dolphin sonar, the signal transmission system. In this chapter we will first discuss the general characteristics of sonar signals used by odontocetes, both in open waters and in enclosed tanks, in order to gain an appreciation of the wide variety of signals that are used. The topic of biosonar (biological sonar) signals will be briefly covered in this chapter, mainly as a prelude to the subject of sound production mechanisms. Characteristics of dolphin sonar signals will be more thoroughly discussed in Chapter 7. Although we are generally concerned with sonar signals used by dolphins, we need to pay some attention to other types of acoustic signals they produce. Bottlenose and other dolphins emit a wide variety of sounds not used for active sonar searches. Sound emissions can be classified into two broad categories of narrow-band frequency-modulated (FM) continuous tonal sounds referred to as whistles and broadband sonar clicks (Evans 1967). Whistles appear to be used for intraspecific communications (Herman and Tavolga 1980). These sounds are generally low frequency emissions between 5 and 30 kHz and may last for several seconds. They are often referred to as squeaks, squawks, and squeals, as well as whistles.

5.1 Preliminary Examination of Biosonar Signals

The most effective way for a dolphin to probe its underwater environment for the purpose of navigation, obstacle and predator avoidance, and prey detection is to use its active and passive sonar capabilities. As mentioned in Chapter 3, acoustic energy propagates in water more efficiently than almost any other form of energy, so that an active sonar capability is ideal for life in an aquatic environment. Electromagnetic, thermal, light, and other forms of energy are severely attenuated in water. The natural habitat of certain dolphin species—shallow bays, inlets, coastal waters, swamps, marshlands, and rivers—is often so murky or turbid that vision is severely limited. These animals must then rely almost exclusively on their auditory skills, including active sonar, for survival.

Evans (1973) presented a review of echolocation by marine delphinids and displayed the waveform and frequency spectrum of a typical sonar signal used by *Tursiops truncatus* in a tank environment. This typical *Tursiops* signal is reproduced in Figure 5.1, with the signal waveform on the left and the frequency spectrum on the

T. Truncatus

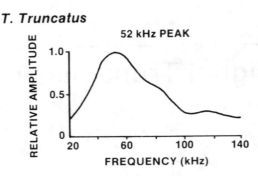

Figure 5.1. Typical waveform and associated spectrum for the echolocation signal of *Tursiops truncatus* in a tank. (From Evans 1973.)

right. Evans (1973) reported that source levels (sound pressure level 1 m from the animal) were typically around 170 dB re 1 μPa. The peak frequency (frequency of maximum energy) in this example was 52 kHz. At the time, most sonar signals of *Tursiops* were measured in tanks, and it was generally believed that peak frequencies of sonar signals were in the vicinity of 30 to 60 kHz. It was not until the study of Au et al. (1974) that certain additional features of biosonar signals used by *Tursiops* and other dolphins were discovered.

Au et al. (1974) measured the sonar signals emitted by two *Tursiops truncatus* during a tar-get detection experiment in the open waters of Kaneohe Bay, Oahu, Hawaii. We discovered that these signals had peak frequencies between 120 and 130 kHz, which was over an octave higher than the peak frequencies reported by Evans (1973). We also measured an average peak-to-peak click level on the order of 220 dB re 1 μPa at 1 m, which represents a level 30 to 50 dB higher than previously measured for *Tursiops*. We attributed the use of high frequency sonar signals to the characteristics of the high ambient noise environment of Kaneohe Bay (see Figs. 1.4 and 1.5). The high source levels were attributed to the absence of boundaries that could cause high

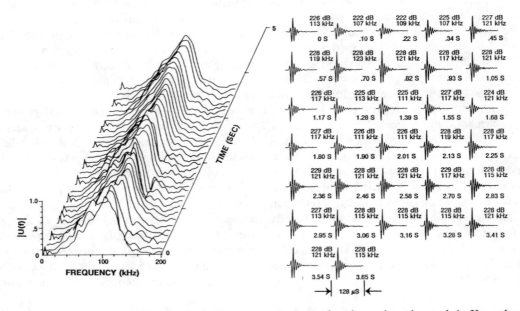

Figure 5.2. A typical sonar click train for a *Tursiops truncatus* performing a detection task in Kaneohe Bay. The peak-to-peak SPL in dB re 1 μPa at 1 m and the peak frequency in kHz is shown to the right of each signal. The time of occurrence of each click relative to the first click is shown below. (From Au 1980.)

reverberation and to the long distances between the animals and the targets. Since then, the use of high frequency, high amplitude signals by *Tursiops* in Kaneohe Bay has been reconfirmed and solidly established in other studies by Au and his colleagues (Au et al. 1978, 1982; Au, 1980).

An example of a typical sonar click train for a *Tursiops* performing a target detection task is shown in Figure 5.2 The frequency spectra plotted as a function of time are shown on the left and the individual click waveform is displayed on the right. The frequency spectra represent the relative magnitude of the fast Fourier transform (FFT) of each click. The peak-to-peak sound pressure level (SPL) in dB re 1 μPa at 1 m along with the peak frequency in kHz is given for each signal. The parameter listed below each signal is the time of occurrence relative to the first click in the click train. The shape of the signals in a click train tends to be very repetitive and stereotypical, and the amount of information in a click train can be considerable and quite unwieldy, especially if many tens of trials and sessions are considered. A simple way to process clicks in a click train is by averaging the digital form of the time and frequency representations of the signals,

as is done with real-time spectrum analyzers and other digital equipment. The average waveform and frequency spectrum of the signals in the click train of Figure 5.2 are shown in Figure 5.3. Note the similarity between the average waveform and the signals in the click train. The higher amplitude signals in most click trains are highly correlated, whereas fluctuations in the waveshape are generally associated with the lower level signals (Au 1980). Therefore, the averaged waveform will tend to resemble the higher level signals. A typical sonar signal resembles an exponentially damped sinusoidal wave with a duration between 40 and 70 μs and with 4 to 10 positive excursions, as shown in Figures. 5.2 and 5.3. A mathematical expression that can be used to approximate dolphin sonar signals consist of the Gabor function (Kamminga and Beitsma, 1990), which is the product of a cos ($2\pi f_0 t$) or sin($2\pi f_0 t$) term and a Gaussian curve as given by

$$s(t) = A \begin{bmatrix} \cos(2\pi f_0 t + \phi) \\ or \\ \sin(2\pi f_0 t + \phi) \end{bmatrix} e^{-\pi^2 \frac{(t-\tau_0)^2}{\Delta\tau^2}} \quad (5\text{-}1)$$

where A = relative amplitude
f_0 = peak frequency
t_0 = centroid of the signal (see 10-29)
$\Delta\tau$ = rms duration of the signal (see 10-28)
ϕ = phase shift

Examples of the shape of the Garbor Function using the sine and cosine terms are shown in Figure 5.4. The Fourier transform of the Gabor Function will also have a Gaussian shape with a peak frequency at f_0. A reasonable approximation to the average waveform in Figure 5.3 can be obtained from (5-1) by letting $\tau_0 = 9.9$ μs and $\Delta\tau = 29.9$ μs and $\phi = 3\pi/4$.

The effects of a noisy environment on the sonar signals used by a beluga, or white whale (*Delphinapterus leucas*), was vividly demonstrated by Au et al. (1985). The sonar signal of the beluga was measured in San Diego Bay, California before the whale was moved to Kaneohe Bay, where noise is between 12 and 17 dB greater than in San Diego Bay (Figs. 1.4 and 1.5). The whale emitted sonar signals with peak frequencies between 40 and 60 kHz and with a maximum average source level (peak-to-peak) of 202 dB re

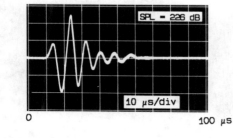

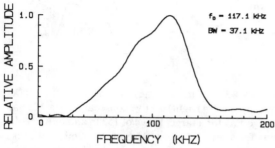

Figure 5.3. Averaged waveform and frequency spectrum for the click train of Fig. 5.2. (From Au 1980.)

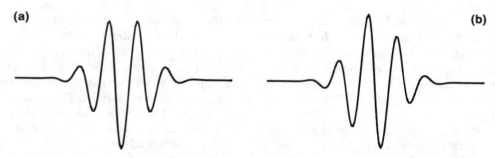

Figure 5.4. Shape of the Gabor Function with (*A*) the sine term and (*B*) the cosine term.

1 μPa in San Diego Bay. After the whale was moved to Kaneohe Bay, it shifted the peak frequency of its sonar signals over an octave higher to 100 and 120 kHz. The source level also increased to over 210 dB re 1 μPa (Au et al. 1985). The averaged time waveforms and frequency spectra for five consecutive trials in San Diego Bay for a target range of 157 m and in Kaneohe Bay for a target range of 16.5 m are shown in Figure 5.5. Au et al. (1985) attributed the beluga's use of a higher peak frequency in Kaneohe Bay to the high ambient noise in Kaneohe Bay. Sonar signals used by belugas in tanks also resemble the low frequency signals shown in the figure (Gurevich and Evans 1976; Kamminga and Wiersma 1981). Turl et al. (1991) measured the sonar signals of a beluga during a task detecting targets in clutter in San Diego Bay and found that the animal was using high frequency (peak frequency above 100 kHz) and high intensity (greater than 210 dB re 1 μPa) signals.

The effects of a highly reverberant environment on the sonar signals of a false killer whale, *Pseudorca crassidens*, can be inferred from two studies by Thomas et al. (1988) and Thomas and Turl (1990). The sonar signals of a *Pseudorca* measured in a tank had peak frequencies between 20 and 60 kHz, with source levels of approximately 180 dB re 1 μPa (Thomas et al. 1988). Most of the sonar signals used by another *Pseudorca* performing a detection task in the open waters of Kaneohe Bay had peak frequencies between 100 and 110 kHz, with source levels between 220 and 225 dB re 1 μPa (Thomas and Turl 1990).

The use of high frequency sonar signals (peak frequencies greater than 90 kHz) by bottlenose dolphins, a beluga, and a false killer whale is

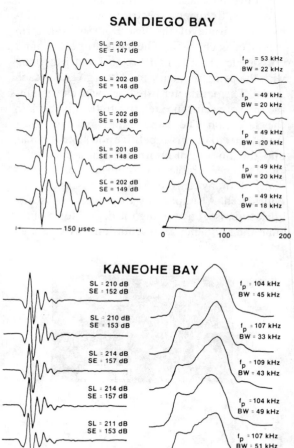

Figure 5.5. Averaged time waveforms (left) and frequency spectra (right) for five consecutive trials in San Diego Bay for a target range of 157 m and in Kaneohe Bay for a target range of 16.5 m. *N*, number of clicks in a trial; SL, peak-to-peak source level in dB re 1 μPa; SE, source energy flux density in dB re 1 μPa^2 s; f_p, peak frequency; BW, 3-dB bandwidth. (From Au et al. 1985.)

definitely connected to the high ambient noise environment of Kaneohe Bay. Most of the noise in Kaneohe Bay is caused by snapping shrimp; as mentioned in Chapter 1, the noise levels (Figs. 1.4 and 1.5) are higher than those reported for any other body of water. In order to overcome the effects of ambient noise, dolphins in Kaneohe Bay typically emit clicks at intensity levels approaching their maximum capability. Au et al. (1985) postulated that the high frequencies were a by-product of producing high intensity clicks. In other words, dolphins can only emit high-level clicks (greater than 210 dB) if they use high frequencies. The animals can probably also emit high frequency clicks at low amplitudes, but cannot produce low frequency clicks at high amplitudes. The data of Moore and Pawloski (1990) seem to support this contention, as do the measurements of Turl et al. (1991). Turl et al. (1991) measured the sonar signals of a beluga in a target-in-clutter detection task in San Diego Bay and found that the animal was using high frequency (peak frequency above 100 kHz) and high intensity (greater than 210 dB re 1 μPa) signals. When a beluga emitted low amplitude signals in San Diego Bay the peak frequency was also low (Au et al. 1985), and when high amplitude signals were used the peak frequency was high (Turl et al. 1991). Additional characteristics of the high intensity, high frequency sonar signals used by dolphins in the open waters of Kaneohe Bay will be considered in Chapter 7.

The dolphin sonar signal measurements performed in Kaneohe Bay indicate that caution must be taken in studying these signals in tanks and highly reverberant environments. The close proximity of walls, which are good acoustic reflectors, might tend to discourage a dolphin from emitting high intensity sonar signals because too much energy would be reflected back to the dolphin. This introduces the possibility that unnatural and sub-optimal signals may actually be used in tanks. There is only one study involving *Tursiops*, in which high frequency signals were measured. Poché et al. (1982) lined a concrete tank with an anechoic material consisting of water-saturated cypress panels and recorded signals emitted by two *Tursiops*; the signals had a major peak at 30 kHz with a secondary peak almost as large as the major peak at 130 kHz. The amount of echo reduction and the transmission loss of the cypress panels increased with frequency (Poché et al. 1977), so that the reflection from the concrete wall of the tank was at least 20 dB lower at a signal frequency of 130 kHz than at 30 kHz. Therefore, it is not surprising that the dolphins shifted their signal to a higher frequency in order to reduce reflections from the concrete wall of the tank. After the panels were removed, Poché et al. (1982) measured signals from one dolphin and found them similar to signals measured with the lined tank. Unfortunately, they did not report on the source levels of the high frequency clicks, nor did they report on signals measured before the cypress panels were inserted.

The results of many studies indicate that dolphins can generate a wide variety of sounds, ranging from frequency-modulated continuous tones called whistles, which can last several seconds, to short broadband pulses or clicks that have durations between 50 and 200 μs. Any discussion of the sound production mechanisms of dolphins must consider the vastly different types of acoustic signals that dolphins can emit. Furthermore, it seems that whistles and pulses can be produced simultaneously, as was observed by Lilly and Miller (1961). The vastly different characteristics of the two general classes of sound along with the fact that they can be produced simultaneously suggest that different generative mechanisms may be involved.

5.2 Sound Production Mechanism

One of the great mysteries concerning the dolphin sonar system is the question of sound production. The seemingly simple topic of the general location of sound production can arouse widely divergent opinions from prominent scientists. Dormer (1979) presented a detailed list of twenty references discussing the site of sound production for different species of dolphins, along with the investigative techniques used. The two most popular proposed sites of sound production are the larynx and the nasal sac system. An excellent review of the various theories concerning sound production was written by Morris (1986).

5.2.1 Theory of Sound Production by the Larynx

The larynx seems to be the most likely location of sound production by dolphins, prima facie at least, since most other mammals produce sounds with their larynx. Therefore many investigators (e.g., Evans and Prescott 1962; Purves 1967; Lilly and Miller 1961; Blevins and Parkins 1973; Schenkkan 1973) have proposed the larynx as the

site of sound production in dolphins. Evans and Prescott (1962), after performing some sound simulation experiments with severed dolphin heads, proposed a dual production theory with whistles being generated in the larynx and echolocation clicks in the nasal sac. The champion and developer of the most detailed theory of sound generation in the larynx is Peter Purves in England (Purves 1967; Purves and Pilleri 1983). He theorized that sounds are generated by vibra-

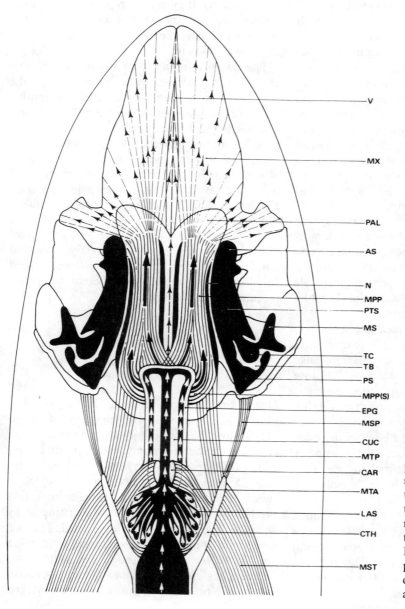

Figure 5.6. Diagram showing the coupling of sound vibration generated by the epiglottic spout to the palatopharyngeal muscles and skull. Black area denote air spaces; white areas denote cartilage or bone. White arrows show direction of air stream; black arrows show propagation of sound waves. AS, air-sac; CAR, arytenoid cartilage; CTH, thyroid cartilage; CUC, cuneiform cartilage; EPG, epiglottis; LAS, laryngeal air sacs; MPP, palatopharyngeus muscle; MPP(S), palatopharyngeal sphincter; MS, middle sinus; MSP, salpingopharyngeal muscle; MST, stenothyroid muscle; MTA, thyroarytenoid muscle; MTP, thyropharyngeus muscle; MX, maxilla; N, naris; PAL, palatine bone; PS, posterior sinus; PTS, pterygoid sinus; TB, tympanic bulla; RC, tympanic cavity; V, vomer. (From Purves and Pilleri 1983.)

tions of the epiglottic cartilages of the larynx at the epiglottic spout of the beaked larynx. Sounds are supposedly transferred from the larynx to the rostrum via the palatopharyngeal muscles, which are in direct contact with the palatine bones at the posterior face of the skull (Fig. 5.6). The acoustic vibrations are then radiated from the bone of the rostrum into the blubber and sea water.

The proponents of a laryngeal sound production mechanism base their arguments mainly on anatomical considerations and experiments with dead animals. Evans and Prescott (1962) used the heads of two *Stenella graffmani and a Tursiops truncatus* and forced air through the larynx, measuring the subsequent sounds with an array of condenser microphones placed 0.6 cm from the surface of the dolphin's head. Depending on the pressure and position of the larynx, sounds similar to barks and whistles were produced. Short duration clicks were generated by air flowing through the nasal sac system of one of the *Stenella*. Purves (1967) inserted a Galton whistle attached to an air line into the trachea of several dolphins to study sound propagation through the head in fresh specimens. *Phocoena, Lagenorhynchus cruciger* and *Tursiops truncatus* (blunt-snouted, short-snouted, and long-snouted cetaceans, respectively) were used by Purves (1967). He found that an air pressure of 25 mm Hg was sufficient to produce an intense and extremely high-pitched, audible tone. He also measured the relative intensity of the vibration on the skull at various locations. The arguments used by Purves (1967) and Purves and Pilleri (1983) to support their theory have been summarized by Morris (1986). Some of the more important ones are as follows: (1) The larynx of dolphins is extremely sophisticated and unlikely to have been developed merely for breathing. It is also more than capable of making all known sounds emitted by dolphins. (2) The larynx is the primary source of sound in every other mammal. (3) The nasal air sac system is used for recycling of air during sound production. The asymmetrical arrangement is related to breathing while asleep. (4) The bony rostrum has a cartilaginous rod, the vomer, which makes an ideal wave guide. It is joined with the larynx via the two lobes of the palatopharyngeal muscle (Fig. 5.6) which are coupled

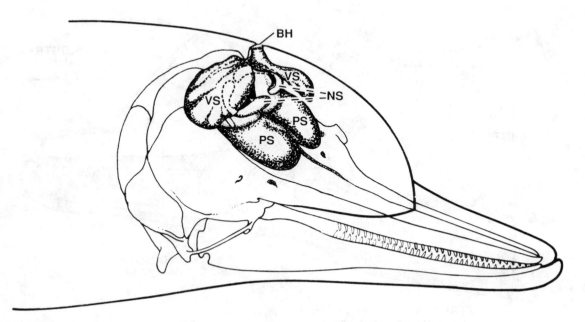

Figure 5.7. Three-dimensional diagram of the air sacs in an inflated condition associated with the recycling of air during echolocation. PS, premaxillary sac; VS, vestibular sac; NS, nasofrontal (tubular) sac; AS, accessory sac. (Adapted from Purves and Pilleri 1983.)

directly to the rostrum. (5) Transmission of sound should be possible along the longitudinal muscle of the vomer. The sound field can emanate from the tip of the rostrum.

5.2.2 Experiments to Localize the Site of Sound Production

Many experiments, using widely different techniques and instrumentation, have been conducted to localize the site of sound generation and to obtain a better understanding of sound production by dolphins. All of these experiments have produced data that seem to support the notion that sounds are produced in the nasal system, in the vicinity of the nasal plugs. Some of the experiments have produced data that seem to specifically rule out the larynx as a source of

sound production. In order to facilitate the discussion of these experiments, a schematic of a dolphin head with a three-dimensional drawing of the air sacs in the nasal system is shown in Figure 5.7. Another schematic that will be helpful is the drawing of a sagittal section through the nasal region of a dolphin shown in Figure 5.8.

Diercks et al. (1971) used an array of contact hydrophones to measure dolphin sonar signals. They obtained results which placed the sound source at a depth of 1.5 to 2.0 cm from the surface, in the vicinity of the nasal plug. Norris et al. (1971) and Dormer (1979) used high-speed X-ray motion pictures to observe movements of the laryngeal and nasal region association with sound production in live phonating *Tursiops truncatus*, *Tursiops gilli*, and *Stenella longirostris*. Hollien et al. (1976) used soft-tissue spot X-rays

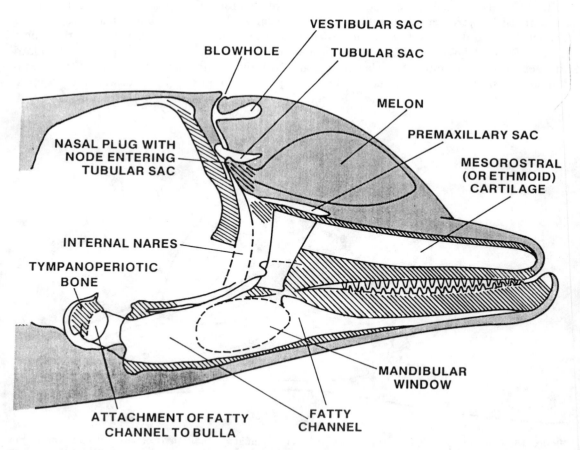

Figure 5.8. Schematic of a dolphin's head showing various structures associated with the theory of sound production in the nasal sac system. (After Norris 1968.)

Figure 5.9. Electromyography and sound and pressure events in the nasal system of *Tursiops*. RAIA, right anterior internus muscle, anterior part; RAIP, right anterior internus muscle, posterior part; LAI, left anterior internus muscle; NPM, nasal plug muscle; W, whistles; B, blows. (From Ridgway et al. 1980.)

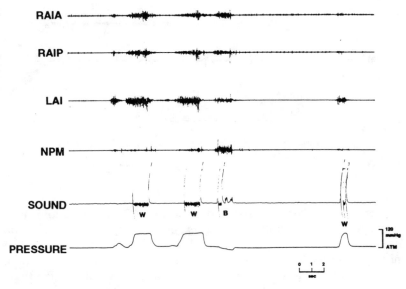

to observe the larynx and melon of two phonating *Tursiops truncatus*. All of these investigators observed movements in the nasal system that correlated with sound production, yet did not see any correlated movements in the larynx. According to Dormer (1979), "the mechanism of sound production in porpoises consist of a series of muscular pumps, valves, and compliant sacs with the sound being generated at the nasal plug". He observed that the movement of the left nasal plug is precisely associated with whistle production and speculated that the right nasal plug is associated with click production. Mackay and Liaw (1981) used an ultrasonic Doppler motion detector to measure vibrational motion in *Tursiops truncatus* and *Delphinus delphis* while the dolphins were phonating. The motion detector emitted a continuous wave of 2-MHz sound and could detect movements smaller than 10 μm. They observed that the nasal plug and the air sacs all vibrated in synchrony with the production of sound. They also observed that the nasal diverticula on the right side vibrated with clicking sounds all the time, but only some of the time on the left side. The vestibular sac was steadily inflated while the animal produced clicks. They suggested that the vibration of the nasal plugs probably is the originator of the clicking or buzz sound. Neither gross movements nor rapid vibrations in the larynx were observed when the

animals emitted clicks and whistles. However, motion in the larynx was detected with each breath.

Ridgway et al. (1980) measured the muscle activities (using electromyographic techniques) and air pressures associated with sound production in five *Tursiops truncatus*. They found that the anterior internus, the posterior internus, the diagonal membrane muscle, and the nasal plug muscles (all associated with the nasal system) fired just before and during the production of sounds, whereas the hyoepiglottal muscle and the intercostal muscles (associated with the larynx) did not. They also found that pressure increased in the nares and premaxillary sacs prior to sound production and dissipated after the sound. There were no pressure changes in the trachea during sound production. An example of muscle activities and pressure events in the nasal system associated with whistles is given in Figure 5.9. An example of muscle activities associated with clicks and whistles is shown in Figure 5.10. Note the simultaneous occurrence of muscle, pressure, and sound events. Amundin and Andersen (1983) measured the muscle activity of the nasal plug muscle and the air pressure in the bony nares of a *Phocoena* and a *Tursiops truncatus* and obtained results similar to Ridgway et al. (1980). Ridgway and Carder (1988) measured the air pressure in the nasal cavities and in the trachea

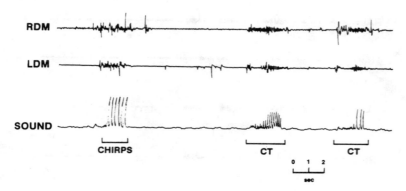

Figure 5.10. Electromyography of the right and left diagonal membrane muscles during pulsed sounds described as chirps and click trains (CT) (From Ridgway et al. 1980.)

adjacent to the larynx associated with sound production in an echolocating *Delphinapterus leucas*. They found that intranasal pressure increased markedly but intratracheal pressure remained unchanged during the emission of echolocation clicks by the beluga. All of the measurements cited in this section suggest that the larynx of dolphins is not involved in the production of either click or tonal sounds.

5.2.3 Theory of Sound Production in the Nasal Sac System

Although experiments have provided rather conclusive evidence that sounds are produced within the nasal sac system of dolphins and not in the larynx, the specific locations and mechanisms involved are not precisely known. Inside the forehead of the dolphin are two muscled nasal plugs (Fig. 5.8) acting as valves in the superior bony nares. Each of the nasal plugs has connective tissue flaps (lips) on its lateral margins. Along the nasal passages are at least six blind-ended air sacs (Fig. 5.7); these airways merge above the nasal plug into a single passage, called the spiracular cavity, which exits at the blowhole. Evans and Prescott (1962) theorized that the production site of echolocation clicks is located where the nasal sac nodes insert into the opening of the tubular sacs. The tubular, or nasofrontal, sacs represent a pair of anteriorly directed tapering horns connected posteromedially to the nasal passage above the nasal plugs (Norris 1964). Evans and Maderson (1973) revised the original theory and suggested that movement of the nasal plugs against the hard edge of the bony nares was responsible for sound production. The alternate

resistance and release of the plugs' movements produced "relaxation oscillations" resulting in the production of acoustic pulses. However, palpations and visual inspections by Amundin and Anderson (1983), through the opened blowhole of both *Phocoena* and *Tursiops*, during click production revealed no movement that would support the "friction–striction" mechanical mechanism proposed by Evans and Maderson (1973). Ridgway et al. (1980) also disagreed with Evans and Maderson (1973) on the role of the nasofrontal sacs in sound production. They suspected that the nasofrontal sacs may indeed be important to sound production, by serving as resonators.

One of the most recent hypotheses of sound production by dolphins was made by Cranford et al. (1987). They examined the heads of several dolphin species using X-ray computer tomography and magnetic resonance imaging and discovered two pairs of small bulbous fatty dorsal projections at the posterior boundary of the melon's shell. These dorsal projections, referred to as dorsal bursae by Cranford (1985), can be seen in Figures 5.11A and 5.11B. The dorsal bursae abut a liplike structure, the *museau de singe*, which is connected to the main nasal passage with interlocking tissue ridges on the opposite surface of each "lip" that may be used to control air flow through the passage. The *museau de singe* is also referred to as "monkey lips"; each pair of dorsal bursae has its own pair of monkey lips associated with it (Cranford 1989). Cranford (1988) theorized that sounds may be produced by passing pressured air bubbles past the monkey lips or between the bursae, a process analogous to the production of glottal pulses in

Figure 5.11. (*A*) X-ray projection of the head of a male spinner dolphin. The melon is the dark gray elliptical structure. Near the posterior end of the melon there are two small bulbous projections (arrow) resembling the ears of a rabbit. (*B*) Magnification of the right dorsal bursae. (From Cranford 1988.)

the human larynx. Cranford et al. (1987) considered their hypothesis similar, even "homologous," to the mechanism proposed by Norris and Harvey (1972) for the generation of sounds by sperm whales.

Unfortunately, there is no direct evidence to support any of these theories of sound production. Acoustic energy can be produced in a dolphin's head by a variety of mechanical and pneumatic mechanisms: Air blowing across a vibrating membrane, air blowing into a resonating chamber or cavity, or air blowing across a tiny orifice can generate sounds. A muscle or ligament rubbing against another muscle or ligament or against a bony structure can be a source of sound. A thin membrane flapping against a hard surface can also be used to generate acoustic energy. Whether any one of these methods of generating acoustic energy is being used by the dolphin is not known and needs to be examined. The physical and mechanical properties of differents parts of the nasal sac system need to be

examined to determine whether they can support the generation of the wide variety of acoustic signals that dolphins are capable of producing.

5.3 Acoustic Propagation in the Dolphin's Head

5.3.1 Role of the Melon

Sounds produced in the vicinity of the nasal plugs will propagate in the forward direction through the melon; the vestibular sacs (Figs. 5.7 and 5.8) will tend to reflect any upward directed sound, the nasal passage and cranial bone will reflect any backward directed sound, and the premaxillary sacs and the rostrum will reflect any downward directed sound. Wood (1964) was one of the first to suggest that the fatty melon of the dolphin forehead may be used to couple sounds from inside the animal's head into the water. The fissue of the melon actually invades the muscles of the nasal plug, allowing for a direct link between a source associated with the nasal plug and the rest of the melon. The melon and jaw fat are composed of a translucent lipid very rich in oil and are completely different from other tissue. Varanasi and Malin (1971), among the first to study the chemical composition of the melon lipid, found that it was composed mainly of triacylglycerol and wax esters. Since then a number of studies have been performed on the chemical, acoustical, and mechanical properties of the melon lipid for a variety of odontocete species (Varanasi and Malin 1971, 1972, 1983; Varanasi et al. 1975; Litchfield et al. 1971, 1973, 1979; Blomberg 1974; Blomberg and Jensen 1976; Blomberg and Lindholm 1976). In some of these studies the oil-rich lipid has been referred to as "acoustic tissue" and the velocity gradient of the lipid, discussed below, has been speculated to focus outgoing acoustic emissions. It is interesting to note that sonar transducers are often constructed with a rho-c rubber housing containing a low viscosity mineral oil which interfaces with the piezoelectric sensing elements (Wilson

1985). The oil, essentially a lossfree acoustic medium, is used to couple acoustic energy from the piezoelectric crystals to the water and vice versa.

Norris and Harvey (1974) measured the sound velocity profile of the tissue in the melon of a just deceased Pacific bottlenose dolphin (*Tursiops* sp.) by dicing the melon into many 1.8-cm × 1.8-cm × 9.5-cm pieces. Each sample was fitted into a holder and immersed in a constant-temperature (22.9°C) water bath. The time interval for an acoustic signal to propagate through the sample was measured, and the sound velocity determined, using transducers on both sides of the sample. The location of each sample and the corresponding sound velocity are shown in Figure 5.12. Norris and Harvey (1974) found a low-velocity core surrounded by a higher-velocity shell, as can be seen in the figure; the low-velocity core consisted of pellucid oily fat, surrounded by denser blubber that was heavily invaded with connective tissue fibers near the surface. They suggested that such a velocity profile could act like a sound channel, causing sounds propagat-

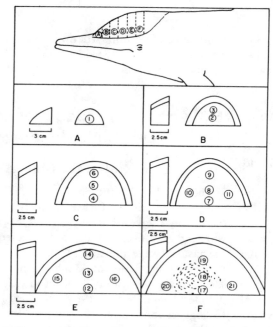

Sample	Time in blubber μ sec	Velocity m/sec
1	17.57	1422
2	17.57	1422
3	16.82	1682
4	18.07	1807
5	18.06	1342
6	17.27	1449
7	19.67	1273
8	19.32	1292
9	18.17	1376
10	17.67	1415
11	17.62	1418
12	19.07	1311
13	19.07	1311
14	18.60	1343
15	17.32	1441
16	17.57	1423
17	17.82	1402
18	18.22	1371
19	16.87	1481
20	17.27	1449
21	17.07	1463

Figure 5.12. Sound velocity measurement locations in melon tissue slices from a dolphin's forehead. The upper diagram shows the positions of the transverse slices, while the other diagrams show the slice thickness and position of velocity measurements. (From Norris and Harvey 1974.)

Figure 5.13. Three-dimensional computer graphic reconstruction of the melon in the dolphin's forehead. The inner set of circles represents the lower-density core which splits and emerges from the posterior shell as two lateral branches ending near the dorsal bursae. (From Cranford 1988.)

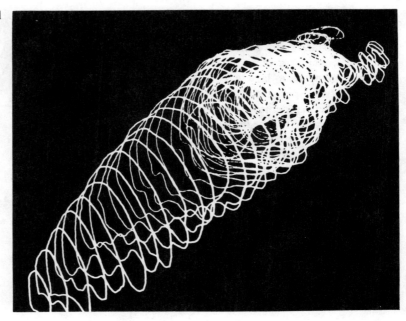

ing in the core to be refracted inward. The sound velocity structure would also provide a gradual impedance transformation from the nasal plug region to the water. The results presented in Figure 5.12 should be adjusted to correspond to the normal body temperature of 37°C for *Tursiops* (Ridgway 1965). Since the sound velocity of melon lipid has a negative gradient of −3.3 m/s per °C (Litchfield et al. 1979), the data should be reduced by approximately 46 m/s. However, if the dolphin is swimming in waters with temperatures different than its normal body temperature, then there should be a temperature gradient very near the surface of the dolphin skin. Unfortunately, little is known about the thickness and temperature profile of this transition region.

Cranford (1988) examined the heads of dolphins using X-ray computer tomography (CT) and combining the serial section images, or tomograms, with high-resolution three-dimensional computer graphics. An example of a three-dimensional image of a melon is shown in Figure 5.13. A low-density core corresponding to the low-velocity core found by Norris and Harvey (1974) can be seen in the reconstructed image. A schematic of

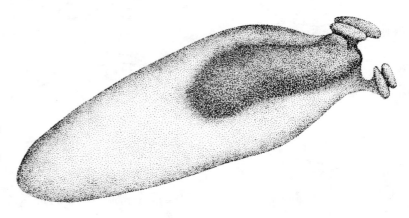

Figure 5.14. A $\frac{3}{4}$-view of the melon emphasizing the relative sizes of the melon's branches and illustrating how the left side appears to be rotated approximately 50° from the midsagittal plane. (From Cranford 1988.)

the melon displaying the low-density core is shown in Figures 5.14.

To understand the concept of impedance matching, consider a plane acoustic wave incident on a plane boundary ($z = 0$) at an angle θ with respect to a perpendicular projection from the boundary, as depicted in Figure 5.15. A portion of the wave will be transmitted across the boundary and a portion will be reflected. The angle of reflection will be the same as the incident angle, and the angle of the transmitted wave will be governed by Snell's Law,

$$\frac{\sin \theta_1}{c_1} = \frac{\sin \theta_2}{c_2} \qquad (5\text{-}2)$$

where c is the velocity of sound in the two media. The specific acoustic impedance of each medium is defined as

$$Z_n = \frac{\rho_n c_n}{\cos \theta_n} \qquad (5\text{-}3)$$

where ρ_n is the density of medium n. The reflection coefficient is the fraction of the acoustic wave

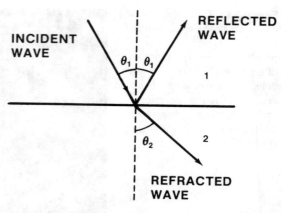

Figure 5.15. Reflection and refraction of a sound wave across a boundary.

that will be reflected at the boundary and is given by the expression

$$R = \frac{Z_2 - Z_1}{Z_2 + Z_1} \qquad (5\text{-}4)$$

The transmission coefficient is given by the ex-

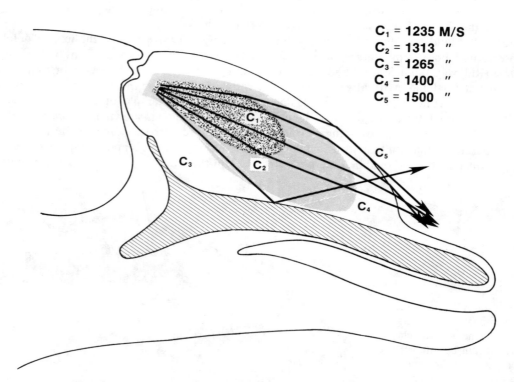

$$C_1 = 1235 \text{ M/S}$$
$$C_2 = 1313 \quad "$$
$$C_3 = 1265 \quad "$$
$$C_4 = 1400 \quad "$$
$$C_5 = 1500 \quad "$$

Figure 5.16. Ray traces through the melon of a dolphin in the vertical plane.

pression

$$T = \frac{2Z_2}{Z_2 + Z_1} \qquad (5\text{-}5)$$

Equation (5-4) indicates that the closer the impedance of medium 2 is to that of medium 1, the smaller will be the reflection at the boundary and the more of the wave will be transmitted across the boundary. In the extreme case in which the specific acoustic impedances of the two media are equal (no-boundary situation), there will be no reflection ($R = 0$) at the "boundary" and all of the acoustic wave is transmitted across ($T = 1$).

The sound velocity and density structure of the melon indicates a specific impedance that gradually increases as one moves from deep within the melon towards the surface. Such an impedance variation will tend to minimize internal reflections as an acoustic signal propagates through the melon. Therefore, the melon can definitely be considered an acoustic matching device with a tapering impedance profile that will enhance the coupling of acoustic energy from the nasal plug area to the water. Matching elements are used in acoustic transducers to provide a graduated impedance transition from the piezoelectric elements to water (Konig et al. 1961; Larson 1983).

The notion of the melon functioning as an acoustic lens to focus outgoing acoustic energy will now be considered in a very gross manner using ray tracing and Snell's Law. When a plane wave strikes the boundary between two media of different sound velocities, as shown in Figure 5.15, the angle of the refracted signal can be determined by Snell's Law, expressed in (5-2). We will approximate the structure of the melon by subdividing it into four distinct volumes of constant sound velocities (Fig. 5.16). The sound velocities of the low-density core, the bottom section of the melon, the outer melon, and the blubber layer next to the surface will be assumed to be 1,235 m/s, 1,265 m/s, 1,313 m/s, and 1,400 m/s, respectively. Ray traces in the vertical plane are shown in Figure 5.16, and in the horizontal plane in Figure 5.17. The ray paths were determined by applying Snell's Law to the various boundaries. The ray traces in both planes show bending or refraction at each boundary, includ-

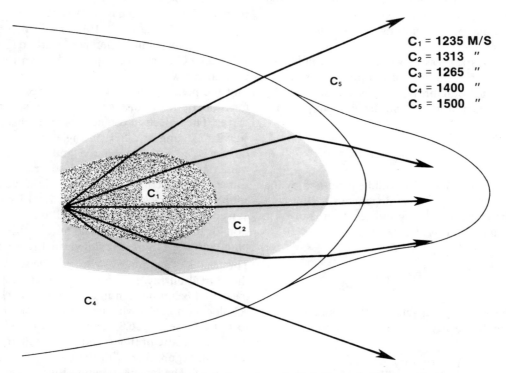

$C_1 = 1235$ M/S
$C_2 = 1313$ "
$C_3 = 1265$ "
$C_4 = 1400$ "
$C_5 = 1500$ "

Figure 5.17. Ray traces through the melon of a dolphin in the horizontal plane.

ing the interface between the melon's surface and the sea water. This type of propagation will cause the outgoing acoustic energy to be focused. The model used in Figures 5.16 and 5.17 is a very crude one since it assumes discrete isovelocity volumes with ray bending at the various interfaces. In actuality the sound velocity will vary continuously, causing the rays to bend continuously as acoustic energy propagates through the melon. The simple model shows that the melon only weakly focuses acoustic energy into the water, and by itself cannot account for the dolphin's directional beam.

5.3.2 Numerical Simulation of Propagation

Aroyan (1990a), performed an elegant simulation of sound propagation in the head of a dolphin by numerically solving the wave equation for an acoustic source within the dolphin's head. His analysis, which dealt with the physics of sound propagation and reflection in both homogeneous and inhomogenous media and with the numerical solution of second-order partial differential equations, is beyond the scope of this book. Only a summary of his analysis and results will be presented.

An acoustic pressure wave represents a fluctuation, or perturbation, of the pressure in a medium. The total pressure p_t is the sum of the acoustic fluctuations p and the ambient pressure p_0:

$$p_t = p_0 + p \qquad (5-6)$$

The pressure fluctuation will be governed by the wave equation for a homogeneous medium in which the sound velocity and density do not vary, and can be expressed as (Kinsler et al. 1982)

$$\nabla^2 p = \frac{1}{c^2} \frac{\partial^2 p}{\partial t^2} \qquad (5-7)$$

where c is the sound velocity of the medium and ∇ is the divergence operator defined as

$$\nabla = \hat{i} \frac{\partial}{\partial x} + \hat{j} \frac{\partial}{\partial y} + \hat{k} \frac{\partial}{\partial z} \qquad (5-8)$$

$$\nabla^2 = \nabla \cdot \nabla = \frac{\partial^2}{\partial x^2} + \frac{\partial}{\partial y^2} + \frac{\partial^2}{\partial z^2} \qquad (5-9)$$

Aroyan considered a two-dimensional case using a CT scan of a parasagittal section of a *Delphinus delphis* provided by Ted Cranford. The CT scan image was digitized with a digitizing tablet. The CT scan and the resulting digitized outline for the skull and skin boundaries are shown in Figure 5.18. For this geometry, there is no dependency on the z-axis (assuming cartesian coordinates in which the horizontal axis is the x-axis, the vertical axis is the y-axis, and the axis perpendicular to the plane of the figure, sticking out of the page, is the z-axis). Therefore, from (5-7) the wave equation in a homogeneous medium can be expressed as

$$\frac{1}{c^2} \frac{\partial^2 p}{\partial t^2} = \frac{\partial^2 p}{\partial x^2} + \frac{\partial^2 p}{\partial y^2} \qquad (5-10)$$

Aroyan (1990a) first assumed a homogeneous medium (melon and water having the same constant density and sound velocity), which was essentially the "skull-only" case shown in Figure 5.18B. A two-dimensional grid of $1,100 \times 1,100$ points (1.5 mm between points) was overlayed on this graph and equation (5-10) was solved numerically using the "finite difference" method (Smith 1985) on a Cray supercomputer (Aroyan 1990b). Cylindrical wave propagation was assumed so that the pressure in the far field decreased as a function of $1/\sqrt{r}$, instead of $1/r$, for spherical waves. The reflection coefficients for an acoustic signal reflecting at different angles of incidence off the skull were calculated and used in the numerical solution. The sound source was assumed to be a continuous harmonic wave (cw) source; frequencies of 50, 100, and 150 kHz were considered. The results of the skull-only simulation for the source at the dorsal bursae, the nasal plug, and the larynx are shown in Figure 5.19. The results of Figure 5.19AB, with the source at the dorsal-bursae location, best matched the measured beam pattern in the vertical plane for echolocation clicks with peak frequencies close to 100 kHz (see Fig. 6.8). Caution must be taken, however, in attempting to directly compare the results in Figure 5.18 with the beam patterns of Figure 6.8. The results shown in Fig. 5.19 are for

Figure 5.18. (*A*) Image of a CT scan of a parasagittal section of a *Delphinus delphis* head. (*B*) Outline of skull and skin boundaries taken from the CT scan, with simulated source position marked. 1, dorsal bursae complex; 2, the nasal plugs just above the nares; 3, the region of the larynx; 4, a point just above the premaxillary bone; 5, a point deep in the melon tissue. (From Aroyan 1990a.)

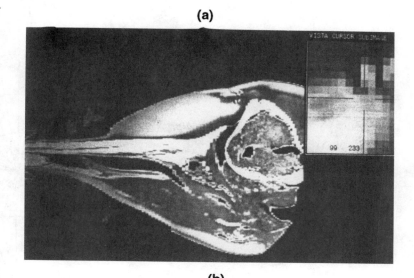

(a)

(b)

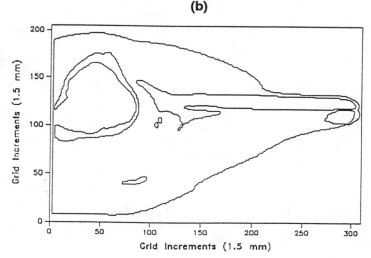

a cw source, for which deep nulls in the beam caused by destructive interferences are evident. A broadband source appropriate to dolphin echolocation signals does not exhibit such deep nulls, as will be shown in Section 6.6. The intensity is plotted on a linear scale here whereas the beams shown in Figure 6.8 are plotted on a log (dB) scale; use of a linear scale will make the beam seem much sharper than use of a log scale.

The results in Figure 5.19 clearly eliminate the larynx as a possible site for the sound generator of sonar signals. Sounds generated at the larynx will be directed downward and away from the front of the dolphin. The results for a hypotheti-

cal source at the larynx are totally incompatible with our understanding of a dolphin's transmitting beam. The results for a source at the nasal plug show a large lobe pointed at about 68° above the horizontal plane, suggesting that the nasal plug location may not be a likely candidate for the actual source site.

Other cases considered by Aroyan (1990a) included the skull with three different arrangements of air sacs, the melon alone, and the full head with skull, air sacs, and the inhomogeneous melon. For the case with an inhomogeneous melon, the appropriate wave equation describing the pressure wave within the melon can be

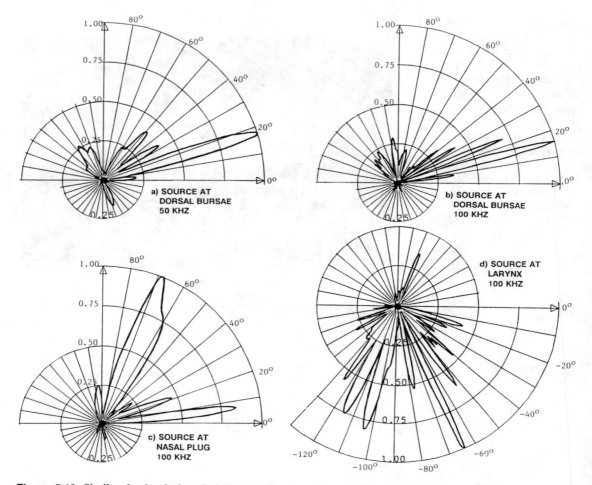

Figure 5.19. Skull-only simulation. Polar plot of acoustic intensity for (A) a 50-kHz cw source at the dorsal bursae; (B) a 100-kHz source at the dorsal bursae; (C) a 100-kHz source at the nasal plug; (D) a 100-kHz source at the larynx. (From Aroyan 1990a.)

expressed as

$$\frac{1}{c^2(x)\rho(x)} \frac{\partial^2 p}{\partial t^2} = \nabla \cdot \left[\frac{1}{\rho(x)} \nabla p \right] \quad (5\text{-}11)$$

where the sound velocity and density vary in the x- and y-direction. Aroyan (1990a) used the CT scan shown in Figure 5.18A to obtain density values within the melon and the sound velocity profile of Norris and Harvey (see Figure 5.12) for *Tursiops*. The simulations with the full head, for cw sources of 50 and 100 kHz, respectively, at the dorsal-bursae location, are shown in Figure 5.20 The results presented are generally consistent with the measured beam in the vertical plane

shown in Figure 6.8. For the melon-only case, Aroyan (1990a) came to the same conclusion arrived at in the previous subsection with ray trace analysis. That is, the velocity profile in the melon is capable of some focusing, but by itself cannot explain the dolphin's directional beam.

5.4 Summary

Dolphins are capable of producing a variety of underwater tonal sounds and sonar pulses with peak frequencies between 30 and 135 kHz. However, the sites of sound production and the

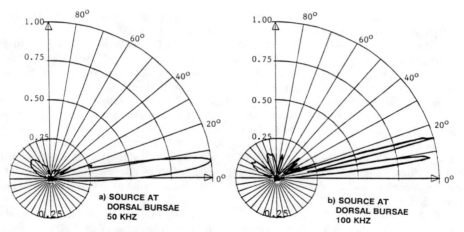

Figure 5.20. Skull, air sacs, and melon simulation. Polar plot of acoustic intensity for a source located at the dorsal-bursae location for a frequency of (A) 50 kHz and (B) 100 kHz. (From Aroyan 1990a.)

mechanisms involved are not known. Recent numerical simulations of sound propagation in the dolphin's head seem to suggest that a likely site of sonar sound production is the dorsal bursae. A variety of different types of measurements suggest that sounds are produced in the skull of the dolphin, in close proximity to the nasal passages. None of the measurements on live phonating animals point to the larynx as a site of sound production. Sonar sounds produced in the skull are focused into the water mainly by the bony structure of the jaw and the various air sacs in the head. The melon has a velocity structure that includes a low-velocity core with a gradual transition to higher velocity towards the skin. The velocity profile of the melon is capable of causing some focusing, but by itself cannot explain the dolphin's directional beam. The melon may function like an impedance transformer, however, coupling the sounds produced deep in the skull into the water.

References

Amundin, M., and Andersen, S.H. (1983). Bony nares air pressure and nasal plug muscle activity during click production in the harbour porpoise, *Phocoena*, and the Bottlenosed Dolphin, *Tursiops truncatus*. J. Exp. Biol. 105: 275–282.

Aroyan, J.L. (1990a). Numerical simulation of dolphin echolocation beam formation. Master of Science Thesis, University of California at Santa Cruz.

Aroyan, J.L. (1990b). Supercomputer modeling of delphinid sonar beam formation. J. Acoust. Soc. Am. 88: S4.

Au, W.W.L. (1980). Echolocation signals of the Atlantic bottlenose dolphin (*Tursiops truncatus*) in open waters. In: R.G. Busnel and J.F. Fish, eds., *Animal Sonar Systems*. New York: Plenum Press, pp. 251–282.

Au, W.W.L., Floyd, R.W., Penner, R.H., and Murchison, A.E. (1974). Measurement of echolocation signals of the Atlantic bottlenose dolphin, *Tursiops truncatus* Montagu, in open waters. J. Acoust. Soc. Am. 56: 1280–1290.

Au, W.W.L., Floyd, R.W. and Haun, J.E. (1978). Propagation of Atlantic bottlenose dolphin echolocation signals, J. Acoust. Soc. Am., 64; 411–422.

Au, W.W.L., Penner, R.H., and Kadane, J. (1982). Acoustic behavior of echolocating Atlantic bottlenose dolphins. J. Acoust. Soc. Am. 70: 687–693.

Au, W.W.L., Carder, D.A., Penner, R.H., and Scronce, B.L. (1985). Demonstration of adaptation in beluga whale echolocation signals. J. Acoust. Soc. Am. 77: 726–730.

Blevins, C., and Parkins, B. (1973). Functional anatomy of the porpoise larynx. Am. J. Anat. 138: 151–

Blomberg, J. (1974). Unusual lipids. II. Head oil of the North Atlantic pilot whale, *Globicephala melaena*. Lipids 11: 461–470.

Blomberg, J., and Jensen, B.N. (1976). Ultrasonic studies of the head oil of the North Atlantic pilot whale (*Globicephala melaena*). J. Acoust. Soc. Am. 60: 755–758.

Blomberg, J., and Lindholm, L.E. (1976). Variations in lipid composition and sound velocity in melon from the North Atlantic pilot whale, *Globicephala melaena*. Lipids 11: 153–156.

Cranford, T.W. (1985). Quantitative morphology of delphinid cephalic anatomy. Sixth Biennial Conf. on the Biol. of Mar. Mamm., Vancouver, British Columbia, Nov. 22–26, (A).

Cranford, T.W. (1988). The anatomy of acoustic structures in the spinner dolphin forehead as shown by X-ray computed tomography and computer graphics. In: P.E. Nachtigall and P.W.B. Moore, eds., *Animal Sonar: Processes and Performances*. New York: Plenum Press, pp. 67–77.

Cranford, T.W. (1989). Comparative morphology of the odontocete nose: anatomy of a sonar signal generator. 8th Biennial Conf. on the Biol. of Mar. Mamm., Pacific Grove, Ca., Dec. 7–11, (A).

Cranford, T.W., Amundin, M., and Bain, D.E. (1987). A unified hypothesis for click production in odontocetes. Seventh Biennial Conf. on the Biol. of Mar. Mamm., Miami, Florida, Dec. 5–9, (A).

Diercks, K.J., Trochta, R.T., Greenlaw, C.F., Evans, W.E. (1971). Recording and analysis of dolphin echolocation signals. J. Acoust. Soc. Am. 49: 1729–17

Dormer, K.J. (1979). Mechanism of sound production and air recycling in delphinids: cineradiographic evidence. J. Acoust. Soc. Am. 65: 229–239.

Evans, W.E. and Prescott, J.H. (1962). Observations of the sound production capabilities of the bottlenose porpoise: a study of whistles and clicks. Zoologica 47: 121–128.

Evans, W.E. (1967). Vocalization among marine animals. In W.N. Tavolga, ed., *Marine Bioacoustics*, Vol. 2, edited by New York: Pergamon Press, pp. 159–186.

Evans, W.E. (1973). Echolocation by marine delphinids and one species of fresh-water dolphin. J. Acoust. Soc. Am. 54: 191–199.

Evans, W.E., and Maderson, P.F.A. (1973). Mechanisms of sound production in delphinid cetaceans: a review and some anatomical consideration, Am. Zool., 13; 1205–1213.

Gurevich, V.S., and Evans, W.E. (1976). Echolocation discrimination of complex planar targets by the beluga whale (*Delphinapterus leucas*). J. Acoust. Soc. Am. 60 (Suppl. 1): S5.

Herman, L.M., and Tavolga, W.N. (1980). The communication systems of cetaceans. In L.H. Herman, ed., *Cetacean Behavior: Mechanisms and Function*. New York: Wiley-Interscience, pp. 149–209.

Hollien, H., Hollien, P., Caldwell, D.K., and Caldwell, M.C., (1976). Sound production by the Atlantic bottlenose dolphin, *Tursiops truncatus*, Cetology, 26: 1–7.

Kamminga, C., and Beitsma, G.R. (1990). Investigations on cetacean sonar IX. Remarks on dominant sonar frequencies from *Tursiops truncatus*. Aquat. Mamm. 16: 14–20.

Kamminga, C., and Wiersma, H. (1981). Investigations of cetacean sonar II. Acoustical similarities and differences in odontocete sonar Signals. Aquat. Mamm. 8: 41–62.

Kinsler, L.E., Frey, A.R., Coppens, A.B., and Sanders, J.V. (1982). *Fundamentals of Acoustics*, 3rd Edition. New York: John Wiley and Sons.

Konig, W.F., Lambert, L.B., and Schilling, D.L. (1961). The bandwidth insertion loss and reflection coefficient of ultrasonic delay lines for backing materials and finite thickness bonds. IRE Int. Conv. Rec. 9: 285–295.

Larson III, J.D. (1983). An acoustic transducer array for medical imaging. Part I. Hewlett-Packard Journal 34: 17–22.

Lilly, J.C., and Miller, A.M. (1961). Sounds emitted by the bottlenosed dolphin. Science 133: 1689–1693.

Litchfield, C., Ackman , R.G., Sipos, J.C., and Eaton, C.A. (1971). Isovaleroyl triglycerides from the blubber and melon oils of the beluga whale (*Delphinapterus leucas*). Lipids 6: 674–681.

Litchfield, C., Karol R., Greenberg, A.J. (1973). Compositional topography of melon lipids in the Atlantic bottlenose dolphin (*Tursiops truncatus*): implications for echolocation. Mar. Biol. 23: 165–169.

Litchfield, C., Karol, R., Mullen, M.E., Dilger, J.P., and Luthi, B. (1979). Physical factors influencing refraction of the echolocative sound beam in delphinid cetaceans. Mar. Biol. 52: 285–290.

Mackay, R.S. and Liaw, C. (1981). Dolphin vocalization mechanisms, Science, 212, 676–678

Moore, P.W.B., and Pawloski, D.A. (1990). Investigations on the control of echolocation pulses in the dolphin (*Tursiops truncatus*). In: J.A. Thomas and R.A. Kastelein, eds., *Sensory Abilities of Cetaceans: Laboratory and Field Evidence*. New York: Plenum Press, pp. 305–316.

Morris, R.J. (1986). The acoustic faculty of dolphins. In: M.M. Bryden and R. Harrison, eds., *Research on Dolphins*. Oxford: Clarendon Press, pp. 369–399.

Norris, K.S. (1964). Some problems of echolocation in cetaceans. In: W. Tavolga, ed. *Marine Bio-Acoustics*. New York: Pergamon Press, pp. 317–336.

Norris, K.S. (1968). The evolution of acoustic mechanisms in odontocete cetaceans. In: E.T. Drake, ed., *Evolution and Environment*. New Haven: Yale University Press, pp. 297–324.

Norris, K.S., Dormer, K.J., Pegg, J., and Liese, G.T. (1971). The mechanism of sound production and air recycling in porpoises: a preliminary report, *In: Proc. VIII Conf. Biol. Sonar Diving Mammals*, Menlo Park, CA.

Norris, K.S., and Harvey, G. W. (1972). A Theory of the function of the spermaceti organ of the sperm whale (*Physeter catodon* L.) NASA SP-262.

Norris, K.S., and Harvey, G.W. (1974). Sound transmission in the porpoise head. J. Acoust. Soc. Am. 56: 659–664.

Poché L.B., Jr., Luker, L.D., and Rogers, P.H. (1977). Improving the acoustic properties of a dolphin tank. J. Acoust. Soc. Am. 62: S89 (A).

Poché, L.B., Jr., Luker, L.D., and Rogers, P.H. (1982). Some observation of echolocation clicks from free-swimming dolphins in a tank. J. Acoust. Soc. Am. 71: 1036–1038.

Purves, P.E. (1967). Anatomical and experimental observations on the cetacean sonar system. In: R.G. Busnel, ed., *Animal Sonar Systems: Biology and Bionics*, Vol. 1. Laboratoire de Physiologie Acoustique, Jouy-en-Jouy, France), pp. 197–270.

Purves, P.E., and Pilleri, G. (1983). *Echolocation in Whales and Dolphins*. London: Academic Press.

Ridgway, S.H. (1965). Medical care of marine mammals. J. Am. Vet. Med. Assoc. 147, 1077–1086.

Ridgway, S.H., and Carder, D.A. (1988). Nasal pressure and sound production in an echolocating white whale, *Delphinapterus leucas*. In: P.E. Nachtigall and P.W. B. Moore, eds., *Animal Sonar: Processes and Performances*. New York: Plenum Press, pp. 53–60.

Ridgway, S.H., Carder, D.A., Green, R.F., Gaunt, A.S., Gaunt, S.L.L., and Evans, W.E. (1980). Electromyographic and pressure events in the nasolaryngeal system of dolphins during sound production. In: R.G. Busnel and J.F. Fish, eds., *Animal Sonar Systems*. New York: Plenum Press, pp. 239–249.

Schenkkan, E.J. (1973). On the comparative anatomy and function of the nasal tract in odontocetes (Mammalia, Cetacea). Bijdragen tot de Dierkunde 43: 127–159.

Smith, G.D. (1985). *Numerical Solution of Partial Differential Equations: Finite Difference Methods*, 3rd Edition. Oxford: Oxford University Press.

Thomas, J.A., and Turl, C.W. (1990). Echolocation characteristics and range detection by a false killer whale (*Pseudorca crassidens*). In: J.A. Thomas and R.A. Kastelein, eds., *Sensory Abilities of Cetaceans: Laboratory and Field Evidence*, New York: Plenum Press, pp. 321–334.

Thomas, J., Stoermer, M., Bowers, C., Anderson, L., and Garver, A. (1988). Detection abilities and signal characteristics of echolocating false killer whale (*Pseudorca crassidens*). In P. Nachtigall, ed., *Animal Sonar Systems II*. New York: Plenum Press, pp. 323–328.

Turl, C.W., Skaar, D.J., and Au, W.W.L. (1991). The echolocation ability of the beluga (*Delphinapterus leucas*) to detect targets in clutter. J. Acoust. Soc. Am. 89: 896–901.

Varanasi, U., and Malin, D.C. (1971). Unique lipids of the porpoise (*Tursiops gilli*): differences in triacylglycerols and wax esters of acoustic (mandibular canal and melon) and blubber tissues. Biochem. Biophys. Acta 231: 415–418.

Varanasi, U., and Malin, D.C. (1972). Triacylglycerols characteristics of porpoise acoustic tissues: molecular structure of diisovaleroylglycerides. Science 176: 926–928.

Varanasi, U., Feldman, H.R., and Malins, D.C. (1975). Molecular basis for formation of lipid sound lens in echolocating cetaceans. Nature 255: 340–343.

Varanasi, U., Markey, D., and Malins, D.C. (1983). Role of isovaleroyl lipids in channeling of sound in the porpoise melon. Chem. and Phys. of Lipids 31: 237–244.

Wilson, O.B. (1985). *An Introduction to the Theory and Design of Sonar Transducers*. Washington D.C.: U.S. Government Printing Office.

Wood, F.G. (1964). Discussion. In: W. Tavolga, ed., *Marine Bio-Acoustics*, Vol. 2. Oxford: Pergamon Press, England), pp. 395–396.

6

Characteristics of the Transmission System

6.1 The Concept of Near and Far Acoustic Fields

As sonar signals propagate out of the dolphin's head, a definite acoustic field is formed that is governed by the size of the animal's head, the peak frequency of the signals, the area of the sound production mechanism, and the amount of focusing within the melon and at the melon–water interface. The acoustic *near field* will form in the vicinity of the dolphin's head as it would near the transmitting surface of any energy radiator. The concept of a near field is often difficult to grasp since many explanations are embroiled in mathematics that seem to confuse rather than clarify the issue. In this section, a simple model will be used to discuss the notion of the near and far acoustic field of a radiator and to provide insights into the physics involved with the propagation of acoustic energy from a dolphin's head.

Consider the circular disc of radius a shown in Figure 6.1, and assume that there are two point sources of acoustic energy spaced a distance a apart, one at the origin and the other at the edge of the disc. Both point sources are being driven by a harmonic or sinusoidal signal generator $(Be^{j\omega t})$. The acoustic pressure at an observation point on the y-axis will be the sum of the pressure from both point sources, which can be expressed as

$$p = A e^{j\omega t} \left(\frac{e^{-jkr}}{r} + \frac{e^{-jkr_s}}{r_s} \right) \qquad (6\text{-}1)$$

where $k = 2\pi/\lambda$, A is an amplitude constant, and r_s is the slant range from point source #2 to the observation point and is equal to

$$r_s = \sqrt{(a^2 + r^2)} \qquad (6\text{-}2)$$

The amplitude of the pressure is the absolute value of the expression in (6-1), which can be expressed as

$$|p(r)|$$
$$= A \sqrt{\left(\frac{\cos k_r}{r} + \frac{\cos k_s}{r_s} \right)^2 + \left(\frac{\sin k_r}{r} + \frac{\sin k_s}{r_s} \right)^2} \qquad (6\text{-}3)$$

The acoustic pressure for a continuous wave (cw) as expressed by (6-3) is plotted as a function of r/a (dashed line) in Figure 6.2. In the near field, the acoustic pressure oscillates with range, exhibiting a number of peaks and nulls. These oscillations are present because the distance (r_s and r) from each point source to the observation point is different so that the sinusoidal waves from the point sources arrive out of phase with one another, causing either constructive or destructive interferences. The phase difference between the waves from the two point sources is related to the range difference ($\Delta r = r_s - r$) and the wavelength (λ) of the signal and can be expressed as

$$\Delta\phi = \Delta r \frac{2\pi}{\lambda} \qquad (6\text{-}4)$$

When Δr equals an integer multiple of $\lambda/2$, the

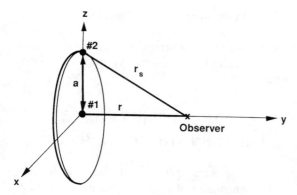

Figure 6.1. Simple model of a disc of radius a with two point sources of acoustic energy separated by the radius of the disc. Each point source is driven by the same sinusoidal signal generator. This model will be used to describe the acoustic near field generated by the two point sources and by the disc transducer.

$$r_s = \sqrt{a^2 + r^2} \approx r \qquad (6\text{-}5)$$

the acoustic pressure of (6-3) can now be expressed as

$$|p(r)| \approx 2\frac{A}{r} \qquad (6\text{-}6)$$

The *far field* is usually defined as starting at the range at which the acoustic pressure begins to decay as a function of $1/r$. This range cannot be defined precisely since the acoustic field makes a gradual transition to the far field. More on this topic will be said in Section 6.6, in the context of modeling the dolphin projection system by an equivalent aperture transducer. The near field is also referred to as the *Fresnel zone* and the far field as the *Fraunhofer zone* in wave propagation studies associated with optics, acoustics and electromagnetics.

The acoustic field created by short duration broadband signals such as the sonar signals of dolphins will be slightly different from that of the cw case just discussed. Assume in Figure 6.1 that the sinusoidal signal generator is replaced by a click generator so that $Ae^{j\omega t}$ can be replaced by $s(t)$, where $s(t)$ is defined by equation (5-1). The acoustic pressure along the y-axis can now be expressed as

$$p = \frac{s(t - \tau)}{r} + \frac{s(t - \tau_s)}{r_s} \qquad (6\text{-}7)$$

where

two waves will be 180° out of phase and will experience the maximum amount of destructive interference, and when Δr equals zero or an integral multiple of λ, the two waves will be in phase and will experience the maximum amount of constructive interference. For short ranges (near the disk), Δr and consequently $\Delta\phi$ vary considerably with range. For large ranges, Δr and $\Delta\phi$ hardly vary at all (approach zero) so that the waves from the point sources arrive nearly in phase. If we consider the case in which $r \gg a$ so that

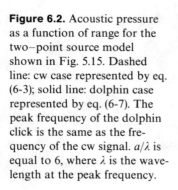

Figure 6.2. Acoustic pressure as a function of range for the two–point source model shown in Fig. 5.15. Dashed line: cw case represented by eq. (6-3); solid line: dolphin case represented by eq. (6-7). The peak frequency of the dolphin click is the same as the frequency of the cw signal. a/λ is equal to 6, where λ is the wavelength at the peak frequency.

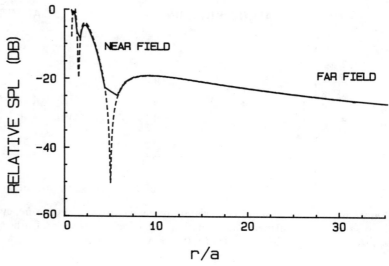

$$\tau = \frac{r}{c} \qquad (6\text{-}8)$$

$$\tau_s = \frac{\sqrt{a^2 + r^2}}{c} \qquad (6\text{-}9)$$

$$s(t - \tau_n) = \begin{cases} s(t) & \text{for } t \geq \tau_n \\ 0 & \text{for } t < \tau_n \end{cases} \qquad (6\text{-}10)$$

The acoustic pressure for a simulated dolphin signal ($\tau_0 = 9.9\ \mu s$ and $\Delta\tau = 29.9\ \mu s$ and $\phi = 3\pi/4$ in (5-1)) plotted as a function of range is depicted in Figure 6.2 (solid curve). The peak frequency of the simulated dolphin signal is the same as the frequency used in the cw case. Notice that in the near field, the pressure does not exhibit large amplitude oscillations as in the cw sinusoidal case. Nevertheless, there still exist a near-field region in which there is an interference effect from the two sources, and a far-field region where the pressure drops off smoothly as a function of range. In the far field, where $r \gg a$, expressions (6-5) and (6-9) reduce to $r_s \approx r$ and, $\tau_s \approx \tau$, respectively, so that

$$p \approx 2\frac{s(t - \tau)}{r} \qquad (6\text{-}11)$$

The acoustic field of a planar aperture can be thought of as being created by a summation of an infinite number of infinitesimally small point sources covering the aperture. As in the simple case of two point sources, fluctuations in the near-field pressure will be caused by signals from the different point sources arriving with different phases and interfering with each other, producing regions of constructive and destructive interference.

6.2 The Dolphin Near- to Far-Field Transition Region

Let us now examine the near-to-far acoustic field transition region of an echolocating dolphin. A series of measurements was conducted by Au et al. (1978) to determine the distance from the tip of a dolphin's rostrum at which the sonar acoustic field began to satisfy the far-field condition (pressure decreasing in amplitude as a function of $1/r$). Miniature hydrophones (Celesco LC-10) were placed at different distances from the tip of a rigid chin-cup station aligned in the direction of the target to measure the variation in the sound pressure level (SPL) as a function of distance away from the animal. The results of our measurements with the hydrophones aligned either approximately 15° below the major axis and along the major axis of the dolphin's transmission beam are plotted in Figure 6.3. The ordinate is the relative SPL at a distance r referenced to the SPL at 1 m from the tip of the dolphin's rostrum. The dashed curve represents an acoustic field decreasing at a rate of $1/r$, with

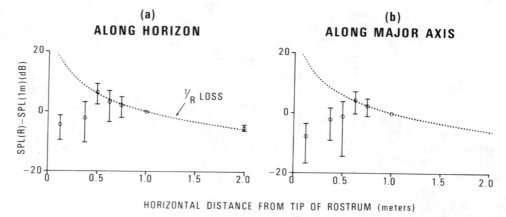

Figure 6.3. Mean and standard deviation of the relative sound pressure level within the dolphin's sonar field across the transitional region between the acoustic near and far fields. Dashed curve: acoustic field with spherical spreading loss, decaying at a rate of $\frac{1}{r}$. (From Au et al. 1978.)

a reference point at 1 m. Since the measurements performed with the hydrophones aligned along the major axis of the vertical beam should more accurately represent the propagation condition of the signals, Figure 6.3b should be used to determine the onset of the far field. The graph shows that the transition region between the near and far fields exists between 0.500 and 0.625 m from the tip of the animal's rostrum for the high frequency sonar signals (peak frequencies between 110 and 130 kHz) emitted by *Tursiops* in Kaneohe Bay. Au et al. (1987) also measured the acoustic field of an echolocating *Delphinapterus leucas* with an array of Brüel and Kjaer 8103 hydrophones. For this measurement, a bite plate was used to keep the whale in a stationary position while performing its sonar searches. Video recordings from an underwater television camera were used to monitor the beluga's position on the bite plate during the acoustic tape recording of the animal's signals. The results of the acoustic measurement made along the major axis of the animal's sonar beam are shown in Figure 6.4. From the diagram we can determine that the transition region between the near and far fields exists between 0.62 and 0.75 m from the tip of the animal's mouth, or approximately 1.03 and 1.15 m from the assumed source, for the high frequency sonar signals (peak frequencies between 100 and 120 kHz) emitted by the beluga in Kaneohe Bay. The head of a *Delphinapterus* is slightly larger than that of a *Tursiops*; therefore, the far field should start at a further distance than

for *Tursiops* given that peak frequencies are similar.

The experimental results displayed in Figures 6.3 and 6.4 show considerable variations. The standard deviations were greatest in the near-field region, which is to be expected. Differences in the animals' position from trial to trial and within a trial probably contributed to the large variations. From Figure 6.2 we can see that slight variations in distance can result in large variations in the measured acoustic pressure.

6.3 The Acoustic Field on a Dolphin's Head

The acoustic field in the immediate vicinity of a dolphin's head should be extremely complex because of the presence of air sacs, cranial bones, and the complex structure of the melon causing reflection and refraction of the acoustic energy generated within the head. Simultaneous measurements of the signals in the near and far fields of a *Tursiops* were made by Au et al. (1978) using a contact hydrophone placed on either the dolphin's melon or rostrum and another hydrophone located 1 m from the tip of the dolphin's rostrum, in line with the longitudinal axis of the chinal. Examples of the averaged signals of two echolocation click trains measured with the contact hydrophone placed on the melon and of two click trains with the contact hydrophone placed on the rostrum are shown in Figure 6.5. Also

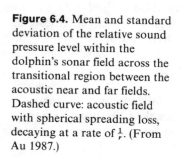

Figure 6.4. Mean and standard deviation of the relative sound pressure level within the dolphin's sonar field across the transitional region between the acoustic near and far fields. Dashed curve: acoustic field with spherical spreading loss, decaying at a rate of $\frac{1}{r}$. (From Au 1987.)

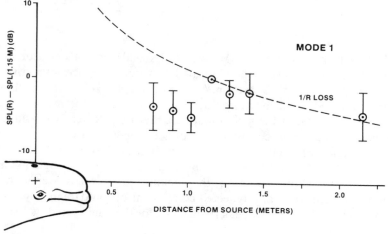

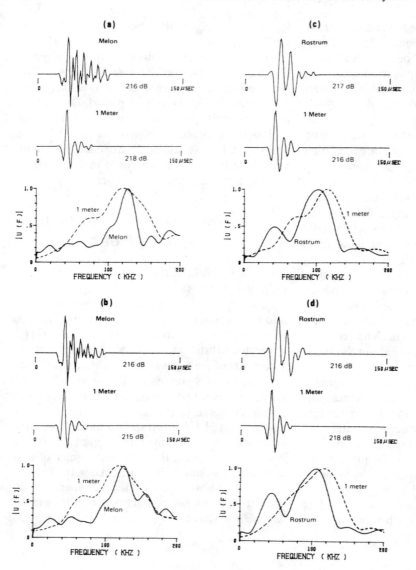

Figure 6.5. Sonar signals of a *Tursiops* measured simultaneously at the melon (A, B) or rostrum (C, D) and at 1 m from the tip of the rostrum. (From Au 1980.)

included in the figure are the averaged signals measured by the hydrophone located 1 m from the tip of the dolphin's rostrum. A comparison of the averaged signal measured at 1 m with the signal measured at the melon indicates that the signal at the melon consisted of a direct signal and internally reflected signals from several locations within the head. Each cusp in the time-domain signal measured at the melon may indicate the arrival of a reflected component of the generated pulse. The peak frequencies for the

signals measured at the melon were 8 and 6 kHz higher, respectively, than the signals at 1 m. The 3-dB bandwidth was also smaller for the signals at the melon. The higher peak frequencies and lower bandwidths were probably due to the many cusps associated with the signals measured at the melon.

Although in all cases the signals measured at 1 m were similar in shape in both the time and frequency domain, the corresponding signals measured at the rostrum and melon were considerably different. The signals measured at the rostrum were not as complex as those at the melon. The waveforms for the rostrum signals seem to suggest that they were formed by the superposition of a direct and a single internally reflected or refracted pulse. The peak frequency of each rostrum signal was 11.7 kHz lower than the signal at 1 m. The signals at the melon seemed to be formed by a large number of internally reflected pulses that arrived very close together soon after the arrival of the direct pulse. This type of propagation is not surprising since the melon region is close to the air sacs and areas of the skull that could reflect acoustic energy into the melon region.

The amplitude of the signals measured by contact hydrophones placed on the melon and rostrum of the dolphin in relationship to the signals measured in the far field was also studied by Au et al. (1978). The dolphin was trained to station in a chin cup that was supported by surgical rubber tubing. A vertical array of three hydrophones was placed 1 m from the animal's rostrum and another hydrophone was placed at 85 m, 5 m beyond the target. Two separate cases were considered, the first with the contact hydrophone placed on the rostrum and the second with the hydrophone on the melon. Since the averaged values of the SPL measured at 1 m were within 0.9 dB for the two cases, the results for both conditions were combined. The results in terms of peak-to-peak SPL are shown in Figure 6.6, with the SPL at 1 m obtained from the hydrophone in the vertical array which detected the largest signal.

The dashed curve represents the variation in SPL as a function of range taking in to account spherical spreading and absorption losses. An absorption coefficient of 0.044 dB/m, corresponding to a typical Kaneohe Bay water temperature 24°C and salinity of 35 ppt, and a peak frequency of 120 kHz were used. There is good agreement between the SPLs at 1 and 85 m after transmission loss has been accounted for. The mean SPLs at 1 m, at the rostrum, and at the melon were 217.7, 217.2, and 215.8 dB re 1 μPa, respectively. Diercks et al. (1973; also reported by Evans 1973), using an array of contact hydrophones, found that the amplitude of a signal

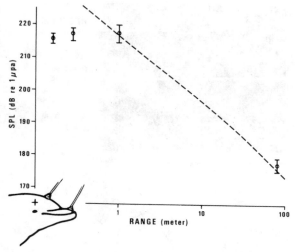

Figure 6.6. Peak-to-peak sound pressure level of projected sonar signals as a function of range. (From Au et al. 1978.)

measured at the melon was approximately equal to that measured at the rostrum. The 1.4-dB difference is in general agreement with their results. The relationship between the signals measured at the dolphin's head and at 1 m is consistent with acoustic propagation within the near field of a projector, as will be seen later in this chapter (Section 6.6).

An example of a sonar signal measured at the rostrum and at ranges of 1 and 85 m from the rostrum was also presented by Au et al. (1978) and is reproduced here in Figure 6.7. The waveshape and corresponding frequency spectrum measured with the contact hydrophone are considerable different from those of the signals measured at 1 and 85 m. This is to be expected, since the contact hydrophone was in the near field of the source, so that signals radiated from different sections of this finite source should tend to interfere with each other. The slight differences in the signals measured at 1 and 85 m are due primarily to the dissimilarity in the hydrophone responses to broadband signals. The effects of frequency-dependent absorption losses on the signal measured at 85 m should be small since there is only a 1.3-dB difference in absorption losses between 50 and 100 kHz at a range of 85 m.

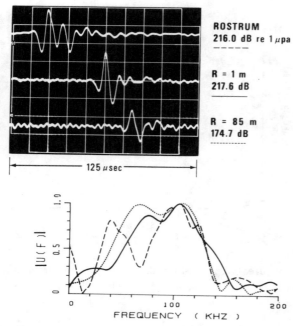

Figure 6.7. Example of a single sonar signal measured by hydrophones located at the rostrum and at ranges of 1 and 85 m. (From Au et al. 1978.)

6.4 Directional Pattern of Biosonar Signals

The outgoing echolocation signals of dolphins and other odontocetes have been shown by a number of investigators to be directional. Early work by Norris et al. (1961) indicated that a blindfolded *Tursiops* could not detect targets below its jaws and at elevation angles greater than 90° above the rostrum. Evans et al. (1964) used a *Stenella longirostris* cadaver and a *Tursiops* skull to study the directional properties of sonar signals. With a cw sound source placed in the region of the nasal sacs, they found that a definite beam was formed, which was directed 15° above the rostrum in the vertical plane and forward in the horizontal plane. The beam was highly dependent on frequency, becoming narrower as the frequency increased. Norris and Evans (1966) measured the directionality of a *Steno bredanensis* sonar signal by systematically moving a single hydrophone placed at a fixed depth to different azimuths as the animal performed sonar searches

on a target. The sound was found to be directed forward in a narrow beam and was highly frequency dependent. Schevill and Watkins (1966) studied the directionality of clicks produced by an *Orcinus orca* and found that the frequency content and amplitude of the clicks varied with the orientation of the animal. As the whale turned away from the hydrophone, the higher frequency components of the clicks would progressively diminish in amplitude. Evans (1973) used an array of hydrophones spaced 5° apart in the horizontal plane to measure the horizontal beam pattern of a restrained *Tursiops*.

The transmission beam patterns in the vertical and horizontal planes for *Tursiops* were measured by Au et al. (1978), Au (1980), and Au et al. (1986) using an array of hydrophones in the vertical and horizontal planes. The results of these three measurements on three different *Tursiops* are shown in Figure 6.8, where the beam patterns represent average values obtained by Au (1980) for 10 trials in each plane (dotted line), by Au et al. (1978) for 9 trials in the vertical plane and 12 trials in the horizontal plane (dashed line), and by Au et al. (1986) for 30 trials in each plane (solid line). The beams represented by the dashed and dotted lines were aligned with the beam represented by the solid line. Using an underwater television monitoring system, and having the dolphin station on a bite plate while resting its tail on a rest bar, Au et al. (1986) were able to control and monitor the position of the dolphin very accurately. They found that the echolocation signals are projected at an elevation angle of 5° above the animal's head in the vertical plane. Au et al. (1978) previously reported that the major axis of the beam was 20° above the rostrum. However, in that study, a chin cup stationing device was used without the animal's position in the chin cup being closely monitored. The tail of the dolphin in the chin cup station may have drooped downward sufficiently to cause the animal's position in the chin cup to be pivoted upwards. The shape and width of the three beams were similar in both the vertical and horizontal planes. The 3-dB and 10-dB beamwidths in the vertical and horizontal planes along with the directivity index obtained by using the beam patterns in equation (3-10) are given in Table 6.1. The 10-dB beamwidths for the three animals

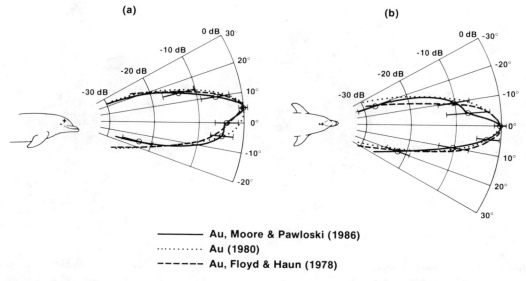

Figure 6.8. Composite broadband transmission beam patterns of three bottlenose dolphins in (A) the vertical plane and (B) the horizontal plane.

Table 6.1. 3-dB and 10-dB beamwidth (BW) and directivity index (DI_T) of the transmitted beams presented in Fig. 6.8

Dolphin	Vertical Beam		Horizontal Beam		Directivity Index
	3-dB BW	10-dB BW	3-dB BW	10-dB BW	DI_T
Sven	10.0°	22.0°	9.8°	21.0°	25.4 dB
Ekahi	12.0°	23.0°	10.7°	22.0°	25.4 dB
Heptuna	8.5°	22.5°	8.5°	20.0°	26.5 dB
Average	10.2°	22.5°	9.7°	21.0°	25.8 dB

differed at most by 1.0° in the vertical plane and by 2.0° in the horizontal plane, indicating the similarity of their beams. Larger variations were found in the 3-dB beamwidth, with the maximum difference being 3.5° in the vertical plane and 2.2° in the horizontal plane. The average 3-dB beamwidths were 10.2° and 9.7° in the vertical and horizontal planes, respectively.

The directivity index for the animals was very similar; the average directivity index was 25.8 dB. In Chapter 3, we saw that the receiving directivity index for *Tursiops* at a frequency of 120 kHz is approximately 20.8 dB. Therefore, the transmitting beam is more directional than the receiving beam by 5 dB. The relative spatial coverage of transmitting and receiving beams is depicted in Figure 6.9, which shows the receiver beam at 120 kHz overlaying the average transmitting beam obtained from Figure 6.8. The polar plot origin used in both figures is only an approximation since the signal is produced in the upper part of the animal's head and signals are received via the lower jaw. Nevertheless, Fig. 6.9 is useful in conveying the relative spatial coverage of the two beams.

Au (1980) also examined the waveform and corresponding frequency spectrum detected by each of the hydrophones in the array. Examples of the averaged waveform and frequency spectrum measured by the array in the vertical plane

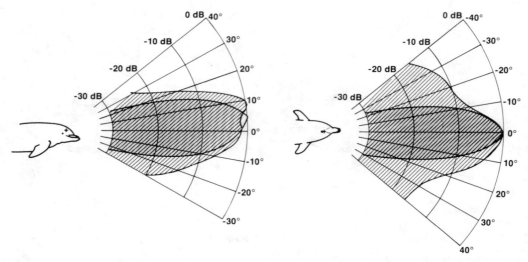

Figure 6.9. Relative spatial coverage of the receiving (for a frequency of 120 kHz) and transmitting beams in the vertical and horizontal planes.

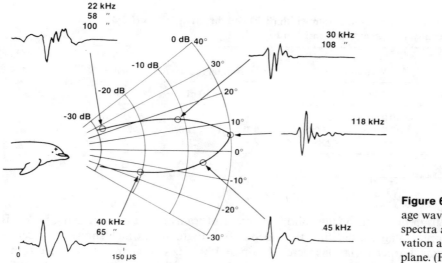

Figure 6.10. Examples of average waveforms and frequency spectra as a function of the elevation angle in the vertical plane. (From Au 1980.)

for a single trial are displayed in Figure 6.10. As the angle departed from the beam axis, the signals in the time domain became progressively distorted relative to the signal on the major axis at +5°. In the frequency domain, the peak frequencies decreased as the hydrophone angle departed from 0°. The presence of multipaths is apparent for the signals measured above the beam axis, at +15° and +25°. The multipaths

may be due to internal reflection and refraction of the signals within the animal's head, to the signal being radiated from different portions of a finite source region, or to combinations of these possibilities. For signals measured above the major axis, there are still considerable amounts of energy at frequencies close to the peak frequency of the signal on the major axis, causing the spectra to have multiple humps or peaks. This

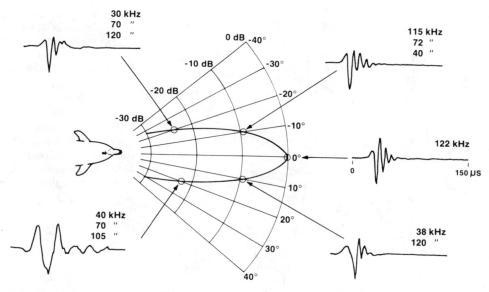

Figure 6.11. Examples of average waveforms at different azimuths in the horizontal plane. The peak frequency of each waveform is shown to the right. For frequency spectra with multiple peaks, the frequencies of the peaks are listed in descending order according to amplitude. (From Au 1980.)

is not true for signals measured below the major axis: there seems to be a general shift to lower peak frequencies, with relatively little energy close to the peak frequency of the signal on the major axis. The signals in the time domain appear to be stretched with few multipath effects.

The average waveforms and their corresponding frequency spectra as detected by the five hydrophones in the horizontal array are displayed for a single trial in Figure 6.11. As the angle departed from the beam axis, the signal became progressively more distorted relative to the signal on the major axis. Signals measured at $-10°$ were the least distorted, having peak frequencies close to those along the beam axis. As the angle departed from the beam axis, the peak frequencies shifted to lower values, with signals measured at $\pm 20°$ exhibiting the greatest amount of shift. The signals were not symmetrical about the beam axis, which is expected since the structure of the skull is not symmetrical about the midline of the animal.

The transmitting beam pattern of a *Delphinapterus* was measured by Au et al. (1987) using a vertical and horizontal array of B&K 8103 hydrophones. The composite vertical beam pattern from 40 trials is shown in Figure 6.12.

The results indicate that the major axis of the vertical beam is elevated at the same angle as for *Tursiops*, 5° above the horizontal axis. The 3-dB and 10-dB beamwidths were approximately 6.5° and 13°, respectively, which is considerably narrower than the average of 10.2° and 22.5° for *Tursiops*.

An example of a single echolocation signal measured by each of the seven hydrophones in the vertical array is also included in Figure 6.12. Hydrophones located away from the major axis detected signals of complex frequency spectra with multiple peak frequencies. Only signals measured close to the beam axis (5° and 10°) were undistorted. Abrupt phase changes in the signals measured away from the beam axis were probably caused by the summation of a direct component with components reflected off air sacs and the skull.

The composite transmitting horizontal beam pattern from 40 trials is shown in Figure 6.13. The results indicate that the major axis of the horizontal beam is pointed directly ahead of the animal, parallel to the longitudinal axis of the whale. The 3-dB beamwidth was the same as for the vertical beam, approximately 6.5°. This beam width was again much smaller than the average

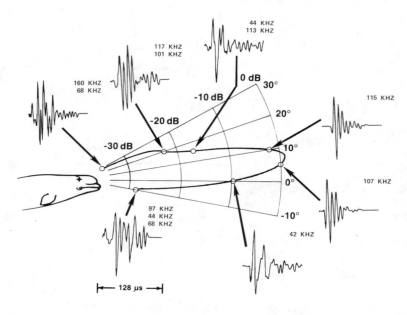

Figure 6.12. Composite vertical beam pattern of 40 trials for a beluga. Also included is an example of the signals measured by the individual hydrophones in the vertical array for a single echolocation signal. Peak frequencies are listed in the upper right of each signal. For signals with multiple peaks in the frequency domain, peaks are listed in descending order according to amplitudes. (From Au et al. 1987.)

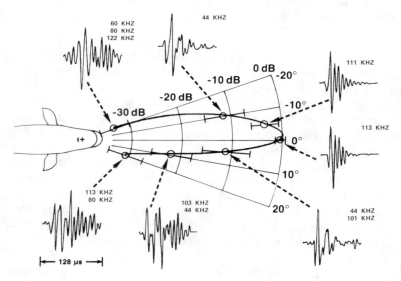

Figure 6.13. Composite horizontal beam pattern of 50 trials, presented as in Fig. 6.12. (From Au et al. 1987.)

of 9.7° for *Tursiops*. The directivity index, calculated from the vertical and horizontal beam patterns in conjunction with (3-21), was found to be 32.1 dB, or 6.3 dB greater than that of *Tursiops*.

An example of the waveforms measured by each of the seven hydrophones in the horizontal array for a single signal is also included in Figure 6.13. As in the vertical beam case, only hydrophones located close to the major axis (0° and

−5°) detected undistorted signals. The effects of interference caused by the arrival of multiple signal components can be seen in the off-axis signals. Their distortion was similar to that of *Tursiops* signals; the amount of distortion increased as the angle from the major axis became greater. The topic of off-axis distortion will be taken up again in the next section after the notion of an equivalent planar aperture has been developed.

6.5 Equivalent Planar Circular Aperture

6.5.1 Transition Region Between Near and Far Fields

The acoustic projection system of the dolphin can be modeled by an equivalent planar transducer that has the same directivity index and the same near-field to far-field transition distance. The goal of this section is to determine the size of an equivalent planar aperture or transducer that would have the same directivity index and near- to far-field transition distance as the dolphin, and to perform some straightforward engineering analysis and modeling. The complex shape of a dolphin's head and the presence of air sacs and a melon with a complex sound velocity profile make it very difficult, if not impossible, to accurately model the propagational characteristics of the sound projection system. Au et al. (1978, 1987) used standard mathematical expressions for the 3-dB beam pattern and the transition distance that are appropriate only for sinusoidal cw signals. Here, we will also consider the effects of using broadband transient signals similar to dolphin sonar signals. While it is not within the scope of this book to perform a detailed derivation of the acoustic field of a planar aperture, some mathematical analysis will be necessary.

Consider a rigid circular piston mounted flush with the surface of an infinite baffle and vibrating with a simple harmonic motion, as shown in Figure 6.14. Assume that the piston is vibrating uniformly with a speed $U_0 e^{j\omega t}$ normal to the baffle. The complex acoustic pressure (see Kinsler et al. 1982, p. 177) on the y-axis can be expressed as

$$p(r, t) = \rho_0 c U_0 e^{j\omega t}(e^{-jkr} - e^{-jk\sqrt{r^2+a^2}}) \quad (6\text{-}12)$$

where ρ_0 is the density and c is the sound velocity of water, $k = 2\pi f/c$ is the wave number, and $\omega = 2\pi f$ is the radial frequency. The amplitude of the acoustic pressure on the y-axis is the magnitude of the above expression, which after some manipulation and use of trigonometric identities involving the products of sin and cos functions can be expressed as

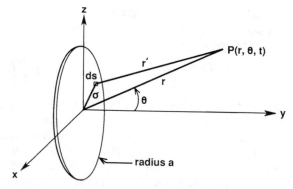

Figure 6.14. Geometry used in deriving the radiation characteristics of a circular piston.

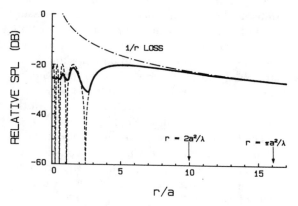

Figure 6.15. Variations in the on-axis acoustic pressure for a circular piston with $\frac{a}{\lambda_0} = 5$. Dashed curve: sinusoidal cw signal; solid curve: simulated dolphin sonar signal. Also included is the $\frac{1}{r}$ spherical spreading loss.

$$p(r) = 2\rho_0 c U_0 \left| \sin\left(\frac{k}{2}[\sqrt{r^2 + a^2} - r]\right) \right| \quad (6\text{-}13)$$

The dashed line in Figure 6.15 shows how the acoustic pressure varies as a function of the range from the circular piston. The pressure varies drastically in the near field, exhibiting a number of oscillations with deep nulls, and decays gradually with range in the far field. The acoustic pressure for a transient signal can be determined by treating (6-12) as a transfer function representing a single spectral component of a generalized piston velocity $U(f)$. Assume that $U(f) = S(f)$, where $S(f)$ is the Fourier transform of the simulated dolphin signal given by (5-1), so that the acoustic pressure for this transient signal can be expressed as

$$p(r,t) = \mathfrak{I}^{-1}[AS(f)(e^{-j2\pi r f/c} - e^{-j2\pi f/c\sqrt{1+(a/r)^2}})]$$

(6-14)

where $\mathfrak{I}^{-1}$ is the symbol for the inverse Fourier Transform, and A is an arbitrary constant. The magnitude of the acoustic pressure is determined by numerically evaluating (6-14) for different ranges r with $\tau_0 = 9.9$ μs, $\Delta\tau = 29.9$ μs and $\phi = 3\pi/4$ in (5-1). The variation in the magnitude of a simulated dolphin sonar signal as a function of range is depicted by the solid curve in Figure 6.15. The magnitude of the dolphin sonar signal does not fluctuate nearly as much as does a sinusoidal signal, and deep nulls in the pressure are not present. The acoustic pressures for both types of signals behave similarly in the transition region between the near and far fields and in the far field. Data of the acoustic field for *Tursiops* and *Delphinapterus leucas* presented in Figures 6.3 and 6.4, respectively, are consistent with the transient signal result of Figure 6.15.

The far field begins, as we said, at the distance from the transducer at which the acoustic pressure decreases as a function of $1/r$. In most physical acoustics texts, this distance is usually defined as

$$r = \pi a^2/\lambda$$

(6-15)

An equivalent aperture size for the dolphin projection system can be calculated with (6-15). From Figure 6.3, the far field for *Tursiops* begins between 0.50 and 0.63 m from the tip of the dolphin's rostrum; we will use a mean value of 0.56 m for our calculation. Au et al. (1978) estimated that the tip of the rostrum is approximately 0.25 m from the acoustic source, so that the far field is established approximately 0.81 m from the source. Therefore, from (6-15) the radius of an equivalent circular aperture is

$$a_{TT} \approx 5.7 \text{ cm}$$

(6-16)

From Figure 6.4, the far field for *Delphinapterus leucas* begins between 1.03 and 1.15 m from the source. Using a mean value of 1.09 m and a frequency of 110 kHz, the radius of an equivalent circular aperture is

$$a_{DL} \approx 7.0 \text{ cm}$$

(6-17)

6.5.2 Directivity Index of the Transmission Beam

The size of a circular aperture that is equivalent to the dolphin's projection system will now be calculated by comparing the directivity index of the dolphin's transmission beam with that of a circular piston. The mathematical expression for the complex pressure in the far field of a circular piston is given by Kinsler et al. (1982, p. 179) as

$$p(r,\theta,t) = jA\frac{k}{r}aU_0e^{j(\omega t - kr)}\frac{2J_1(ka\sin\theta)}{ka\sin\theta}$$

(6-18)

where A is an arbitrary constant, U_0 is the uniform surface velocity of the piston, and J_1 is the Bessel function of the first kind. The angular dependency is in $J(ka\sin\theta)/ka\sin\theta$ term, which goes to unity as θ goes to zero. Thus the beam pattern pattern can be expressed as

$$H(\theta) = \left|\frac{2J_1(ka\sin\theta)}{ka\sin\theta}\right|$$

(6-19)

where $H(\theta)$ is the beam pattern. The beam pattern for a circular piston, calculated from (6-19) for a cw signal with $a/\lambda = 2.3$, is shown in Figure 6.16 (thin solid line).

The beam pattern of a circular piston for broadband transient signals can be determined with the transfer function technique of Mazzola and Raff (1977). The acoustic pressure in the far field of a piston projecting a transient signal can be expressed as

$$p(r,\theta,f) = H(r,\theta,f)U_n(f)$$

(6-20)

where $p(r,\theta,t)$ is the Fourier transform of the acoustic pressure in the time domain, $H(r,\theta,f)$ is the transfer function of the circular piston, and U_n is the Fourier transform of a broadband vibrational velocity on the face of the piston. The expression in (6-20) represents a single spectral component of a generalized piston velocity and can therefore be thought of as the transfer function

$$H(r,\theta,f) = j\frac{A'}{r}f\left[2\frac{J_1(2\pi fa/c\sin\theta)}{2\pi fa/c\sin\theta}\right]e^{-\frac{j2\pi fr}{c}}$$

(6-21)

On the beam axis (y-axis of Fig. 6.16) the acoustic pressure from (6-20) can be expressed as

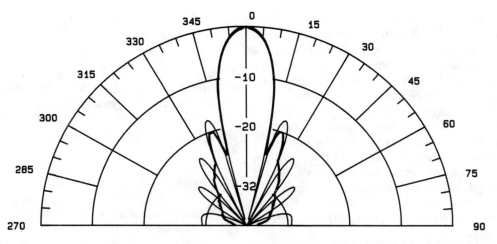

Figure 6.16. Beam pattern of a circular piston for a cw signal (thin line) and a broadband dolphin signal (thick line); $\frac{a}{\lambda} = 2.3$.

$$p(r, 0, f) = jA'\frac{f}{r}e^{-\frac{j2\pi fr}{c}}U(f) \qquad (6\text{-}22)$$

Let the on-axis acoustic pressure equal the simulated dolphin signal expressed by (5-1), so that

$$U(f) = -j\frac{r}{fA'}e^{\frac{j2\pi fr}{c}}S(f) \qquad (6\text{-}23)$$

where S is the Fourier transform of the transient signal $s(t)$ of (5-1). Inserting (6-21) and (6-23) into (6-20) and taking the inverse Fourier transform of the result, we obtain for the acoustic pressure the expression

$$p(r, \theta, t) = \text{Real}\left\{\mathfrak{I}^{-1}\left[\frac{2J_1(2\pi fa/c\sin\theta)}{2\pi fa/c\sin\theta}S(f)\right]\right\}$$
$$(6\text{-}24)$$

The beam pattern is defined as

$$H(\theta) = \frac{\max|p(r, \theta, t)|}{\max|p(r, 0, t)|} \qquad (6\text{-}25)$$

and can be calculated from (6-25). The thick solid line in Figure 6.16 is the beam of a circular disk projecting a simulated dolphin sonar signal. The major lobe of the beams for the transient and cw signals are essentially the same. The beam for the transient signal does not have as much side lobe structure as the cw signal. Therefore, the biosonar beams shown in Figures 6.8, 6.12 and 6.13 seem reasonable in light of this analysis.

Since the major lobe of the beam is essentially the same for transient and cw signals when the peak frequency of the transient signal is equal to the cw frequency, the expression for the directivity index derived for the cw signal can be used. The directivity d for a circular piston is given by Kinsler et al. (1982, p. 185) as

$$d = \frac{(ka)^2}{1 - J_1(2ka)/ka} \qquad (6\text{-}26)$$

The directivity index is the directivity expressed in dB as

$$\text{DI} = 10\log(d) \qquad (6\text{-}27)$$

Equation (6-26) cannot be solved directly for the radius, but it can be used to calculate the directivity at a given peak frequency for different radii, and an interpolation technique is then used to determine the radius for a given directivity. The results presented in Table 6.1 indicate that the directivity index for the transmission beam of *Tursiops* is approximately 25.8 dB. From (6-26) and (6-27), the radius of an equivalent circular aperture is

$$a_{\text{TT}} \approx 4 \text{ cm} \qquad (6\text{-}28)$$

The directivity index calculated for the beluga was 32.1 dB. Therefore, the radius of an equivalent aperture is

$$a_{\text{DL}} \approx 8.9 \text{ cm} \qquad (6\text{-}29)$$

There is a 29% difference for *Tursiops* and a 21% difference for *Delphinapterus* between the radius of an equivalent aperture calculated from the transition field and from the directivity data, respectively. The results obtained with the directivity data should be more representative: transition field measurements are subject to greater variation than beam pattern measurements because large fluctuations in pressure can occur with small changes in distance in the transition region of a projector. It seems safe to conclude that the dolphin sonar projection system can be modeled by a circular transducer with a diameter of approximately 4 cm for *Tursiops* and 8.9 cm for *Delphinapterus*.

6.5.3 Off-Axis Distortion of Biosonar Signals

It was mentioned earlier that distortions in the off-axis signals are probably due to the signal radiating from different portions of a finite aperture and to internal reflection of the signal off air sacs and bones. To gain insight into the finite aperture effect, consider the simple two point-source model of Figure 6.17, where each source is at the edge of a circular disk of radius a. There is an observer on the x–y plane at a distance r from the center of the disk at an azimuth ϕ, where $r \gg a$. From the geometry (see also equations (A-3) and (A-4 of Chapter 3), the distances from the observer to the two point sources can be expressed as

$$r_1 \approx r - a \sin \phi \qquad (6\text{-}30)$$

and

$$r_2 \approx r + a \sin \phi \qquad (6\text{-}31)$$

If both point sources emit a click signal $s(t)$ simultaneously, the signal from point source #1 will arrive at the observer first, followed by the signal from point source #2 arriving with a delay of

$$\Delta t = \frac{r_2 - r_1}{c} = \frac{2a}{c} \sin \phi \qquad (6\text{-}32)$$

The acoustic pressure at the observer position can be expressed as

$$p(\phi) \approx \frac{s(t)}{r} + \frac{s(t - \Delta t)}{r} \qquad (6\text{-}33)$$

Using (5-1) to approximate a dolphin sonar signal, the acoustic pressure as a function of the azimuthal angle from the y-axis is shown in Figure 6.18 for angles of 0° to 20°, in 5° increments. At 0° both signals arrive at the observer at the same instance. As the azimuthal angle increases, the signals from the two sources arrive with a relative delay and interfere with each other, causing the total acoustic pressure to be distorted. The situation for a planar aperture can be approximated by covering the aperture with point sources emitting signals that will interfere with each other as the azimuth increases from 0°. Therefore, off-axis signals from a finite aperture radiating broadband transient-like signals will

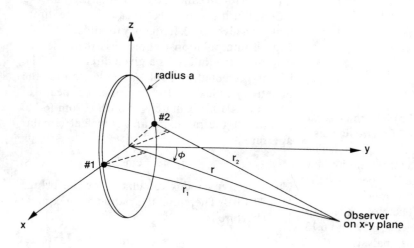

Figure 6.17. Two–point source model to demonstrate off-axis distortion of broadband transient-like signals.

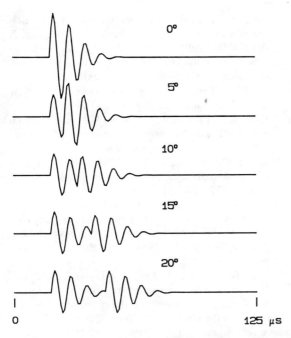

Figure 6.18. Acoustic pressure from the point sources in Fig. 6.17 as a function of the azimuthal angle of the observer, for $\frac{a}{\lambda_0} = 5$.

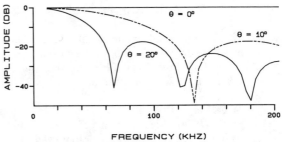

Figure 6.19. Relative magnitude of the transfer function of a circular aperture for $\theta = 0°$, $10°$, and $20°$.

appear distorted because the signals will be radiated simultaneously from different portions of the aperture. For an observer in the far field along the y-axis, the signals from each point source will essentially arrive together, producing the maximum acoustic pressure. The sonar projection system of a dolphin can be considered an aperture of unknown shape; the distortions in off-axis signals are due in part to the aperture having a finite size rather than being a point source.

Off-axis distortion of signals can also be examined by considering the transfer function of the circular aperture expressed in (6-20) and (6-21). It is a function of range, azimuth, and frequency. Let $r = 1$ and let us numerically evaluate $H(r, \theta, f)$ as a function of f for $\theta = 0°$, $10°$ and $20°$. The results are plotted in Figure 6.19. The transfer function for θ not equal to zero resembles the transfer function of a low-pass filter. As the angle θ increases, the cutoff frequency of the low-pass filter decreases, resulting in greater distortions to the radiated signal in the far field.

6.6 Summary

The acoustic field established in the vicinity of a dolphin's head by an outwardly traveling sonar signal does not have sharp nulls in either the near-field pattern and the beam pattern, as would a linear transducer which projects pure tone signals. This characteristic derives from the fact that short-duration broadband clicks do not experience large destructive and constructive interferences from multipaths the way pure tone signals do. The click projection system of dolphins is directional, with a 3-dB beamwidth of approximately 10° in both the vertical and horizontal planes. In the horizontal plane, the beam is pointed directly ahead of the dolphin parallel to the longitudinal axis of the animal. In the vertical plane, the beam is directed between 5° and 10° above the longitudinal axis of the animal. The directivity index for the transmitter beam of *Tursiops* is approximately 26 dB. Beluga whales produce a beam of approximately 6.5°, which is narrower than that of *Tursiops*, in both the horizontal and vertical planes. The corresponding directivity index is about 32 dB, 6.3 dB greater than that of *Tursiops*. The signal projection system of dolphins can be modeled with a circular transducer or aperture. A circular transducer of a radius between 4 and 5.7 cm is equivalent to the signal projection system of *Tursiops*. For a beluga whale, the equivalent circular transducer has a radius between 7 and 8.9 cm.

References

Au, W.W.L. (1980). Echolocation signals of the Atlantic bottlenose dolphin (*Tursiops truncatus*) in open waters. In: R.G. Busnel and J.F. Fish eds., *Animal Sonar Systems*. New York: Plenum Press, pp. 251–282.

Au, W.W.L., Floyd, R.W., Penner, R.H., and Murchison, A.E. (1974). Measurement of echolocation signals of the Atlantic bottlenose dolphin, *Tursiops truncatus* Montagu, in open waters. J. Acoust. Soc. Am. 56: 1280–1290.

Au, W.W.L., Floyd, R.W., and Haun, J.E. (1978). Propagation of Atlantic bottlenose dolphin echolocation signals. J. Acoust. Soc. Am. 64: 411–422.

Au, W.W.L., Moore, P.W.B., and Pawloski, D. (1986). Echolocation Transmitting beam of the Atlantic bottlenose dolphin. J. Acoust. Soc. Am. 80: 688–691.

Au, W.W.L., Penner, R.H., and Turl, C.W. (1987). Propagation of beluga echolocation signals. J. Acoust. Soc. Am. 83: 807–813.

Diercks, K.J., Trochta, R.T., and Evans, W.E. (1973). Delphinid sonar: measurement and analysis. J. Acoust. Soc. Am. 54: 200–204.

Evans, W.E. (1973). Echolocation by marine delphinids and one species of fresh-water dolphin. J. Acoust. Soc. Am. 54: 493–503.

Evans, W.E., Sutherland, W.W., and Beil, R.G., (1964). The directional characteristics of delphinid sounds. In: W.N. Tavolga, ed., *Marine Bioacoustics*. New York: Pergamon Press, pp. 353–372.

Kinsler, L.E., Frey, A.R., Coppens, A.B., and Sanders, J.V. (1982). *Fundamentals of Acoustics*, 3rd Edition. New York: John Wiley and Sons.

Mazzola, C.J., and Raff, A.I. (1977). On the generation of transient acoustic pulses in water. J. Sound and Vib. 53: 375–388.

Norris, K.S., and Evans, W.E. (1966). Directionality of echolocation clicks in the rough-tooth porpoise, *Steno bredanensis* (Lesson). In: W. Tavolga, ed., *Marine Bioacoustics*. New York: Pergamon Press, pp. 305–316.

Norris, K.S., Prescott, J.H., Asa-Dorian, P.V., and Perkins, P. (1961) Experimental demonstration of echolocation behavior in the porpoise *Tursiops truncatus* (Montagu). Biol. Bull. 120: 163–176.

Schevill, W.E., and Watkins, W.A. (1966). Sound structure and directionality in *Orcinus* (killer whale). Zoologica 51: 71–76.

7

Characteristics of Dolphin Sonar Signals

The sonar task for dolphins perceiving their environment involves detection, localization, discrimination, and recognition of objects of interest. Target information such as range, azimuth, direction of movement, speed, and size should also be of interest. The ability of dolphins to accurately perceive their environment and to perform difficult recognition and discrimination tasks depends to a large extent on the characteristics of their sonar signals and how these signals are emitted. The signals must have sufficient energy to detect small targets at large ranges. They must also have sufficient information-carrying capacity so that fine features and characteristics of objects and targets can be determined by analyzing their sonar echoes. At a minimum, a dolphin sonar system should be able to detect and recognize prey, obstacles, and predators. The sonar task is usually performed in a noisy or highly reverberant environment associated with shallow waters, or during searches near the bottom or in the presence of many obstacles.

7.1 Click Intervals

When searching for distant targets, dolphins often emit bursts of clicks normally referred to as a click train. The number of clicks and the time intervals between clicks depend on the a variety of factors such as the distance of interest, the

difficulty of detecting a target, the presence or absence of a target of interest, and on the animal's expectation of finding a specific target. The expression *click interval* will be used instead of *click repetition rate* in describing the sequence in a click train because the time interval often changes from click to click, especially for a moving dolphin.

Click interval studies with free-swimming *Tursiops* have been performed by Johnson (1967), Morozov et al. (1972), and Evans and Powell (1967) as the dolphins echolocated and closed in on a target. All three studies indicated that the dolphins operated in a pulse mode, sending out a click and receiving the target echo before sending out another click after a specific lag time. Lag time is defined as the time difference between the click interval and the two-way transit time an acoustic signal requires to travel from the dolphin to the target and back to the dolphin. Evans and Powell (1967), using a simultaneous audio-video monitoring system, found an almost constant mean lag time of 15.4 ms for target ranges from 1.4 m to 0.4 m. As the animal approached within 0.4 to 0.03 m of the target, the mean lag time decreased to a minimum of 2.5 ms. Morozov et al. (1972) reported a mean lag time of 20 ms, which was fairly constant over a distance of 40 to 4 m as the animal swam toward the target.

A number of experiments have been performed at the Naval Ocean Systems Center, Hawaii Laboratory, in which *Tursiops truncatus, Del-*

phinapterus leucas, and *Pseudorca crassidens* performed sonar discrimination and detection tasks at relatively fixed target ranges varying from 1 m to 120 m. Au et al. (1974) examined the sonar signals associated with two *Tursiops* performing a target detection task between ranges of 55 and 73 m. Two examples showing the variation in the click interval on a pulse-to-pulse basis, for a target range of 64 m, are shown in Figure 7.1. One of the most interesting features of this diagram is the variation in click interval from pulse to pulse. The variation seems to be fairly regular, with the click interval increasing to a peak, then decreasing, and once again increasing, forming a cyclic

pattern of maxima and minima. This pattern was typical of all the click trains examined by Au et al. (1974). Another interesting feature consists of the click intervals for the target-absent cases also being greater than the two-way transit time had a target been present. This implies that the animals had a general expectation of hearing the target echo at a specific time and may have been performing "mental time-gating" of the echoes.

The average click interval as a function of target range for four different experiments with *Tursiops* conducted at our laboratory is shown in Figure 7.2. Also included in the figure is the two-way transit time. Lag times associated with

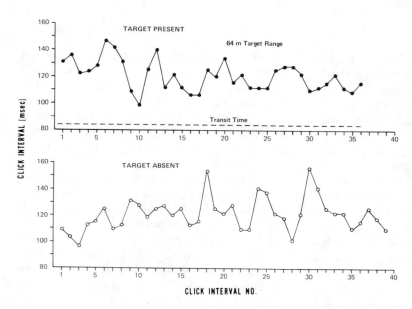

Figure 7.1. Typical variation in click interval during two sonar searches, one with target present and the other with target absent. Two-way transit time is also shown for the target present case. (From Au et al. 1974.)

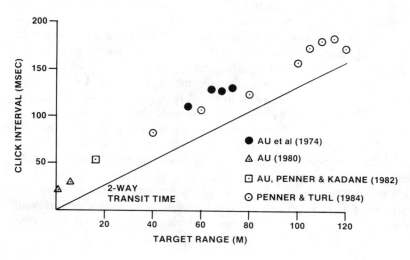

Figure 7.2. Click interval as a function of target range for the bottlenose dolphin. The value of the two-way transit time for any target range can be read off the click interval scale.

Figure 7.2 and those measured for free-swimming dolphins, except for very small target ranges (less than 0.4 m), tend to have values between 19 and 45 ms. If dolphins process received sonar echoes before emitting the next click, then the click interval data suggest a processing time between 19 and 45 ms. At very close ranges (less than 0.4 m), when the lag time decreases to 2.5 ms (Evans and Powell 1967), the dolphin may be processing several echoes at a time.

The target ranges used to obtain the data shown in Figure 7.2 were either fixed for an entire session or were moved after successive ten-trial blocks, in each of which the range was constant. Penner (1988) performed an experiment in which the target range was randomly chosen from trial to trial to be between 40 and 120 m, at 20-m increments . Half of the trials in a session were target-absent trials. During target-absent trials, the *Tursiops* emitted signals with click intervals that were greater than the two-way transit time for the longest target range (120 m), indicating that the animal was expecting a target as far out as 120 m. In target-present trials, the animal quickly narrowed its click interval to match the target range.

The click interval data shown in Figures 7.1 and 7.2 and the results of Penner's (1988) study indicate that dolphins have a certain degree of control over click intervals, since most click intervals observed were greater than the two-way transit time. Yet the dolphin's control over the click interval is not fine enough to produce a more constant click interval for a specific target range of interest, as evidenced by the presence of cyclic variations in click interval patterns.

The false killer whale (*Pseudorca crassidens*) emits sonar signals with similar click interval patterns as *Tursiops*; again, the interval between each pair of clicks is greater than the two-way transit time (Thomas and Turl 1990). The same cannot be said for the beluga whale (*Delphinapterus leucas*). Au et al. (1987) measured the sonar signal of a *Delphinapterus* while the animal was performing a target detection task with the target at 80 m. They observed three different patterns or modes of click intervals. Mode 1 signals had click intervals that were greater than the two-way transit time. The amplitudes of mode 1 signals were always high. Mode 2 signals had click intervals that were less than the two-

way transit time but greater than 5 ms. These signals had mixed amplitudes; some were almost as high as mode 1 signals, while other were about 12 dB below. Mode 3 signals had click intervals less than 5 ms, typically between 1 and 2 ms. The amplitudes of these signals were always low, on the order of 12 dB below mode 1 signals. Mode 3 signals typically occurred toward the beginning or ending portions of click trains. They were probably not used in the sonar search process since they occurred in only half of the trials and the amplitudes were probably too low to provide useful target information.

Turl and Penner (1989) examined the click interval patterns used by *Delphinapterus leucas* for target ranges from 40 to 120 m. They found three distinct patterns of clicks. Pattern I started with an initial series of clicks having click intervals of about 41 ms, followed by packets of two to three clicks each. The time intervals between packets were greater than the two-way transit time, while the interval between clicks in each packed remained fairly constant at about 41 ms. An example of a pattern I click train is shown in Figure 7.3. This pattern did not emerge until after the target range increased to 100 m and more. Pattern II consisted of an initial series of clicks with click intervals less than the two-way transit time (mode 2 and 3 signals according to the nomenclature of Au et al. 1987), followed by a number of clicks with click intervals greater than the two-way transit time. Pattern II clicks were similar to those observed by Au et al. (1987). Pattern III click trains were those with all click intervals less than the two-way transit time. Only about 5 % of the click trains fell into pattern III. For correct detection trials, the beluga emitted pattern I click trains about 80% and pattern II clicks about 18% of the time. For correct rejection trials, about 60% of the click trains were of pattern II and 30% of pattern I.

The beluga seemed to have a strong preference for click intervals between 40 and 50 ms. Whenever the click intervals were below the two-way transit time, whether for a whole click train (pattern III) or part of a click train (patterns I and II), the click intervals were in the vicinity of 45 ms (Au et al. 1987; Turl and Penner 1989). Clicks that belonged to packets (pattern I signals) also tended to be separated by approximately 45 ms. It is not clear why the beluga had

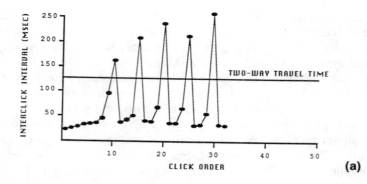

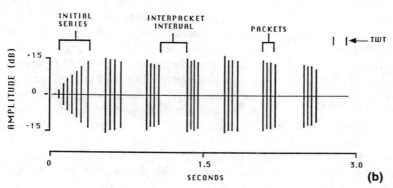

Figure 7.3. Example of a beluga pattern I click train for a target at 100 m. (*A*) Click interval versus click no.; (*B*) amplitude versus time. (From Turl and Penner 1989.)

this strong click interval preference. The different patterns of sonar emissions by the beluga may be related (in a manner yet unknown) to the animal's arctic habitat. In the arctic, belugas are often seen swimming in, around, and under pack ice, which implies that they are able to operate their sonars in the highly reverberant under-ice acoustic environment. On the other hand, click interval patterns have only been studied with one beluga whale, and it is possible that this animal may have had a peculiar echolocation pattern. Research needs to be performed with other beluga whales in order to determine, or confirm, typical echolocation behavior.

7.2 Frequency Characteristics

7.2.1 Peak Frequency and 3-dB Bandwidth

The frequency spectra of typical clicks emitted by dolphins in the open waters of Kaneohe Bay were given in Figures 5.2 and 5.3. A sonar signal in the frequency domain can best be characterized by

its *peak frequency* (frequency of maximum energy) and by its *bandwidth*. There are several ways in which bandwidth can be defined; one of the simplest definitions is that of the 3-dB bandwidth, which is the frequency width between the halfpower points of the spectrum. If we let p_{max} be the maximum amplitude of the frequency spectrum of a signal, and take into account that acoustic power is proportional to the square of pressure, the amplitude at which the power of the signal is half that of the peak is given by

$$p_{1/2\,power} = 10\log(p_{max}^2/2)$$
$$= 20\log(p_{max}) - 3\,dB \qquad (7\text{-}1)$$

Therefore, the amplitude associated with the halfpower points of a pressure pulse is 3 dB below the maximum value. On a linear scale, the pressure at a halfpower point is given by

$$p_{1/2\,power} = 0.707\,p_{max} \qquad (7\text{-}2)$$

Histograms of the peak frequency and the 3-dB bandwidth of the signals used by four *Tursiops* performing different sonar tasks in Kaneohe Bay

Figure 7.4. Histograms of peak frequency (*A*) and bandwidth (*B*) for four *Tursiops truncatus* performing different sonar tasks in Kaneohe Bay. (From Au 1980.)

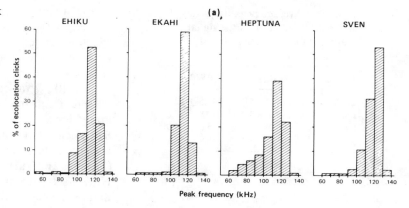

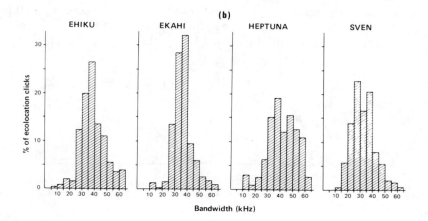

were reported by Au (1980) and are reproduced here in Figure 7.4. The dolphins Ehiku and Heptuna were involved in a target detection task at ranges greater than 50 m, with a 7.62-cm-diameter water-filled sphere as the target. The signals used by these two dolphins were measured with a hydrophone located 5 m behind the target, 73 m from the animals. Ekahi was involved in a shape discrimination study at a target range of 6 m. Sven was involved in a target recognition and discrimination task, also at a range of 6 m. The same hydrophone was used in all measurements of signals emitted by these four dolphins.

The frequency histograms indicate that with the exception of the dolphin Heptuna, at least 70% of the clicks of each animal had peak frequencies between 110 and 130 kHz. Only 60% of Heptuna's clicks had these peak frequencies. The peak frequencies most favored by Ehiku,

Ekahi, and Heptuna fell into the interval between 110 and 120 kHz. Sven showed a preference for the frequency range between 120 and 130 kHz. As stated in Chapter 5, these peak frequencies are over an octave higher than peak frequencies reported for *Tursiops* (animals housed in tanks) prior to 1973.

The bandwidth histograms indicate that at least 75% of the signals had a bandwidth greater than 25 kHz, with the frequency interval between 30 and 40 kHz being the most common. Heptuna had a tendency to use signals of wider bandwidth than the other three dolphins. These bandwidths are comparable to the critical bandwidth (at 120 kHz) for *Tursiops* discussed in Section 3.2.

The averaged peak frequency and bandwidth of all signals used to obtain the histograms are shown in Table 7.1. These averaged values correspond well with the histograms of Figure 7.4. Signals that are representative of the average

Table 7.1. Average peak frequency and 3-dB bandwidth of the signals evaluated for the histograms of Fig. 7.4 (From Au 1980)

Dolphin	f_p(kHz)	BW(kHz)
Ekahi	114.9 ± 5.0	41.5 ± 12.4
Ehiku	115.0 ± 3.9	40.8 ± 3.6
Heptuna	116.2 ± 5.4	45.6 ± 6.9
Sven	121.3 ± 3.2	38.0 ± 12.2

signal used by each dolphin are shown in Figure 7.5. These signals have peak frequencies and bandwidths that are closest to the averaged peak frequency and bandwidth in Table 7.1 The rela-

tive width of the frequency spectra can be described by the Q or quality factor of the signal, which is defined by equation (3-4) as $Q = f_p/\Delta f$, where Δf is the bandwidth of the signal. For the values of the averaged peak frequencies and bandwidth, Q varied from 2.6 to 3.2.

Peak frequency and 3-dB bandwidth histograms for sonar signals emitted by a beluga in San Diego and Kaneohe Bays (Au et al. 1985) and by a false killer whale in Kaneohe Bay (Thomas and Turl 1990) are shown in Figure 7.6. Ninety-six percent of the signals measured in San Diego had peak frequencies between 40 and 60 kHz, whereas 70% of the signals in Kaneohe Bay (same animal) had peak frequencies above 90 kHz. Approximately 91% of the clicks emitted by the *Pseudorca* had peak frequencies equal to

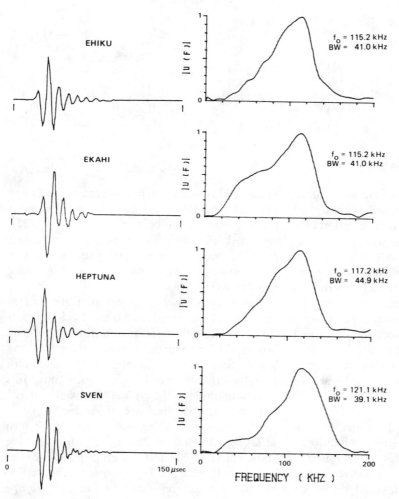

Figure 7.5. Representative signals have peak frequencies and bandwidths approximately similar to the averaged values of Table 7.1. (From Au 1980.)

Figure 7.6. Peak frequency and 3-dB bandwidth histograms for sonar signals used by a beluga in San Diego and Kaneohe Bays (from Au et al. 1985) and by a false killer whale in Kaneohe Bay (from Thomas and Turl 1990).

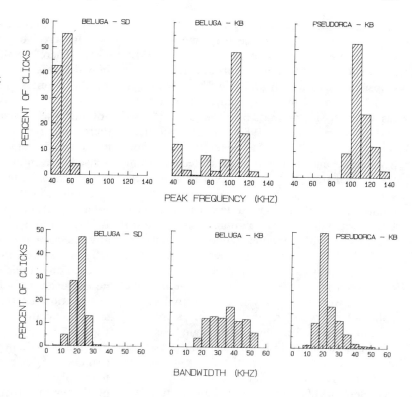

or greater than 100 kHz. The most preferred peak frequency for both the *Delphinapterus* and *Pseudorca* in Kaneohe Bay was between 100 and 110 kHz, about 10 kHz lower than for *Tursiops*.

The bandwidth histograms indicate that the signals used by the beluga in Kaneohe Bay tended to be wider than in San Diego Bay. Eighty-seven percent of the signals in San Diego Bay had bandwidths between 15 and 30 kHz, whereas 62% of the signals in Kaneohe Bay had bandwidths between 30 and 65 kHz. On the other hand, the false killer whale used relatively narrow frequency signals: most of its signals had bandwidths close to 20 kHz. The bandwidths of the signals used by the beluga in Kaneohe Bay were quite similar to those of the bottlenose dolphin shown in Figure 7.4.

7.2.2 Spectral Adaptation

Spectral adaptation of sonar signals used by dolphins in Kaneohe Bay to the ambient noise condition has been reported by Au et al. (1974). The data in Table 7.1 indicate that dolphins typically use signals with peak frequencies be-

tween 100 and 130 kHz, over an octave higher than previously reported for *Tursiops* (cf. Chapter 5). Therefore, it is correct to state that dolphins can and will make gross adjustments in the frequency content of their sonar signals. In Chapter 5 we said that dolphins in Kaneohe Bay typically emit clicks at levels approaching their maximum capability. Au et al. (1985) postulated that these high frequencies are a by-product of producing high intensity clicks. In other words, dolphins can only emit high intensity clicks (greater than 210 dB) if they use high frequencies. The animals can probably also emit high frequency clicks at low amplitudes, but cannot produce low frequency clicks at high amplitudes. The data of Moore and Pawloski (1990) are consistent with this contention. They never observed low frequency clicks with high amplitudes. The high amplitude clicks tended to be slightly bimodal, with a major peak frequency above 100 kHz and a secondary peak frequency between 35 and 60 kHz.

The notion that dolphins make fine adjustments to the spectrum and shape of their sonar signals in order to better examine different targets seems attractive and appealing. However, there is very little evidence from meticulously conducted experiments to support this idea. In fact, available data seem to contradict the notion. Consider the case of the dolphin Ekahi, who was initially involved in an experiment to determine his ability to discriminate between foam spheres and cylinders (Au et al. 1980). In any given trial, one target out of a set of three foam spheres and five foam cylinders would be presented 6-m from the animal stationed in a hoop. The six targets were designed to have similar target strengths, to eliminate target strength differences as a cue. Ekahi was later used in a sonar discrimination experiment with metallic cylinders as targets. Typical mean peak frequencies of the sonar signals used by the dolphin in both experiments are shown in Figure 7.7. Although the metallic cylinders were completely different from the foam spheres and cylinders, there was no significant shift in the peak frequency of the signals used by Ekahi. Also shown in Figure 7.7 are the average peak frequencies used by the dolphin Sven in performing a sonar discrimination task involving cylinders of different material composition but identical outer diameter and length. Again, there was no significant difference between the peak frequencies used for the different targets. It is important to realize that the data of Thompson and Herman (1975) indicate that for frequencies

between 100 and 130 kHz, a bottlenose dolphin cannot perceive pitch differences of less than 8 kHz. Therefore, the small frequency differences shown in Figure 7.7 should not be perceivable for the animals. The data in this figure seem to support the contention that dolphins do not purposefully adjust the peak frequency of their sonar signals to "match" the targets being investigated. Bel'kovich and Dubrovskiy (1977) arrived at a similar conclusion for dolphins kept in tanks at their facility.

Changes in the spectral composition of biosonar signals in a pulse train do occur. However, Bel'kovich and Dubrovskiy (1977) argued that any attachment of significance to these changes reflects incorrect interpretation of the results or is based on insufficient statistics. To illustrate the need to examine dolphin signals from a statistical basis, consider the spectral history plots for six echolocation trials by the dolphin Ekahi in Figure 7.8. The signals were measured with the dolphin performing a shape discrimination task (Au et al. 1980); on each trial, the same 10.2-cm foam sphere was presented to the animal. In cases (a) through (c), there are obvious changes in the spectral composition of the signals: observations based on a single or a few trials may lead to the conclusion that the dolphin was adaptively adjusting its sonar pulses to maximize the information available in the echoes. Upon closer look,

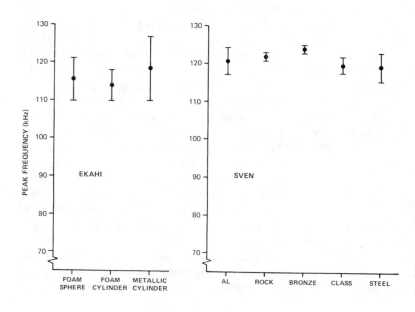

Figure 7.7. Average peak frequencies of sonar signals used by the bottlenose dolphins Ekahi and Sven while performing in different target discrimination experiments. (From Au 1980.)

Figure 7.8. Spectral history plots of Ekahi's signals for six different trials with the same target present. (From Au 1980.)

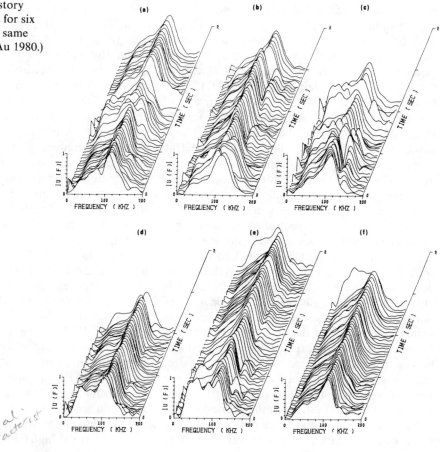

however, plots (a) through (c) do not seem to show a consistent pattern in which the signals were being altered. Furthermore, the dolphin also emitted signals that were fairly consistent and steady, as can be been seen in plots (d) through (f). Yet all of these spectral history plots involve the same target. The spectral changes may have been due to the dolphin turning momentarily away from the target, to internal steering of the beam even when its rostrum was pointed to the target, to fluctuations in the click generation mechanisms, or to lapses in the animal's attention during a trial.

It is possible that spectral adaptations were not observed for the animals in the studies above because of the dominant influence of the ambient noise condition in Kaneohe Bay, where the experiments were conducted. Since the dolphins preferred to use signals with peak frequencies between 110 and 130 kHz, any changes or frequency adjustments within this range probably had insignificant effects on the echoes. In an environment where a dolphin is not masked by noise and can achieve approximately the same signal-to-noise ratio per unit of effort at any frequency between 30 and 130 kHz, spectral adaptation may possibly take place, although it seems unlikely (Bel'kovich and Dubrovskiy 1977) since it is difficult to think of a situation where spectral adaptation would help a dolphin better detect or discriminate targets.

Dziedzic and Alcuri (1977) examined the sonar emission of a *Tursiops* performing a form discrimination study. They used a sophisticated array of hydrophones to track the dolphin's position and a television camera 1 m above the water to monitor the animal's orientation. The dolphin was required to use its sonar to localize the position of a standard ring from different polygonal shapes, with two targets always presented simultaneously, separated by 40 cm. An array of six hydrophones positioned 1 m behind an imaginary line connecting both targets was used to measure the dolphin's sonar signals as it echolocated and swam toward the target of choice (Dziedzic 1978). Dziedzic and Alcuri (1977) found that the sonar signals were relatively invariant when the dolphin was more than 4 m from the target. However, when the animal–target distance was less than 4 m, they found a spectral spreading of the signal accompanying an increase in the difficulty of the discrimination task.

Although the experiment of Dziedzic and Alcuri (1977) seems to support or favor the notion of some sort of spectral adaptation, especially as a dolphin closes in on targets, their data are difficult to interpret. With the hydrophones spread out 1 m behind the targets it is difficult to accurately ascertain the orientation of the dolphin with respect to any one of the hydrophones during any given sonar emission. Spectral changes recorded from any of the hydrophones may not have been produced purposefully by the dolphin, but may instead be due to changes in the dolphin–hydrophone orientation. Variations in the signal waveform as a function of the angle around a dolphin's head (discussed in the previous chapter) compound the difficulties of obtaining reliable sonar emission data on fine spectral adaptation produced for the purpose of optimizing target discrimination and recognition. To study spectral changes in a target discrimination task, contact hydrophones placed on the head of a dolphin should be used. This would eliminate the animal–hydrophone orientation problem. Although dolphins probably do not tailor their signals to targets being investigated, we do know that dolphins can make broad changes in the spectral content of the emitted signals.

A general impression of the degree to which a dolphin can control the characteristics of its sonar signals can be gained by considering the study of Moore and Pawloski (1990) performed at our laboratory in Kaneohe Bay. They succeeded in placing the amplitude and frequency structure of the clicks emitted by a *Tursiops truncatus* under stimulus control while the dolphin was performing a target detection task. An Apple IIe computer interfaced with a real-time spectral analyzer consisting of eight contiguous 15-kHz bandpass filters with center frequencies from 30 to 135 kHz, in 15 kHz increments, was used to analyze each emitted click detected by a miniature hydrophone. The dolphin was required to station on a bite plate/tail rest assembly to ensure the correct dolphin–hydrophone geometry. The animal was trained to emit low frequency clicks (peak frequency < 60 kHz) when a specific audio-cue was given, and high frequency clicks (peak frequency > 105 kHz) when another audio-cue was presented. The normalized energy distribution of the eight-channel analyzer for 200 randomly cued high and low frequency clicks is shown in Figure 7.9. Both click types exhibited a bimodal structure: the low frequency clicks had maximum energy at 60 kHz and a secondary energy peak at 135 kHz, whereas the high frequency clicks had maximum energy at 120 kHz and a secondary energy peak at 60 kHz. A clear distinction in the amplitudes of the low and high frequency clicks was observed by Moore and Pawloski (1990). The amplitudes of the low frequency clicks had an average SPL of approximately 197 dB, and the high frequency clicks had an average SPL of approximately 209 dB. This amplitude–frequency relationship is consistent with our contention expressed earlier in this section that dolphins can only emit high-level clicks if they use high frequencies, and can probably emit high frequency clicks at low amplitudes but may not be able to produce low frequency clicks at high amplitudes (> 210 dB).

7.3 Click Source Levels

7.3.1 Variations in Clicks Source Levels

Dolphins can vary the amplitude of their sonar emissions over a very large dynamic range. Peak-to-peak *source levels* as low as 150–160 μPa and as high as 230 μPa have been measured (Evans 1973; Au et al. 1974). The lower source levels are usually measured in tanks and the higher source levels in open waters. Even in a given situation, when a dolphin is performing a specific sonar task over and over again, there can be large fluctuations in the amplitude of the sonar signals emitted. In most of the sonar experiments conducted at our laboratory, dolphins in floating

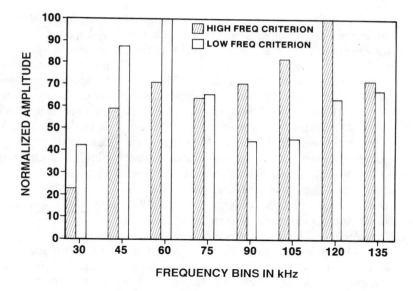

Figure 7.9. Normalized energy across frequency bins for 200 randomly cued frequency control trials. Both low and high frequency clicks exhibit a bimodal distribution. (From Moore and Pawloski 1990.)

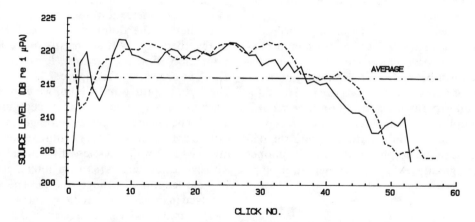

Figure 7.10. Variation in source level of sonar emissions of a *Tursiops truncatus*. Two click trains with the same average source level are shown.

enclosures are required to scan or search for targets that are outside the enclosures at some standoff range. The dolphins in these experiments usually emit sonar signals with typical amplitude fluctuations as shown in Figure 7.10, where two click trains with the same average source level are diagramed. In this example a 17-dB variation is shown, with source levels as high as 222 dB and as low as 205 dB. Variations as much as 20 to 25 dB are not uncommon. In a typical click train, the amplitude is usually low at the beginning, then rises to a peak—with some fluctuations on the order of about 5 dB during the middle portion—and finally decreases at the end of the sonar search. The click train depicted by the solid line in Figure 7.10 and the cases shown by Au and Pawloski (1989) are examples of this general amplitude pattern. However, there can be any number of variations from this general pattern. Signals can start off with relatively high amplitudes and decrease to a local minimum before increasing to a peak, as depicted by the dashed curve in Figure 7.10. In some cases, usually with short click trains, the signals can start off relatively high and remain high throughout the sonar search, as was shown in Figure 5.2.

These source level variations are to be expected and are probably not consciously controlled by the animal but are associated with the natural process of producing sonar clicks.

Despite the presence of large amplitude variations in typical dolphin sonar emissions and the different types of amplitude patterns for click trains, there seems to be a relatively orderly relationship between the largest amplitude and the average amplitude of most click trains. Figure 7.11 shows the mean difference in dB between the largest amplitude and the average amplitude value of a click train for each of five different *Tursiops* performing different sonar tasks in Kaneohe Bay. The mean difference was approximately 4.5 dB for four of the dolphins and 5.4 dB for the fifth dolphin. At least twenty trials per dolphin are represented in the data. The similar relationship between the largest amplitude and the average amplitude may be due to the mechanisms involved or the manner in which dolphins generate clicks. However, since the exact mechanisms of click production are not known, this question must be left for future investigation.

Large amplitude fluctuations can make it difficult for an experimenter to obtain tape recordings of dolphin sonar signals over complete sonar searches, especially in situations where the distance between the dolphin and the hydrophones can vary greatly. Most instrumentation tape

recorders on direct record have a dynamic range of approximately 30 dB (no matter what their advertised specifications may be) so that an experimenter must cope with variations caused by spreading losses and by the animal. A common solution to this problem is to use several channels simultaneously, with each channel having different gain, so that low amplitude signals can be detected on one channel and very high amplitude clicks can be recorded on another channel.

7.3.2 Factors Affecting Click Source Levels

The amplitude of sonar signals emitted by a dolphin will be influenced by several factors, such as the background noise environment, target size (target strength) and target range, the specific task (detection versus discrimination), and whether the dolphin is masked by noise or reverberation. Variations in source level as a function of masking noise level for several *Tursiops* are shown in Figure 7.12. Two of the *Tursiops*, Heptuna and Ehiku, were in a noise-masking-target detection experiment involving a 7.62-cm sphere target at a range of 16.5 m (Au and Penner 1981). The third dolphin was involved in a wall thickness discrimination task with cylindrical targets (3.81 cm outer diameter × 12.7 cm length) at a range of 8 m. The animals involved in target detection tasks did not increase their source levels with increases in the masking noise. The source levels used by Heptuna and Ehiku at masking noise levels of 82 and 87 dB were adversely influenced by the dolphins' somewhat erratic performance. Both animals seemed to have partially abandoned the detection task at these noise levels and during many of the trials mimicked echolocation by going through normal sonar scans without emitting any detectable signals (Au and Penner 1981). The detection threshold occurred at approximately 75 dB for Ehiku and 77 dB for Heptuna. One possible reason that Ehiku and Heptuna did not increase their signal level as the noise increased may have been the fact that the dolphins were already producing relatively high intensity signals in order to overcome the effects of the ambient noise

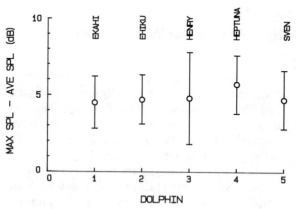

Figure 7.11. Mean difference in dB between the largest amplitude and the average amplitude for the click trains of each of five different *Tursiops* performing different sonar tasks in Kaneohe Bay.

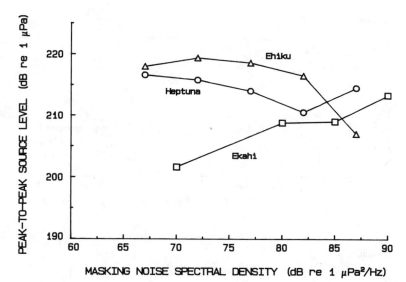

Figure 7.12. Peak-to-peak source levels for *Tursiops* as a function of the masking noise level for a biosonar detection task (Ehiku and Heptuna; from Au and Penner 1981) and for a biosonar discrimination task (Ekahi; data from the study of Au and Pawloski 1992).

and could not significantly further increase the power of their projected signals. Turl et al. (1987) obtained similar results for an experiment on target detection in artificial noise in which three different target ranges (16.5, 40, and 80 m) were used. Their *Tursiops* did not vary its signal levels as the masking noise increased over a 12-dB range.

The dolphin Ekahi, however, which was performing the discrimination task, did increase the amplitude of its emitted signals as the masking noise increased. Ekahi did not need to use high-level signals for the low-noise conditions since the cylindrical targets had target strengths that were approximately 8 to 10 dB larger than the target strengths of the spheres used in the detection tasks and since the two-way transmission loss at the 8-m range was 13 dB less than at the 16.5-m range used in the detection studies. Therefore, Ekahi had considerably more flexibility in adjusting its signal amplitude to compensate for the increasing masking noise level. The results with Ekahi are consistent with the results of Babkin and Dubrovskiy (1971), who reported an almost linear increase in the amplitude of emitted signals with increasing noise level.

The variation of source levels as a function of target range for two Atlantic bottlenose dolphins (Au et al. 1974), a beluga whale (Au et al. 1985), and a false killer whale (Thomas and Turl 1990)

is shown in Figure 7.13. The data for the bottlenose dolphins and the false killer whale were obtained in Kaneohe Bay, while the data for the beluga whale were obtained in San Diego Bay. The data indicate that for all four animals, the source levels increased only slightly as the target range increased. The small increases in source level with range were not nearly enough to compensate for increased propagation loss with range. For instance, the difference in two-way transmission loss between target ranges of 40 and 120 m was approximately 26 dB, yet the *Pseudorca* increased its source level only by about 5 dB for this range difference. Therefore, over the distances considered in Figure 7.13, differences in target range had very little effect on source levels. The same can be concluded from the data on the averaged maximum energy flux density per trial used by *Tursiops* and *Delphinapterus* for three target ranges (16.5, 40, and 80 m).

We will now consider the combined effect of target size and target range on source levels used by *Tursiops*. The amplitude of the echoes from a target (*echo level*) is dependent on the source level of the projected signal, the target range, and the reflectivity of the target (*target strength*), according to the simple relationship expressed in dB

$$EL = SL - 2TL + TS \qquad (7\text{-}3)$$

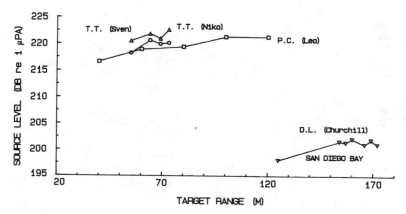

Figure 7.13. Averaged peak-to-peak source level as a function of target range for *Tursiops truncatus* (from Au et al. 1974), *Delphinapterus leucas* (from Au et al. 1985), and *Pseudorca crassidens* (from Thomas and Turl 1990).

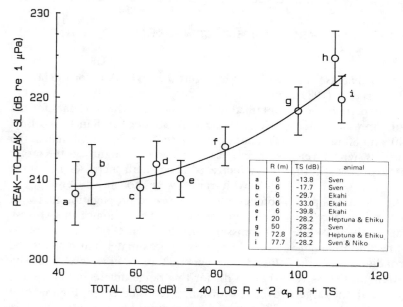

	R (m)	TS (dB)	animal
a	6	-13.8	Sven
b	6	-17.7	Sven
c	6	-29.7	Ekahi
d	6	-33.0	Ekahi
e	6	-39.8	Ekahi
f	20	-28.2	Heptuna & Ehiku
g	50	-28.2	Sven
h	72.8	-28.2	Heptuna & Ehiku
i	77.7	-28.2	Sven & Niko

Figure 7.14. Peak-to-peak source level as a function of total acoustic energy loss for five different *Tursiops* performing different echolocation tasks in Kaneohe Bay. (From Au 1980.)

where EL = peak-to-peak echo level in dB re 1 μPa

SL = peak-to-peak source level in dB re 1 μPa

TL = transmission loss = $20 \log R + \alpha_p R$

R = target range in meters

α_p = absorption coefficient at the peak frequency of the signal

TS = target strength in dB

Equation (7-3) simply states that the amplitude of the return echo is equal to the amplitude of the outgoing signal minus the total loss, which is the sum of losses due to propagation and the reflection. The peak-to-peak source level as a function of total loss for five different *Tursiops* under different circumstances was presented by Au (1980) and is reproduced here in Figure 7.14; The data fell into two distinct categories: one associated with total loss less than approximately 70 dB, and the other associated with total loss greater than 70 dB. A least-square second-order polynomial curve is fitted to the data. For a total loss of less than 70 dB, the source levels were fairly constant. As total loss increased above 70 dB, source levels increased almost linearly with loss. However, the source levels did not increase at the same rate as the total loss. This indicates that the dolphins seem to prefer to operate at a high signal-to-noise ratio. The decrease in loss from case (i) to case (a) is over 62 dB, and the corresponding decrease in source level is only 12 dB.

7.3.3 Maximum Source Levels

The source levels reported for sonar experiments performed in Kaneohe Bay are considerably higher than any levels previously reported for odontocetes. The maximum average peak-to-peak source level for *Tursiops* was recorded at 227.6 dB in one of the trials in case (h) of Figure 7.14. The largest single click measured was 230 dB, emitted by Heptuna in case (h). The largest single click measured for *Delphinapterus* was approximately 225 dB (Au et al. 1987) and for *Pseudorca*, 228 dB (Thomas and Turl 1990). Mohl et al. (1990) measured high intensity clicks from a narwhal (*Monodon monoceros*) and found amplitude levels similar to those emitted by odontocetes in Kaneohe Bay.

These source levels may seem inordinately high, especially when compared to conventional man-made sonars, and considering the amount of energy required to project high-level signals into the water. One must realize, however, that these are peak-to-peak level signals, that the signals are very short, and that the beam is relatively narrow. The rms pressure for these biosonar signals is approximately 15 dB lower than the peak-to-peak amplitude. In comparing transient signals similar to dolphin sonar signals, it is more meaningful to refer to the energy flux density of the acoustic wave than to the sound pressure level (Urick 1983). For signals reported in case (h) of Figure 7.14, the average energy flux density was 167 dB re 1 $(\mu Pa)^2s$. The energy flux density (E) of an acoustic signal is defined as the product of intensity (I) and time duration (T) of the signal

$$E = I \times T \qquad (7-4)$$

The intensity in dB can be expressed as

$$10 \log(I) = 10 \log(E) - 10 \log(T) \qquad (7-5)$$

However, the rms pressure in dB is related to intensity by

$$SPL(dB) = 20 \log(p) = 10 \log(I) \qquad (7-6)$$

where p is the rms pressure. Therefore, the rms pressure of a cw pulse tone that would have an energy flux density equivalent to that of a high intensity dolphin sonar signal can be expressed as

$$SPL(dB) = 167 - 10 \log(T) \qquad (7-7)$$

A 10-ms cw pulse tone with an rms SPL of 187 dB will have the same energy flux density as the dolphin signal of case (h) in Figure 7.14. For a 1-ms cw pulse tone, the equivalent SPL will be 197 dB. Therefore, the amount of acoustic energy emitted by a dolphin is well below the maximum levels of man-made sonars; pulse sonars typically use sound pressure levels in the vicinity of 210 to 220 dB.

Another way to consider what the high peak-to-peak source levels of dolphin sonar signals represent is to calculate the *acoustic power* in a click. An expression for the radiated acoustic power in Watts/m² can be derived by considering the relationship between the rms intensity and rms pressure,

$$I_{rms} = \frac{p_{rms}^2}{\rho c} \times 10^{-12} \qquad (7-8)$$

where p_{rms} is in units of μPa, ρ is the density of water (1,060 kg/m³) and c is the sound velocity of water (1,500 m/s). The factor of 10^{-12} is a conversion factor allowing the acoustic pressure to be expressed in μPa. The power emitted by an ommidirectional source was given in equation (3-7), repeated here:

$$P_0 = 4\pi r^2 I_{rms} \qquad (7-9)$$

The radiated power for a directional source is related to that of an omnidirectional source by the directivity index, expressed in (3-6) as

$$DI = 10 \log\left(\frac{P_0}{P_D}\right) = 10 \log(di) \qquad (7-10)$$

where $di = P_0/P_D$ is the directivity index of the source. Therefore, the radiated power from a directional source of directivity di can be expressed in Watts as

$$P_D = \frac{4\pi r^2 I_{rms}}{di} \qquad (7-11)$$

The radiated power should be evaluated at a

distance 1 m from the source, which can be done by letting $r = 1$ in (7-11). Substituting (7-8) into (7-11), then taking the log of the result and using (7-10), we arrive at

$$10 \log P_D = SL_{rms} - DI - 171 \quad (7\text{-}12)$$

where SL_{rms} is the rms source level (sound pressure level at 1 m). The rms pressure is related to the instantaneous acoustic pressure $p(t)$ by the relationship found in equation (1-6), repeated here:

$$p_{rms} = \sqrt{\frac{1}{T} \int_0^T p^2(t)\, dt} \quad (7\text{-}13)$$

Let us represent a dolphin sonar signal such as the one shown in Figure 5.3 by letting

$$p(t) = As(t) \quad (7\text{-}14)$$

where A is the peak amplitude and $s(t)$ is the normalized waveform function that describes the shape of the click and has a maximum value of unity as described by equation (5-1). The rms source level ($20 \log p_{rms}$) is related to the peak-to-peak source level by

$$SL_{rms} = 20 \log \left(\sqrt{\frac{A^2}{T} \int_0^T s^2(t)\, dt} \right)$$

$$= SL_{pp} - 6 + 20 \log \left(\sqrt{\frac{1}{T} \int_0^T s^2(t)\, dt} \right) \quad (7\text{-}15)$$

where $SL_{pp} = 20 \log(2A)$. If we substitute the digitized form of the signal shown in Figure 5.3 into (7-9) and numerically evaluate the integral, the rms source level will be equal to the peak-to-peak source level minus 15.5 dB:

$$SL_{rms} = SL_{pp} - 15.5 \text{ dB} \quad (7\text{-}16)$$

A value for T of 45 μs was used in (7-9); T was determined as the time at which $\int p^2(t)\, dt$ increased no more than 1% as t increased. From Table 6.1 the averaged transmission directivity index for three *Tursiops* is 25.8 dB. Inserting (7-10) and the directivity index value into (7-9), we obtain

$$10 \log P = SL_{pp} - 212.3 \text{ dB re 1 Watt} \quad (7\text{-}17)$$

The radiated acoustic power in watts as a function of the peak-to-peak source level for a signal having the waveform shown in Figure 5.3 is given

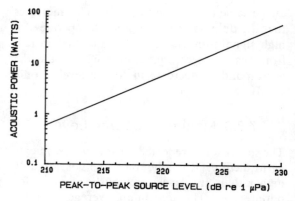

Figure 7.15. Acoustic power of a dolphin's sonar signal as a function of the peak-to-peak source level.

in Figure 7.15. The diagram indicates that the signal represents a source level of 23 watts. The largest amplitude click measured while obtaining the data shown in Figure 7.15 was 230 dB, which represents 59 watts of acoustic power. These acoustic power levels are not very high when compared to some man-made sonar.

The high peak-to-peak source levels, between 225 and 230 dB re 1 μPa, emitted by dolphins will not cause the water to cavitate. The pressure change required for an acoustic signal to cause cavitation is a function of frequency, duration, repetition rate, and depth. Although the cavitation threshold near the sea surface is only 220 dB for a cw signal at 1 kHz, this threshold value increases rapidly as a function of frequency, to over 237 dB at 100 kHz (Urick 1983). The cavitation threshold also increases as the pulse duration decreases. A 500-μs pulse will have a cavitation threshold approximately 5 dB greater than the cw value (Urick 1983).

7.4 Number of Clicks and Response Latencies

The number of clicks used by dolphins to perform a sonar task is a highly variable parameter. A typical count of the clicks emitted per trial by a *Tursiops* involved in a target detection task, for ten consecutive trials with the target range held constant, is shown in Figure 7.16. The dolphin's performance in detecting the presence or absence of the target was 100% correct. The number of

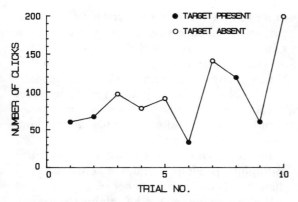

Figure 7.16. Number of clicks emitted per trial for ten consecutive trials by a *Tursiops* performing a target detection task. The target was at a fixed range and the dolphin's performance was perfect for these ten trials.

clicks per trial typically varies over a wide range and in an unpredictable manner from trial to trial. In Figure 7.16, for example, the number of clicks emitted ranged from 33 to 199. Consequently, when the average number of clicks per trial is reported for biosonar experiments, the standard deviation for the data will tend to be high. A satisfactory explanation for the high variability in the number of clicks used by dolphins has yet to be suggested.

There is a tendency for dolphins to emit more clicks as a sonar task becomes progressively more difficult. However, in order to observe this trend, data over many trials and sessions must be obtained because of the large variability in the number of clicks emitted per trial. The number of clicks per trial for four experiments involving different target detection tasks is presented in Figure 7.17. Note the large variations in the data for all four experiments. (The results of these experiments will be explained in detail in Chapter 8.) In Figure 7.17A, two *Tursiops* were involved in a target detection task in which the target was at a fixed range and artificial white noise was used to mask the dolphins (Au and Penner 1981). As the noise level increased above ambient levels, the number of clicks used per trial also increased, until the noise level reached 77 dB. Beyond a noise level of 77 dB, the task became unsolvable (performance was below 58% correct) and the dolphins ceased to increase their effort at detecting the targets (Au and Penner 1981). The data in Figure 7.17B were obtained in a target detection experiment using an electronic phantom target, with an attenuator controlling the level of the echo returned to the *Tursiops* (Au and Pawloski 1988). As the attenuation increased in value, the level of the echo decreased, making the task more difficult; the number of clicks emitted by the animal also increased. The data of Figure 7.17C were obtained in a target detection experiment in which three cylindrical targets were moved progressively closer to a clutter screen (Au and Turl 1983). The task became more

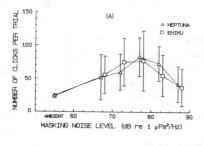

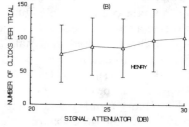

Figure 7.17. Average and standard deviation of the number of clicks per trial used by dolphins in four different target detection experiments.

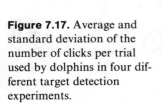

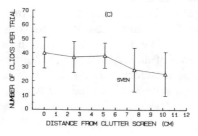

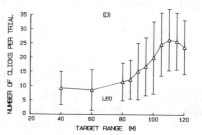

difficult the closer the targets were placed to the clutter screen. Figure 7.17C indicates a progressive increase in the number of clicks emitted per trial as the distance between the target and the clutter screen became smaller. Figure 7.17D shows the number of clicks per trial emitted by a *Pseudorca* in a target detection experiment in which the range of the target was varied (Thomas and Turl 1990). As the target range increased, the number of clicks used by the whale also increased, to a peak at 115 m.

The data in Figure 7.17 were not separated into categories according to the contingency table of Figure 1.8. There seem to be individual-animal differences in the number of clicks used in correct detection and correct rejection trials. Au and Penner (1981) found that one dolphin (Heptuna) emitted the same number of clicks during both correct detection and correct rejection trials. Another dolphin (Ehiku) emitted more clicks for correct detection trials than for correct rejection trials. Au and Turl (1983) also found their dolphin (Sven) emitting more clicks in correct detection trials than in correct rejection trials. The opposite was true for the study of Thomas and Turl (1990) whose animal emitted less clicks for correct detection trials. Therefore, the data from these different studies indicate that there is no general tendency for dolphins to systematically vary the number of clicks used in correct detection versus correct rejection trials, but that there are individual differences in the strategies employed to solve a sonar problem.

The data presented in Figure 7.17 may also indicate a species-specific difference in the number of clicks emitted per trial. The *Pseudorca* emitted considerably fewer clicks per trial than the other four *Tursiops*. The dolphins Ehiku and Heptuna had peaks of 70 to 80 clicks per trial, while Henry's and Sven's peaks were approximately 100 and 40 clicks per trial, respectively. The *Pseudorca* Leo had a peak of only 26 clicks per trial. Leo's minimum number of clicks per trial was 9, compared to about 24 for Ehiku and Heptuna, 75 for Henry, and 40 for Sven. We do not know why dolphins emit the number of clicks they do in performing a sonar task and therefore cannot address the question of why a *Pseudorca* would tend to use fewer clicks than a *Tursiops*. More research is certainly needed in this area in order to obtain a better understanding of what controls or influences an echolocating dolphin in selecting the number of clicks to emit. It would also be interesting to study how dolphins utilize the echoes from each click. Does an animal make a series of independent judgments of the target situation as it receives each echo, or does it make a judgment only after receiving a minimum number of echo returns?

In a biosonar experiment, the amount of time a dolphin spends performing a sonar search is directly related to the number of clicks emitted and the time required for the dolphin to respond. The *response latency* is defined as the time between the emission of the last click and the moment a response paddle is activated to end a trial. This parameter, although not a characteristic of dolphin sonar signals, is discussed here to provide a better understanding of the sonar process from a dolphin's perspective. Au and Penner (1981) in a noise masking experiment and Au and Turl (1983) in a reverberation masking experiment measured the response latencies of dolphins in making correct detection and correct rejection responses. A detailed discussion of both experiments will be given in Chapter 8.

The response latencies for two dolphins each are shown in Figure 7.18A as a function of masking noise level (Au et al. 1983) and in Figure 7.18B as a function of the separation distance between a clutter screen and a target. In all three situations covered by Figure 7.18, the response latencies for correct rejection trials were greater than for correct detection trials. The differences were significant at the 0.01 level. In both experiments, the response paddles were located equidistant from a hoop station so that differences in response latencies could be attributed to different swim distances to the paddles. It seems that the dolphins may have had some kind of intrinsic internal reinforcement associated with the detection of the target. In captivity, dolphins are taught or conditioned to respond to "target-absent" conditions. In the wild, target-absent is not a condition or state—it's just the absence of a target. Natural echolocation is most likely "one-sided" as far as the animal's response is concerned. Consequently, dolphins probably attach greater value to the presence of acoustic returns from objects of interest than to the absence of

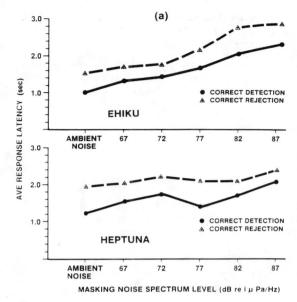

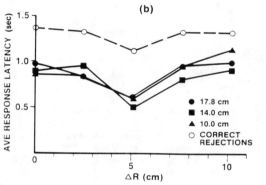

Figure 7.18. Average response latencies per trial for (*A*) a noise masking experiment of Au and Penner (1981; from Au et al. 1982), and (*B*) a reverberation masking experiment (from Au and Turl 1983).

target echoes. It may be a natural tendency for them to respond faster to a target-present case than to a target-absent case, even though the animals in the studies above did not exhibit a strong tendency to emit more clicks in correct detection trials than in correct rejection trials.

7.5 Signals from Other Species

There are approximately 65 species of small toothed whales and dolphins (Ellis 1989), and most of them probably possess active sonar

capabilities. A list of species with demonstrated sonar capabilities was given in Section 1.1. However, despite the demonstration of active sonar capability in various odontocetes, there are very few statistical data on biosonar signals except for the three species discussed in this chapter (*Tursiops truncatus, Delphinapterus leucas, Pseudorca crassidens*). Some of the earlier acoustic measurements were made with band-limited instrumentation, so the results are of limited value. Nevertheless, there have been acoustic measurements performed in tanks and in the field on untrained animals that can provide some information about the types of biosonar signals used by different species. Such measurements are usually obtained by tossing fish in front of a hydrophone and recording the dolphin's signals as the animal echoranges and swims toward the fish. Another popular technique is to suddenly dip a hydrophone into the water, taking advantage of dolphins' natural inclination to examine strange objects introduced into their environment. Unfortunately, these techniques have several serious disadvantages. Dolphins often swim very rapidly toward a fish bait or hydrophone, making it very difficult to obtain accurate ranges for the emitted signals from which signal source levels can be computed. The dynamic range of the instrumentation system must be fairly large since the signal level can change considerably as an animal swims and approaches a hydrophone. The signal level measured will often be relatively low, especially if an animal is within several meters of the hydrophone. From Figure 7.14, we saw that *Tursiops* will not increase its source level unless the total acoustic loss is greater than some particular value. The same threshold effect is probably observable for most other echolocating dolphins. The source level used for a target at a long range (i.e., beyond 20 m) will probably be much higher than the source level for a close target (i.e., 1 to 2 m). Finally, the hydrophone and the animals are often close to the surface (less than 1 m) so that a surface-reflected component of the signal may not be detectable, or may be mistaken for a double-pulsed signal. Some of the characteristics of sonar signals used by other dolphin species are tabulated in Table 7.2. Both marine and freshwater dolphins are included in the table.

In Chapter 5, we have seen that low intensity

Table 7.2. Properties of biosonar signals of different dolphin species. The parameters are peak frequency (f_p), 3-dB bandwidth (BW), peak-to-peak source level (SL), signal duration (τ)

Species	f_p(kHz)	BW(kHz)	$\tau(\mu s)$	SL(dB)	Cond	Reference
Cephalorhynchus commersonii (Commerson's dolphin)	120–134	17–22	180–600	160	Tank	Kamminga and Wiersma (1982) Evans et al. (1988)
Cephalorhynchus hectori (Hector's dolphin)	112–130	≈14	≈140	151	Sea	Dawson (1988)
Delphinapterus leucas (Beluga whale)	100–115	30–60	50–80	225	Bay	Au et al. (1985) Au et al. (1987)
Delphinus delphis (Common dolphin)	23–67	17–45	50–150	—	Sea	Dziedzic (1978)
Globicephala melaena (Pilot whale)	30–60	—	—	180	Tank	Evans (1973)
Grampus griseus (Risso's dolphin)	65	72	40–100	~120	Bay	Unpublished data measured by author in Kaneohe Bay
Inia geofrensis (Amazon River dolphin)	95–105	—	200–250	—	River	Kamminga et al. (1989)
Lagenorhynchus obliguidens (Pacific white-sided dolphin)	30–60	—	—	180	Tank	Evans (1973)
Lipotes vexillifer (Chinese river dolphin)	100–120	37	—	156	Tank	Youfu and Rongcai (1989)
Monodon monoceros (Narwhal)	40	27	250	218	Sea	Mohl et al. (1990)
Neophocaena phocaenoides (finless porpoise)	128	11	127	—	Tank	Kamminga (1988)
Orcaella brevirostris (Irrawaddy dolphin)	50–60	≈22	150–170	—	Tank	Kamminga et al. (1983)
Orcinus Orca (Killer whale)	14–20	≈4	210	178	Tank	Evans (1973)
Phocoena phocoena (Harbor porpoise)	120–140	10–15	130–260	162	Tank	Mohl and Andersen (1973) Kamminga and Wiersma (1981) Hatakeyama et al. (1988)
Phocoenoides dalli (Dall's porpoise)	120–160	11–20	180–400	170	Sea Tank/ Sea	Awbrey et al. (1979) Hatakeyama and Soeda (1990)
Platanista gangetica (Ganges River dolphin)	15–60	—	—	—	Tank	Herald et al. (1969)
Pseudorca crassidens (False killer whale)	100–130	15–40	100–120	228	Bay	Thomas and Turl (1990)
Steno bredanensis (Rough-toothed dolphin)	5–32	—	—	—	Tank	Norris and Evans (1966)
Sotalia fluvatilis (Tucuxi)	95–100	~40	120–200	—	Tank River	Wiersma (1982) Kamminga et al. (1989)
Tursiops Truncatus (Atlantic bottlenose dolphin)	110–130	30–60	50–80	228	Bay	Au (1980)
Trusiops gilli (Pacific bottlenose dolphin)	110–130	30–60	50–80	228	Bay	Unpublished data measured by author in Kaneohe Bay

biosonar signals of *Tursiops truncatus*, *Delphinapterus leucas*, and *Pseudorca crassidens* measured in tanks have low peak frequencies (30–60 kHz) and that high intensity signals measured in open waters have high peak frequencies (100–130 kHz). Odontocetes from the family *Phocoenidae* (harbor porpoise *Phocoena*, finless porpoise *Neophocaena phocaenoides*, Dall's porpoise *Phocoenoides dalli*) and from the genus *Cephalorhynchus* (Commerson's dolphin *Cephalorhynchus commersonii* and Hector's dolphin *Cephalorhynchus hectori*) all seem to emit high frequency yet low intensity signals, which may not differ whether measured in a tank or the open ocean. The signals

Figure 7.19. Examples of sonar signals of (*A*) harbor porpoise (from Kamminga and Wiersma 1981), (*B*) finless porpoise (from Kamminga 1988), (*C*) Dall's porpoise (from Hatakeyama and Soeda 1990), (*D*) Commerson's dolphin (from Kamminga and Wiersma 1981), and (*E*) Hector's dolphin (from Dawson 1988).

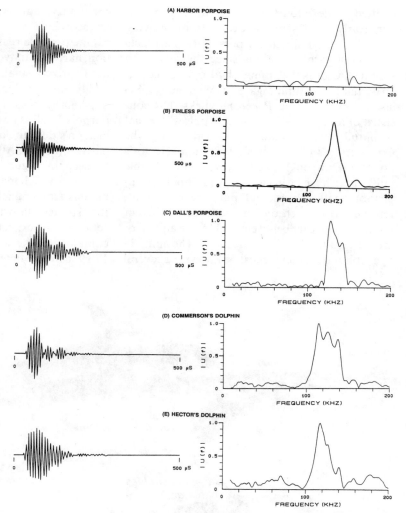

of the harbor porpoise (Mohl and Andersen 1973; Kamminga and Wiersma 1981; Hatakeyama et al. 1988) and the finless porpoise (Kamminga 1988) are similar in shape and time duration, with peak frequencies between 120 and 140 kHz. Dall's porpoise sonar signals with peak frequencies of 90 to 115 kHz have been measured in a tank (Hatakayama and Soeda 1990) and signals with peak frequencies of 120 to 160 kHz in the open sea (Awbrey et al. 1979; Hatakayama and Soeda 1990). Kamminga and Wiersma (1981) measured peak frequencies of about 124 kHz for a Commerson's dolphin in a tank and Evans et al. (1988) measured peak frequencies of 130 to 140 kHz in the open sea. Examples of sonar signals used by five of these species are shown in Figure 7.19. The duration of these signals is much longer and the spectra are much narrower (bandwidths are less than half) than for the sonar signals used by *Tursiops*, *Delphinapterus*, and *Pseudorca*. The signals of the species depicted in Figure 7.19 often have multiple components, as can be seen in (*C*) for the Dall's porpoise and in (*D*) for the Commerson's dolphin. Kamminga and Wiersma (1981) showed the waveform of 24 consecutive clicks measured from a swimming Commerson's dolphin. All 24 signals had nearly identical waveforms indicating that the multiple clicks were not the result of multipath propagation or some other artifact but were actually emitted by the dolphin. The reception of consecutive, nearly identical multiple clicks from a moving animal would eliminate the possibility of artifacts from multipath reflections since the multipath would change as the animal–hydrophone geometry changes. Kamminga and Wiersma (1981) also presented some *Phocoena* clicks that consisted of multiple pulses. Double pulses of equal amplitudes were commonly recorded from the Commerson's (Evans et al. 1988) and Hector's dolphin (Dawson 1988). However, these double pulses may have been caused by multipath propagation, with the second pulse being surface reflected. It is often difficult to discern whether multiple pulses are being emitted by the animal or are the results of surface reflections. The animal–hydrophone geometry must be known, or several

consecutive signals from a moving animal must be recorded.

One possible reason for the use of long duration, narrow bandwidth signals may be related to the relatively small size of the porpoises in the family *Phocoenidae* and the dolphins in the genus *Cephalorhynchus*. The relative size of the harbor porpoise, finless porpoise, Dall's porpoise, Commerson's dolphin, and Hector's dolphin in comparison with the Atlantic bottlenose dolphin can be seen in Figure 7.20. Except for the Dall's porpoise, which may be relatively short in length but has a larger girth, and the bottlenose dolphin, the species shown are among the smallest cetaceans in existence. The energy flux density contained of the waveforms displayed in Figure 17.19 can be determined by letting the time

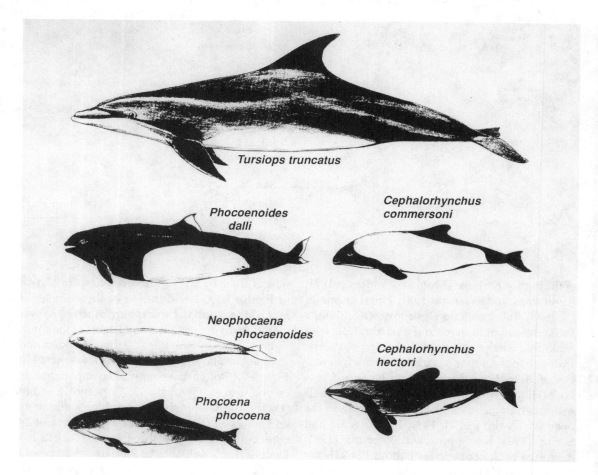

Figure 7.20. Relative sizes of some of the odontocetes belonging to the family *phocoenidae* and the genus *Cephalorhynchus*, compared with a *Tursiops truncatus*.

Table 7.3. Peak-to-peak source level (SL), relative energy in the normalized form function (E_N), and energy flux density (E) of the biosonar signals in Figure 7.19 and a *Tursiops* signal

Species	SL (dB re 1 μPa)	E_N	10 Log(E_N) (dB)	E (dB re 1 μPa^{2}s)
Tursiops t.	220	6×10^{-6}	-52	162
Phocoena p.	162	15×10^{-6}	-48	108
Neophocoena p.	—	16×10^{-6}	-48	—
Phocoenoides d.	170	19×10^{-6}	-47	117
Cephalorhynchus c.	160	14×10^{-6}	-49	105
Cephalorhynchus h.	151	19×10^{-6}	-47	98

waveform be written as

$$p(t) = As(t) \tag{7-18}$$

where A is the peak amplitude and $s(t)$ is the normalized form factor describing the shape of the signal in the time domain, $|s(t)|_{max} \leq 1$. The energy flux density, E, of a 1 μPa plane wave taken over an interval of 1 second is simply the integral of the instantaneous pressure squared (Urick, 1983), and can be expressed as

$$E = A^2 \int_0^T s(t)^2 dt = A^2 E_N \tag{7-19}$$

where E_N is the energy of the normalized form function. Values of E_N for the *Tursiops* signal shown in Figure 5.3 and for the narrow band signals of Figure 7.19 are given in Table 7.3 in the linear and log form, along with the peak-to-peak source level (equal to $2A$ in dB) and energy flux density. Note that the signals of Figure 7.19 contain approximately three to four times more energy than the *Tursiops* signal. Therefore, the smaller animals may have some peak amplitude limitations stemming from their size and must compensate for this limitation by emitting longer, narrower-band signals than *Tursiops* and other larger odontocetes. The use of double pulses is also helpful in projecting acoustic signals with greater overall energy ($\approx$ 3 dB more energy) per emission into the water.

7.6 Summary

Dolphins emit sonar signals in an adaptive manner, varying click intervals, amplitude, and waveform according to the environment and specific sonar task at hand. Signals are emitted with intervals that are longer by 20 to 40 ms than the time required for a signal to travel to an expected target and back. However, the click interval is never constant but often varies with time in a cyclic manner. The average peak frequency and 3-dB bandwidth of signals used by different *Tursiops* in Kaneohe Bay are relatively consistent, varying from 115 to 121 kHz for peak frequency and from 38 to 46 kHz for bandwidth. Other species, such as *Delphinapterus leucas* and *Pseudorca crassidens*, emit high amplitude (> 220 dB re 1 μPa) and high frequency signals (> 100 kHz) in Kaneohe Bay. There is no evidence of any spectral adaptation in which the spectral content of the signals is deliberately varied to "match" specific targets. The amplitude of the sonar signals is affected by various factors such as transmission loss (target range), masking noise level, target size, and difficulty of the task. As target range or masking noise increases, as target size decreases, and as the difficulty of a discrimination task increases, dolphins will tend to increase the amplitude of their signals. Although the peak-to-peak amplitudes of signals projected by dolphins can be relatively high (as high as 230 dB re 1 μPa), the short duration, exponential decaying property of the signals, and directivity of the transmission beam result in the acoustic power being relatively low, not exceeding 60 watts. The signal intensity in a click train can vary by as much as 20 dB; however, the variation is not random but orderly. Click intensity usually starts off low and raises to a maximum, remains relatively high during the midportion of a click train, and then gradually decreases toward the end of the train. The average source level is usually between 4.5 and 5.5 dB below the maximum source level in a click train. Statistical data on the properties of sonar signals exist only for a few species, such as *Tursiops truncatus, Delphinapterus leucas, Pseudorca crassidens, Phocoena phocoena, Delphinus delphis,* and *Cephalorhynchus hectori.* Only "snap shot" data exist for other species of odontocetes.

References

Au, W.W.L. (1980). Echolocation signals of the Atlantic bottlenose dolphin (*Tursiops truncatus*) in open waters. In: R.G. Busnel and J.F. Fish, eds., *Animal Sonar Systems.* New York: Plenum Press, pp. 251–282.

Au, W.W.L., and Pawloski, D.A. (1988). Detection of complex echoes in noise by an echolocating dolphin. J. Acoust. Soc. Am. 83: 662–668.

Au, W.W.L., and Penner, R.H. (1981). Target detection in noise by echolocating Atlantic bottlenose dolphins. J. Acoust. Soc. Am. 70: 251–282.

Au, W.W.L., and Turl, C.W. (1983). Target detection in reverberation by an echolocating Atlantic bottlenose dolphin (*Tursiops truncatus*). J. Acoust. Soc. Am. 73: 1676–1681.

Au, W.W.L., Floyd, R.W., Penner, R.H., and Murchison, A.E. (1974). Measurement of echolocation signals of the Atlantic bottlenose dolphin, *Tursiops truncatus* Montagu, in open waters. J. Acoust. Soc. Am. 54: 1280–1290.

Au, W.W.L., Schusterman, R.J., and Kersting, D.A. (1980). Sphere-cylinder discrimination via echolocation by *Tursiops truncatus.* In: R.G. Busnel and J.F. Fish, eds., *Animal Sonar Systems,* edited by New York: Plenum Press, pp. 859–862.

Au, W.W.L., Penner, R.H., and Kadane, J. (1982). Acoustic behavior of echolocating Atlantic bottlenose dolphin. J. Acoust. Soc. Am. 71: 1269–1275.

Au, W.W.L., Carder, D.A., Penner, R.H., and Scronce, B.L. (1985). Demonstration of adaptation in beluga whale echolocation signals. J. Acoust. Soc. Am. 77: 726–730.

Au, W.W.L., Penner, R.H., and Turl, C.W. (1987). Propagation of beluga echolocation signals. J. Acoust. Soc. Am. 82: 807–813.

Awbrey, F.T., Norris, J.C., Hubbard, A.B., and Evans, W.E. (1979). The bioacoustics of the Dall's porpoise-salmon drift net interaction. H/SWRI Techn. Rep. 79–120, pp. 79–120.

Babkin, V.P., and Dubrovskiy, N.A. (1971). Range of action and noise stability of the echolocation system of the bottlenose dolphin in detection of various targets. Tr. Akust. Inst. 17: 29–42.

Bel'kovich, V.M., and Dubrovskiy, N.A. (1977). Sensory bases of cetacean orientation. U.S. Joint Publication Research Service JPRSL/7157, May 27, 1977.

Dawson, S.M. (1988). The high frequency sounds of free-ranging Hector's Dolphin, *Cephalorhynchus hectori.* Rep. Int. Whal. Commn. (Spec. Iss. 9), 339–341.

Dziedzic, Z.-A. (1978). Etude experimentale des emissions sonar de certains delphinides et notamment de *D. Delphis* et *T. Truncatus.* These de Doctorat D'Etat Es-Sciences Appliquees, l'Universite de Paris VII.

Dziedzic, A., and Alcuri, G. (1977). Reconnaissance acoustique des formes et caracteristiques des signaux sonars chez *Tursiops truncatus,* famille des delphinides. C.R. Acad. Sc. Paris 285, Series D, 981–984.

Ellis, R. (1989). *Dophins and Porpoises,* New York: Alfred Knopf.

Evans, W.E. (1973). Echolocation by marine delphinids and one species of fresh-water dolphin. J. Acoust. Soc. Am. 54: 191–199.

Evans, W.W., and Powell, B.A. (1967). Discrimination of different metallic plates by an echolocating delphinid. In: R.G. Busnel, ed., *Animal Sonar Systems: Biology and Bionics.* Laboratoire de Physiologie Acoustique, Jouy-en-Josas, France, pp. 363–382.

Evans, W.E., Aubrey, F.T., and Hackbarth, H. (1988). High frequency pulse produced by free ranging Commerson's dolphin (*Cephalorhynchus commersonii*) compared to those of phocoenids. Rep. Int. Whal. Commn. (Spec. Iss. 9), 173–181.

Hatakeyama, Y., Ishii, K., Soeda, H., and Shimamura, T. (1988). Observation of harbor porpoise's behavior to salmon gillnet, (Document submitted to the International North Pacific Fisheries Commission.), 17 p. Fisheries Agency of Japan, Tokyo, Japan.

Hatakeyama, Y., and Soeda, H. (1990). Studies on echolocation of porpoises taken in salmon gillnet fisheries. In: J.A. Thomas and R. Kastelein, eds., *Sensory Abilities of Cetaceans.* New York: Plenum Press, pp. 269–281.

Herald, E.S., Brownell, R.L., Jr., Frye, F.L., Morris, E.J., Evans, W.E., and Scott, A.B. (1969). Blind river dolphin: first side-swimming cetacean. Science 166: 1408–1410.

Johnson, C.S. (1967). Discussion. In: R.G. Busnel ed., *Animal Sonar Systems: Biology and Bionics.* Laboratoire de Physiologie Acoustique, Jouy-en-Josas, France, pp. 384–398.

Kamminga, C. (1988). Echolocation signal types of odontocetes. In: P.E. Nachtigall and P.W.B. Moore, eds., *Animal Sonar: Processes and Performance.* New York: Plenum Press, pp. 9–22.

Kamminga, C., and Wiersma, H. (1981). Investigations on cetacean sonar II. Acoustical similarities and differences in odontocete sonar signals. Aquatic Mamml. 8: 41–62.

Kamminga, C., and Wiersma, H. (1982). Investigations on cetacean sonar V. The true nature of the sonar sound of *Cephaloryncus Commersonii*. Aquatic Mamml. 9: 95–104.

Kamminga, C., Dudok van Hell, W.H., and Tas'an, G. (1983). Investigations on cetacean sonar VI. Sonar sounds in *Orcaella Brevirostris* of the Makaham River, East Kalimanta, Indonesia; first descriptions of acoustic behaviour. Aquatic Mamml. 10: 83–104.

Kamminga, C., Engelsma, F.J., and Terry, R.P. (1989). Acoustic observations and comparison on wild, captive and open water Sotalia, and riverine Inia. 8th Biennial Conf. Biol. of Mar. Mamm., Pacific Grove, Cal.

Mohl, B., and Andersen, S. (1973). Echolocation: high frequency component in the click of the harbour porpoise (*Phocoena phocoena L.*), J. Acoust. Soc. Am. 54: 1368–1372.

Mohl, B., Surlykke, A., and Miller, L.A. (1990). High intensity narwhal clicks. In: J.A. Thomas and R.A. Kastelein, eds., *Sensory Abilities of Cetaceans*. New York: Plenum Press, pp. 295–303.

Moore, P.W.B., and Pawloski, D. (1990). Investigation of the control of echolocation pulses in the dolphin (*Tursiops truncatus*). In: J.A. Thomas and R.A. Kasterlein, eds., *Cetacean Sensory Systems: Field and Laboratory Evidences*. New York: Plenum Press, pp. 305–316.

Morozov, B.P., Akapiam, A.E., Burdin, V.I., Zaitseva, K.A., and Y.A. Solovykh. (1972). Tracking frequency of the location signals of dolphins as a function of distance to the target. Biofiika 17: 139–145.

Norris, K.S., and Evans, W.E. (1966). Directionality of echolocation clicks in the rough-tooth porpoise, *Steno Bredanensis* (Lesson). In: W.N. Tavolga, ed., *Marine Bio-Acoustics*. New York: Pergamon Press, pp. 305–324.

Penner, R.H. (1988). Attention and detection in dolphin echolocation. In: P.E. Nachtigall and P.W.B. Moore, eds., *Animal Sonar: Processes and Performance*. New York: Plenum Press, pp. 707–713.

Thomas, J.A., and Turl, C.W. (1990). Echolocation characteristics and range detection by a false killer whale (*Pseudorca crassidens*). In: J.A. thomas and R. Kasterlein, eds., *Cetacean Sensory Systems: Field and Laboratory Evidences*. New York: Plenum Press, pp. 321–334.

Thompson, R.K.R., and Herman, L.M. (1975). Underwater frequency discrimination in the bottlenose dolphin (1–140 kHz) and the human (1–8 kHz). J. Acoust. Soc. Am. 57: 943–948.

Turl, C.W., Penner, R.H., and Au, W.W.L. (1987). Comparison of target detection capabilities of the beluga and bottlenose dolphin. J. Acoust. Soc. Am. 82: 1487–1491.

Turl, C.W., and Penner, R.H. (1989). Differences in echolocation click patterns of the beluga (*Delphinapterus leucas*) and the bottlenose dolphin (*Tursiops truncatus*). J. Acoust. Soc. Am. 68: 497–502.

Urick, R.J. (1983). *Principles of underwater sound*. New York: McGraw-Hill.

Wiersma, H. (1982). Investigations on cetacean sonar IV, a comparison of wave shapes of odontocete sonar signls. Aquat. Mamm. 9: 57–67.

Youfu, X., and Rongcai, J. (1989). Underwater acoustic signals of the baiji, *Lipotes vexillifer*. In: W.R. Perrin et al., eds., *Biology and Conservation of the River Dolphins*. IUCN Species Survival Commission, Gland, Switzerland, pp. 129–136.

8

Target Detection Capability of the Active Sonar System

In this chapter, we will begin to examine the active sonar capabilities of dolphins by considering their target detection capabilities. However, it will be helpful to first take a short excursion and consider the reflection of acoustic energy by underwater targets and the application of the sonar equation to dolphin target detection. The subject of acoustic scattering by targets under water is an extremely complicated one, with physical processes often described by extremely complex mathematical functions and operations. This topic has been the subject of much research since the end of World War II. A good review of some of the modern research on acoustic scattering has been written by Neubauer (1986). Even after much intensive research, only problems involving simple geometrical shapes such as spheres, cylinders, spheroids, and plates are well understood and have lent themselves to some form of analytical solution. In the next section we will examine the acoustic backscatter from spheres and cylinders in order to gain an appreciation of the complex nature of echoes from targets.

8.1 The Physics of Acoustic Reflection by Targets

Acoustic energy emitted by a sonar will propagate outward in the form of a *longitudinal* or *compressional* wave until it strikes an object or a boundary and scatters. Longitudinal waves are associated with molecular vibrations directed along the path of propagation and produce alternating regions of compression and rarefaction. *Transverse* or *shear waves* are associated with molecular vibrations that are directed perpendicular or transverse to the path of propagation. Liquid media such as water will only support the longitudinal mode of vibration. Most solid media will support both longitudinal and transverse waves. When an acoustic wave strikes a target, a small percentage of the energy will be scattered back toward the source; this energy is often referred to as the acoustic backscatter, acoustic reflection, or target echo. A portion of the incident energy will penetrate the target and propagate in the form of both longitudinal and transverse waves. These waves may undergo further scattering at various interfaces within a target and can contribute to the structure of the backscattered energy. If the duration of the incident signal is relatively short, the results of different scattering processes can be resolved, as components of a complex *echo structure*.

Let us first consider the acoustic reflection from a sphere. An example of a simulated dolphin sonar signal and the echoes from a 7.62-cm diam. water-filled stainless steel sphere and a 2.54-cm diam. solid steel sphere are shown in Figure 8.1. Although the incident signal is rather simple, the echo from the sphere is quite complex, consisting of many *highlights* or *echo components*. A highlight is considered to be any individually recognizable component in the time waveform within the structure of the echo. The echo from the sphere seems to consist of the superposition of many differently delayed and attenuated versions of the incident signal; the duration of the echo is considerably longer than that of the incident signal. The presence of multiple highlights will cause the spectrum of the echo to exhibit an oscillatory behavior with local maxima and minima (see Fig. 8.1 B) in a way similar to the summation of a broadband signal with its delayed replica results in a rippled spectrum (discussed in Chapter 4). The scattering process for a cylinder will be very similar to that of a sphere if the incident signal is perpendicular to the axis of the cylinder.

Two major scattering processes will influence the echo structure. The first process consists of specular reflection from the front of the target and the transmitted longitudinal and transverse waves which undergo multiple internal reflections at the boundaries of the target. Shirley and Diercks (1970) experimentally identified two of the major interior sound paths for a water-filled sphere or cylinder. Echo highlights due to bound-ary reflections are produced at the front boundary and by acoustic energy propagating along two interior paths, the "central path" and the "square path" shown in Figure 8.2A. The central path is along the diameter that is parallel with the incident plane wave. Energy penetrates the front surface and travels along this path until the back boundary is met, where a portion of the energy is transmitted through the back surface and another portion is reflected. Part of this reflected energy will penetrate the front surface and part will be re-reflected towards the back boundary, and so on. The square path consists of three sides of a square circumscribed by the interior circumference, plus the extension of each leg to the reference plane. Multiple traverses along this path will also occur. The second process involves the propagation of various *circumferential waves* that travel around the target and back toward the source. Two types of circumferential waves may be excited: highly attenuated Franz waves with velocities that are slower than the speed of sound (denoted as c) in the surrounding fluid, and very slightly attenuated Rayleigh waves with velocities that are faster than c (Neubauer 1986). These circumferential waves continuously circumnavigate a sphere

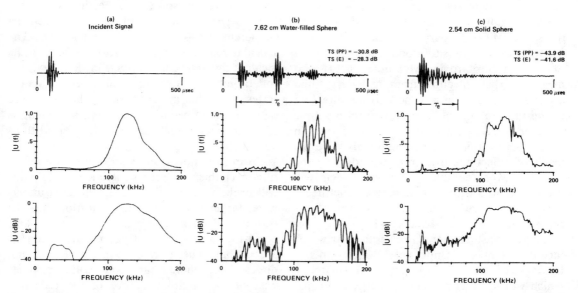

Figure 8.1. Example of a simulated dolphin sonar signal (*A*) and the echo from a 7.62-cm diam. water-filled stainless steel sphere (*B*) and a 2.54-cm solid steel sphere (*C*). (From Au and Snyder, 1980.)

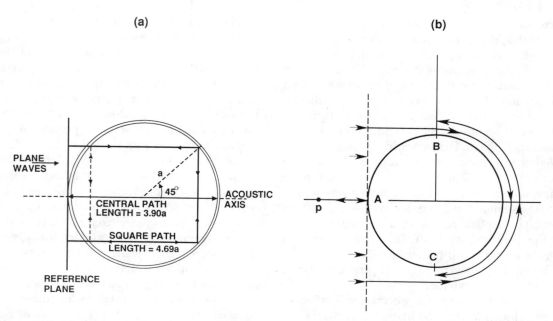

Figure 8.2. (*A*) Schematic diagram of principal interior sound paths in a water-filled sphere and cylinder at normal incidence (from Shirley and Diercks 1970); (*B*) schematic of circumferential waves contributing to the echo structure (from Neubauer 1986).

or cylinder and radiate energy into the water, producing periodic echo highlights of decreasing amplitudes. The two scattering processes distinguished above will also be in effect for solid elastic spheres and cylinders.

Echo highlights can also be caused by reflections from different parts of a target. A simple example of such a situation can be found in the study of Barnard and McKinney (1961), who measured the reflection from cylinders tilted at different angles with respect to the direction of the incident signal. The echo from a simple cylinder tilted at an angle of 45° is shown in Figure 8.3. In this case, the highlights were the result of reflections from the two edges of the cylinder. More complexly shaped objects will produce echoes with more complex structures.

So far we have considered the reflection processes associated with spheres and cylinders constructed of elastic material (a classification which includes all metals). Underwater targets can also be constructed of acoustically "soft" material, which will not allow acoustic energy to propagate into it (i.e., acoustic energy does not penetrate into the material). Targets constructed out of foam, cork, corprene (a cork-neoprene mix-

ture), and closed-cell neoprene (material used for some wet suits) are examples of acoustically soft targets. The structure of the echoes from a soft sphere or cylinder will be relatively simple, consisting of a specular reflection from the front surface of the target plus an attenuated circumferential Franz wave (Neubauer 1986). The specular reflection is 180° out of phase with the incident signal. Metallic targets in air behave like soft targets in water: no acoustic energy penetrates into the material since the difference between the impedance of air and any metal is so great. Such targets are usually referred to as acoustically "rigid" targets.

From this short discussion of acoustic reflection mechanisms we can conclude that dolphins will probably receive echoes (from most underwater targets) of much longer duration than the transmitted click. The echoes will most likely consist of many highlights, and these highlights are probably used by dolphins to detect and discriminate different targets. Finally, an awareness of the complex nature of the reflection mechanism is important for the design of biosonar experiments that address specific issues in an unambiguous manner.

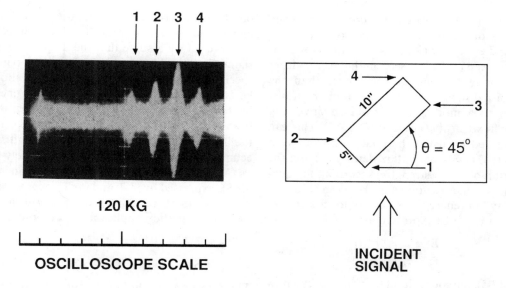

Figure 8.3. Back-scattered pulses from a brass-wall hollow cylinder. Pulse length, 50 μs; sweep time, 100 μs per division; cylinder diameter, 5 in.; cylinder length, 10 in. (From Barnard and McKinney 1961.)

8.2 Noise-Limited Form of the Sonar Equation

The *noise-limited* form of the *sonar equation* can be used to describe the target detection performance of any sonar. It equates the detection threshold (DT) with the signal-to-noise ratio of the echo at a point where the receiver can just detect the target. In its simplest form, it can be expressed as the difference in dB between the echo level and the received noise spectral density

$$DT = EL - NR \qquad (8-1)$$

where   EL = intensity of the echo
NR = received noise spectral density

The intensity of the echo from a target is equal to the source level of the transmitted signal minus the transmission loss of going to the target and back, plus the target strength, which is related to the fraction of energy reflected by the target back to the source (usually a negative quantity). It was expressed in equation (7-3) as

$$EL = SL - 2TL + TS \qquad (8-2)$$

where   EL = intensity of the echo
SL = source level or intensity at 1 m from source

TL = one-way transmission loss
TS = target strength

The background noise is assumed to be isotropic, which means that the noise arriving at the receiver is the same from all directions in three-dimensional space. If the receiver has a directional receiving beam, then only noise that is within the beam will be received, so that the received noise is equal to the isotropic noise minus the directivity index of the receiver, or

$$NR = NL - DI \qquad (8-3)$$

where   NL = isotropic noise spectral density
DI = receiving directivity index

The receiving directivity index for a *Tursiops* was calculated by Au and Moore (1984), and was discussed in Chapter 3. The receiving directivity index as a function of frequency was plotted in Figure 3.18, and the linear equation describing its variation with frequency was expressed in (3-12) as

$$DI(dB) = 16.9 \log f(kHz) - 14.5 \qquad (8-4)$$

Substituting expressions (8-2) and (8-3) into (8-1), the sonar equation in dB can be expressed as (Urick 1983)

$$DT = SL - 2TL + TS - (NL - DI) \qquad (8-5)$$

Equation (8-5) is written in terms of acoustic intensity, or the average acoustic power per unit area, and each term is expressed in dB. However, we have seen in the last section that echoes from targets may be many times longer than the transmitted signal, so that (8-5) should be transformed to a more generalized form involving energy flux density (Urick 1962, 1983). This can be achieved by using the relationship between intensity (I) and energy flux density (E) and the definition of target strength. According to equation (7-4), the relationship between acoustic intensity and energy flux density for a signal of duration τ_i can be expressed as

$$I = \frac{E}{\tau_i} = \frac{1}{\tau_i} \int_0^{\tau_i} \frac{p^2(t)}{\rho c} dt \qquad (8\text{-}6)$$

From (7-5), the source level term in (8-5) can now be expressed as

$$SL = 10 \log I = 10 \log E - 10 \log \tau_i \quad (8\text{-}7)$$

where τ_i is the duration of the transmitted signal. E is the energy flux density of the source at 1 m and is measured in units of the energy flux density of a 1-μPa plane wave taken over a 1-second interval (Urick 1983). The source energy flux density in dB (SE) can be written as

$$SE = 10 \operatorname{Log} E = 10 \log \left[\int_0^{\tau_i} p^2(t) dt \right] \quad (8\text{-}8)$$

The target strength of a target is defined as the ratio in dB of the echo intensity measured 1 m from the target to the intensity of the incident signal at the location of the target:

$$TS =$$

$$10 \log \left(\frac{\text{echo intensity 1 m from target}}{\text{incident intensity}} \right) \quad (8\text{-}9)$$

From the relationship between intensity and energy in (8-6), equation (8-9) can be expressed as

$$TS = TS_E - 10 \operatorname{Log}(\tau_e/\tau_i) \qquad (8\text{-}10)$$

where

$$TS_E =$$

$$10 \operatorname{Log} \left(\frac{\text{echo energy 1 m from target}}{\text{incident energy}} \right) \quad (8\text{-}11)$$

The expression TS_E, is the target strength based

on the ratio of the energy in the echo to the incident energy, and τ_e is the duration of the echo. However, if we assume that the dolphin detects signals in noise like an energy detector having a specific integration time of τ_{int}, then τ_{int} should be used in place of τ_e in (8-10) and the echo energy in (8-11) should be determined by integrating from 0 to τ_{int}. We will see in Section 8.4 that the integration time for *Tursiops* for broadband echolocation–type signals is approximately 264 μs (Au et al. 1988). Substituting the relationship for SL expressed by (8-7) and for TS expressed by (8-10) into (8-5), we arrive at the *transient form* of the sonar equation applicable to a dolphin:

$$DT_E = DT + 10 \log \tau_{int}$$

$$= SE - 2TL + TS_E - (NL - DI) \quad (8\text{-}12)$$

The detection threshold DT_E corresponds to the energy-to-noise ratio commonly used in human psychophysics:

$$DT_E = 10 \log(E_e/N_0) \qquad (8\text{-}13)$$

where E_e is the echo energy flux density and N_o is the noise spectral density or the rms noise level in a 1-Hz band. DT is the signal-to-noise ratio commonly used in sonar engineering and in electronics.

In Chapter 7, we saw that the transmitted signals are usually expressed in terms of the peak-to-peak sound pressure level (SPL_{pp}) rather than energy flux density. A simple relationship between E and SPL_{pp} can be derived by letting the acoustic signal equal $A s(t)$, where A is the peak amplitude and $s(t)$ is the normalized waveform function such that $|s(t)| \leq 1$ (equation (5-1) was essentially written in this form). We can now write

$$SPL_{pp}(dB) = 20 \log 2A = 20 \operatorname{Log} A + 6 \quad (8\text{-}14)$$

and

$$E(dB) = 10 \log \left[A^2 \int_0^T s^2(t) dt \right]$$

$$= 20 \log A + 10 \log \left[\int_0^T s^2(t) dt \right] \quad (8\text{-}15)$$

The log(integral) term in (8-15) does not vary much for dolphin signals in Kaneohe Bay; it is approximately -52 ± 1 dB. Therefore, substituting (8-14) into (8-15) we obtain for dolphin signals measured in Kaneohe Bay

$$E(dB) \approx SPL_{pp}(dB) - 58 \qquad (8\text{-}16)$$

Expression (8-16) should be valid for the high frequency signals of the bottlenose dolphin, beluga whale, and false killer whale. If dolphins use a longer duration, lower frequency signal in a different environment, then the log(integral) term must be evaluated. This will also be true for the narrow-band, high frequency signals used by the dolphins in Figure 7.17.

8.3 Biosonar Detection Capabilities

The target detection capability of any sonar system, man-made or biological, is limited by interfering noise and reverberation. The target detection sensitivity of a sonar can be measured by a variety of equivalent methods: (1) the range of a target of known target strength can be increased until the target can no longer be detected; (2) a fixed target range can be used and the size of the target can be reduced continuously until the target can no longer be detected; (3) a fixed target range can be used with a specific target and the amount of interfering noise or reverberation be gradually increased until the target cannot be detected; (4) conversely, the interfering noise or reverberation can be held constant and the target size reduced until the detection threshold is reached. Whatever method is used, source levels, target strength, and noise or reverberation levels should be measured so that the echo-to-noise ratio expressed in (8-12) or the echo-to-reverberation ratio (E/R) at threshold can be determined.

8.3.1 Maximum Detection Range

A variety of biosonar experiments using three of the four equivalent methods mentioned in the preceding paragraph have been performed in Kaneohe Bay to determine the sonar detection capability of *Tursiops truncatus*. Murchison (1980a,b) performed a maximum range detection experiment with two *Tursiops*, using a 2.54-cm-diam. solid steel sphere and a 7.62-cm-diam.

stainless steel water-filled sphere as targets. The catenary suspension system shown in Figure 8.4A was used to vary the target range. A thin monofilament line was used to lower the target to a depth of 1.2 m or raise it out of the water. Both animals performed the test from a floating pen as shown in Figure 8.4B. During the experiment, a 0.9-m × 3-m gate in front of the pen was lowered to provide an unobstructed aperture through which the animals could project their sonar signals. During a trial, both animals maintained station by facing away from the target toward their respective response paddles. A two-alternative forced-choice response paradigm was used. When an audio-cue was presented to one of the animals, that dolphin would turn and swim to the front of the pen and perform a sonar search while the other animal maintained station (Fig. 8.5). After performing his sonar search, the dolphin would return to his station and touch one of two paddles to indicate "target present" or "target absent." The target was then either raised or left submerged and the other animal was cued to begin a sonar search.

The target range was varied in 10-trial blocks according to the method of constant stimuli. The composite 50% correct detection thresholds were at ranges of 72 and 77 m for the 2.54-cm and 7.62-cm spheres, respectively. However, a bottom ridge at approximately 73 m (see Fig. 8.4a) limited the dolphins' ability to detect the 7.62-cm target beyond 73 m. The animals' performance degraded rapidly when the target was in the vicinity of the ridge, suggesting that the dolphins were probably limited by reverberation from the bottom at ranges beyond 73 m for tests with the 7.62-cm sphere. However, the limit for the 2.54-cm sphere seemed valid, because of its lower target strength.

Au and Snyder (1980) remeasured the maximum detection range in a different part of Kaneohe Bay using one of the same dolphins (Sven), the same 7.62-cm spherical target, and the same response paradigm and method of constant stimuli as Murchison (1980a,b). Sven's target detection performances for the 2.54-cm sphere (Murchison 1980a,b) and the 7.62-cm sphere (Au and Snyder 1980) are plotted in Figure 8.8 as a function of range. The 50% correct detection threshold for the 7.62-cm sphere occurred at

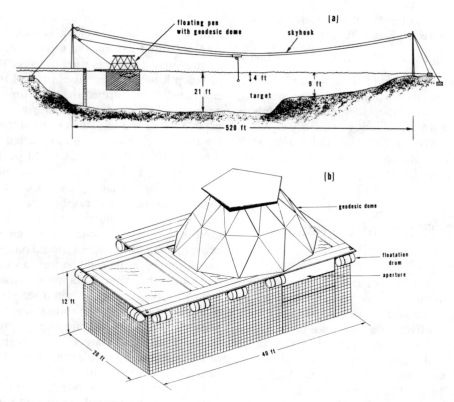

Figure 8.4. (*A*) Cross-sectional view of the biosonar range; (*B*) floating pen with geodesic dome. (From Au et al. 1974.)

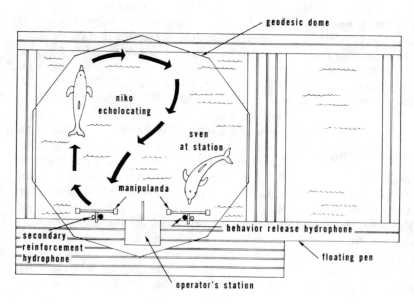

Figure 8.5. Example of an echolocation trial sequence. For conceptual purposes, the diagram is drawn with a transparent geodesic dome. (From Au et al. 1974.)

Figure 8.6. Target detection performance of a *Tursiops truncatus* as a function of range for two different spherical targets.

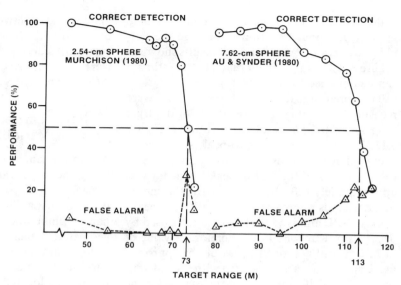

113 m, a considerably longer range than the 76.6 m reported by Murchison, verifying the suggestion that the underwater ridge at 73 m affected the dolphins' performance in Murchison's experiment.

The results shown in Figure 8.6 are very specific to the ambient noise condition of Kaneohe Bay. In order to make the results more general and useful, the detection performance should be plotted as a function of the estimated received signal-to-noise ratio. The transient form of the sonar equation for a noise-limited situation, expressed by equation (8-12), can be used to analyze the dolphins' performance plotted in Figure 8.6 in terms of the echo energy-to-noise ratio. From (8-12), values for the parameters SE, TL, TS_E, NL, and DI_R are needed to determine the dolphin's detection threshold. The transmission loss is merely the spherical spreading loss plus an absorption term, as expressed in (7-3). The target strengths for the 7.62-cm and 2.54-cm spheres are shown in Figure 8.2. The TS_E of -28.3 dB for the 7.62-cm sphere was obtained by integrating over 512 μs instead of the integration time of *Tursiops*, which is 264 μs. A 264 μs integration time would have reduced the target strength by only 0.3 dB, an insignificant difference. The target strength for the 2.54-cm sphere was measured to be -41.6 dB. The ambient background noise of Kaneohe Bay is shown in

Figure 1.4. At a frequency of 120 kHz, NL $\approx$ 54 dB re 1 μPa^2/Hz. The receiving directivity index for *Tursiops truncatus* is given by (8-4); for a peak frequency of 120 kHz, $DI_R = 20.2$ dB. We now have only the source energy flux density to consider in (8-12) in order to calculate the detection threshold.

During a sonar search, dolphins typically vary the amplitude of their sonar signals over a range of as much as 20 dB (Fig. 7.9). Therefore, it is difficult to estimate the detection threshold accurately. If the average value of the source energy flux density is used, the estimate for the detection threshold may be too low. Au and Penner (1981) approached this problem by using the maximum source energy flux density per trial, which will lead to a conservative or best-case (highest S/N available) estimate of the dolphin's detection threshold. In an experiment that will be discussed in Section 8.5.2, Au and Pawloski (1989) determined that the average detection threshold of their *Tursiops* corresponded to an echo level that was approximately 2.9 dB below the maximum echo level produced by the maximum source level emitted by the dolphin during a trial. Therefore, a more accurate estimate of E_e/N_0 can be obtained by the relationship

$$(E_e/N_0)_{cor} \approx (E_e/N_0)_{max} - 3\,dB \quad (8\text{-}17)$$

The subscript "cor" stands for "corrected estimate". The maximum energy flux density will be used in the sonar equation to analyze the dolphin's detection performance for results shown in Figure 8.6, bearing in mind that a better estimate can be obtained by using (8-17).

The sonar signals of the dolphin Sven were measured in the study of Au et al. (1974) for target ranges of 59 to 77 m; results were shown in Figure 7.13. The maximum peak-to-peak source level averaged over 12 trials at the 77-m range was 225 dB re 1 μPa, and the typical peak frequency was approximately 120 kHz. From (8-16), the energy flux density is approximately equal to the peak-to-peak SPL minus 58 dB for the signals used by *Tursiops* in Kaneohe Bay. Therefore, a value of 167 dB re 1 μPa^{2}s for SE_{max} would be appropriate. Using the values associated with the different parameters of the sonar equation discussed in the last three paragraphs, the target detection results of Figure 8.6 are replotted as a function of the echo energy-to-noise ratio in Figure 8.7. The graph shows that the dolphin's performance was consistent for the two studies. The 75% correct thresholds were at an $(E_e/N_0)_{max}$ of 10.4 dB for the 2.54-cm sphere and of 12.7 dB for the 7.62-cm sphere. From (8-17), the corresponding $(E_e/N_0)_{cor}$ is 7.4 dB for the 2.54-cm sphere and 9.7 dB for the 7.62-cm sphere. The difference of approxi-

mately 2 dB between the threshold estimates obtained with the two different-sized spheres is small, considering the fact that the two studies were done approximately two years apart, in different areas of Kaneohe Bay.

8.3.2 Target Detection in Noise: Variable Noise Level

Au and Penner (1981) used the technique of fixing the target range and varying the level of a masking noise source to determine the target detection capabilities of two *Tursiops truncatus* named Ehiku and Heptuna. Their experimental geometry, depicted in Figure 8.8, consisted of a hoop station on one side of the pen, with a 7.62-cm O.D. stainless steel water-filled sphere as the target, located 16.5 m from the hoop. An acoustically opaque screen consisting of a sheet of aluminum plate with a piece of corprene (an acoustically soft material; cf. Section 8.1) glued on it, was placed directly in front of the hoop to prevent a dolphin stationed in the hoop from performing a sonar search until the screen was lowered out of the way. A spherical transducer was placed 4 m from the hoop on a direct path to the target. The dolphin was required to station in the hoop and commence its sonar search as soon as the acoustic screen was lowered. After completing the sonar search, the dolphin would back out of the hoop and respond by striking one of two paddles to indicate "target present" or "target absent" in a two-alternative forced-choice paradigm. The noise transducer was equalized with a 30-kHz two-pole low-pass filter to make its transmission response constant with frequency, so that the noise projected at the hoop had a flat frequency spectrum between 40 and 160 kHz.

Masking noise levels between 67 and 87 dB re 1 μPa^2/Hz in 5-dB increments were randomly used in blocks of 10 trials for a 100-trial session. The dolphins' performance as a function of the masking noise level is shown in Figure 8.9. Each point represents the average performance for 200 trials. The results indicate that the dolphins' accuracy decreased monotonically as the noise level increased.

The dolphins' signals themselves were also measured by Au and Penner (1981) so that the animals' performance as a function of the echo

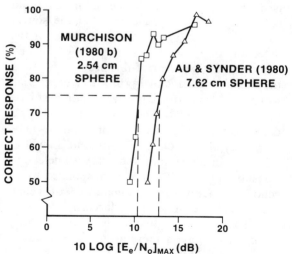

Figure 8.7. Target detection performance of a *Tursiops* as a function of the echo energy-to-noise ratio for the range detection data of Fig. 8.6. (From Au 1988.)

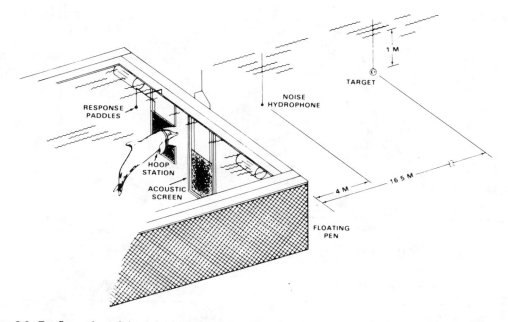

Figure 8.8. Configuration of the dolphin sonar target detection experiment in the presence of artificial masking noise. (From Au and Penner 1981.)

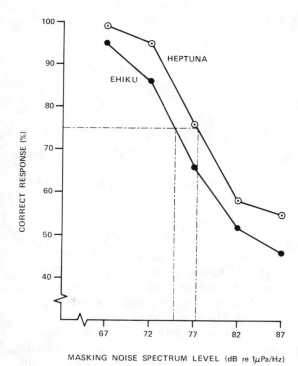

Figure 8.9. *Tursiops* target detection performance as a function of the masking noise level. (From Au and Penner 1981.)

energy-to-noise ratio could be determined. The dolphins' performance plotted as a function of $(E_e/N_0)_{max}$ is shown in Figure 8.10. In the calculations the average value of the maximum source energy flux density per trial was used. The results of Turl et al. (1987), who used the same technique as Au and Penner (1981) to compare the detection capability of a *Tursiops truncatus* with a *Delphinapterus leucas*, are also included in Figure 8.10. A 7.62-cm sphere was used at ranges of 16.5 and 40 m and a 22.86-cm sphere at a range of 80 m. The noise source was located 5 m from the hoop station between the target and the hoop. The experimental procedure was similar to that of Au and Penner (1981) except that a smaller noise increment of 3 dB was used. The 75% correct response threshold occurred at $(E_e/N_0)_{max}$ values of 7 and 12 dB in the study of Au and Penner (1981) and at approximately 10 dB in the study of Turl et al. (1987). The corresponding values for $(E_e/N_0)_{cor}$ were 4 and 9 dB for the study of Au and Penner and 7 dB in the study of Turl et al. The results of Figure 8.10 indicate good agreement and consistency between the three *Tursiops* involved, with less than 5 dB difference in target detection ability between

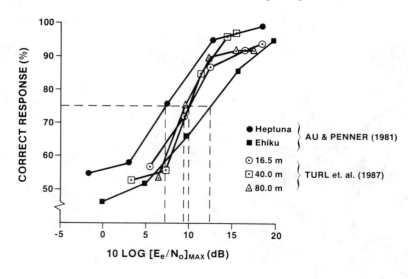

Figure 8.10. Target detection in noise performance results as a function of the echo energy-to-noise ratio. (From Au and Penner 1981; Turl et al. 1987.)

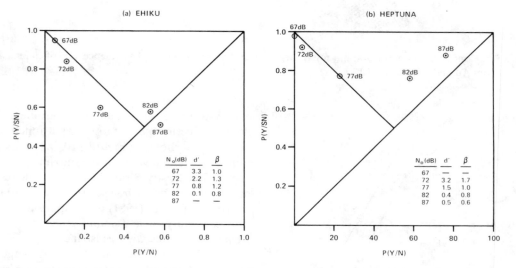

Figure 8.11. Performance data of two dolphins plotted in a receiver operating characteristic (ROC) format. (From Au and Penner 1981.)

animals. At the two highest noise levels of the Au and Penner study, probably both dolphins began to guess at the target condition. At the 82- and 87-dB noise levels, Ehiku did not emit any clicks detected by the click detection system in 20% and 41% of the trials, respectively. Heptuna did not emit any detectable signals in 14% of the trials at the 87-dB noise level. Therefore, the averages of the maximum signal per trial for the noise levels between 67 and 77 dB were used to calculate $(E_e/N_0)_{max}$. The animal used by Turl et al. did not stop emitting signals at any noise level tested,

probably because the highest noise level was 10 dB lower than that of Au and Penner.

Au and Penner (1981) also plotted the dolphins' performance in an ROC format, with the probability of detection plotted against the probability of false alarm, as shown in Figure 8.11. Values of d' (sensitivity) and β (response bias) as defined in Section 1.5 are included in the figure. As the noise level increased from 67 to 82 dB, Ehiku's performance followed the minor diagonal, indicating very little response bias of the animal. Heptuna also exhibited no response bias for noise levels between 67 and 77 dB. However, as the noise level increased to 82 and 87 dB, Heptuna adopted a liberal response bias. This caused $P(Y/sn)$ to remain high (above 0.75), but with a corresponding increase in $P(Y/n)$. The ROC plot suggests that for noise levels of 77 dB and lower, both dolphins were relatively unbiased. At the two highest masking noise levels, the dolphins' performances were close to chance, supporting the notion that the animals may have done a great deal of guessing for the two highest noise levels.

8.3.3 Target Detection in Noise: Variable Target Size

The third technique, fixing the target range and reducing the target size, was used by Au et al. (1988) to measure the target detection capability of a *Tursiops*. An electronic transponder system was used in conjunction with a Franklin Ace-1000 PC so that the effective echo strength could be varied by adjusting the level of the simulated echoes. A detailed description of the phantom electronic target simulator can be found in Au et al. (1987). The experimental configuration, depicted in Fig. 8.12, shows a dolphin in a hoop station with three hydrophones directly in front of the animal. An acoustic screen was located between the hoop and the hydrophones. When raised, the screen blocked the animal's sonar signals from the hydrophones.

The dolphin's sonar signals were detected by the first hydrophone (Celesco LC-10), which triggered an 8-bit analog-to-digital (A/D) converter to digitize the signal received by the second hydrophone (B&K 8103), located 1.9 m from the hoop. The A/D converter operated at a 1-mHz rate, digitized 128 points per signal, and stored the data in a static random access memory (RAM). The output of the first hydrophone also triggered an external delay generator and flagged the computer that a signal had been emitted. After an appropriate delay corresponding to the simulated target range, the delay generator flagged the computer to project the signal stored in RAM. The number of highlights, the time separation between highlights, and the amplitude of each highlight were controlled by the computer. Masking noise was mixed with the RAM signal and projected from the third transducer (Naval Research Laboratory F-42D), located 2.4 m from the hoop. The projector was driven by an equalization circuit that flattened the transmit response of the projector, allowing it to transmit broadband noise and simulated dolphin

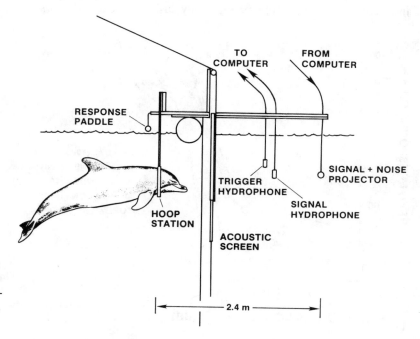

Figure 8.12. Experimental configuration with dolphin in hoop station and transducers used in electronic target simulation system. (From Au et al. 1988.)

signals with a minimum of distortion (Au et al. 1987).

A trial started when the dolphin was directed to swim into the hoop station; the acoustic screen was in raised position. Then the masking noise was projected and the acoustic screen lowered, cuing the dolphin to echolocate. A go/no-go response procedure was used: the dolphin was required to leave the hoop and strike a paddle for a "targets present" response, or remain in the hoop until the trial ended for a "target-absent" response. The dolphin's target detection capability was determined by playing masking noise at a fixed level and varying the intensity of the simulated echo played back to the animal with an attenuator. The signal attenuator was randomly varied in 10-trial blocks by increments of 2 dB. Each phantom target consisted of two clicks separated by 200 μs, played back to the animal at a time delay corresponding to a 20-m target range. The time delay between detecting the outgoing signal and transmitting the simulated echo was approximately 10.5 ms. The dolphin was required to station in a hoop, echolocate, and report whether the "target" was present or "absent".

The phantom target results are shown as the solid line in Figure 8.13 overlaying the maximum range results (from Fig. 8.7) and the noise masking results (from Fig. 8.10) represented by dashed curves. The abscissa represents the average of the maximum E_e/N_0 per trial at each of the five attenuator settings. The phantom target results were consistent with the real target results. The dolphin responded to the phantom echo as if it originated from a real target. At the 75% correct response level, $(E_e/N_0)_{max}$ was 10 dB and $(E_e/N_0)_{cor}$ was 7 dB. The results of the various studies represented in Figure 8.13 agree extremely well, considering the differences in animals, time periods, and experimental procedures. Murchison (1980a,b) and Au and Snyder (1980) varied the target range in small increments in terms of the resultant E_e/N_0. Au and Penner (1981) and Turl et al. (1987) used a constant target range and randomly varied the masking noise level. Relatively large increments of 5 dB were used by Au and Penner (1981) and smaller 3-dB increments were used by Turl et al. (1987). Au et al. (1988) used a fixed phantom target range and noise level and randomly varied the amplitude of the target echoes in 2-dB increments. The shallower slope of the performance curves in the Au and Penner (1981) study were probably the result of using a larger noise increment.

A comparison of the detection sensitivity of humans and dolphins is interesting, although we need to keep in mind that the detectability of a signal in noise is a function of the signal duration, frequency, and bandwidth, as well as the psycho-

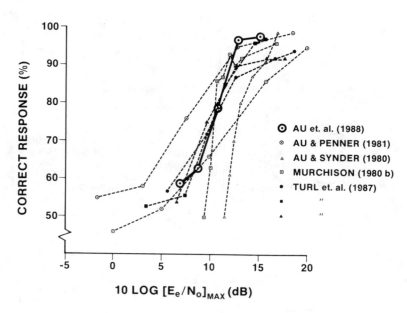

Figure 8.13. Phantom target detection performance of a *Tursiops* as function of the echo energy-to-noise ratio.

physical testing procedure. Results of human listening experiments using the two-alternative forced-choice procedure, in which the subject must specify which one of two intervals in a trial contains the signal, have resulted in S/N_{75} (the S/N at the 75% correct response level) that vary from 7.9 dB (Green et al. 1957) to 5 dB (Jeffress 1968). McFadden (1966), using a yes/no paradigm, obtained a S/N_{75} of 10 dB for a 400-Hz, 125-ms signal in continuous and burst noise. Although the dolphin data obtained with broadband echolocation signals cannot be directly compared with the human data obtained with narrow-band tonal signals, $(E_e/N_0)_{cor}$ values between 4 and 9 dB obtained with the bottlenose dolphin are certainly compatible with S/N_{75} values between 5 and 10 dB for humans. This brief comparison of signal detection in noise between dolphins and humans is consistent with the discussion in Chapter 3 pointing out the similarities of the two auditory systems.

8.3.4 Target Detection in Noise: Forward and Backward Masking

The target detection in noise experiments just discussed involved the use of continuous noise so that the target echoes were always received by the dolphins in the presence of masking noise. Bel'kovich and Dubrovskiy (1976) used a broadband click masker to interfere with two echolocating dolphins that were attempting to detect

the presence of a 40-mm spherical target. Strictly speaking, the click masker was not white noise, but it was also not like reverberation since no effort was made to make the masking click resemble the outgoing sonar signal of the animal. Backscatter from volume or surface reverberation usually resembles the projected signals. However, the masking click can be considered a broadband transient noise interference in the target detection process. The projection of the click masker was time-locked to the emission of sonar signals by the animals so that by changing its delay it was possible to control the spatial and temporal relationship of the masker and the echo from the sphere. The dolphins' performance in detecting the sphere for various delays between the masker signal and the echo from the sphere is presented in Figure 8.14. The masking signal had an amplitude that was 50 ± 3.5 dB greater than the sphere echo. At $\tau < -300$ μs and $\tau > 500$ μs, the dolphins were able to detect the sphere very reliably. The performance curves were steeper when the masking click followed the target echo (backward masking). The dolphins' performance improved more smoothly when the masking click preceded the target echo (forward masking). Dolphin 1 experienced a sharp rise followed by a sharp drop in its performance near a zero delay between the masker and the target echo. Bel'kovich and Dubrovskiy (1976) were of the opinion that this behavior was due to the dolphin making a transition from detection of

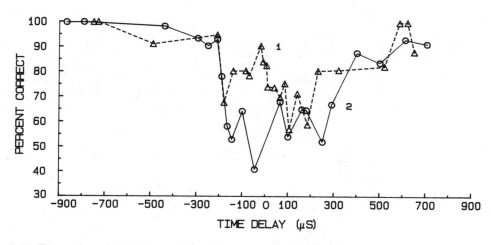

Figure 8.14. Target detection performance as a function of the time delay between the click masker and the echo from a spherical target for two dolphins denoted as 1 and 2. (From Bel'kovich and Dubrovskiy 1976.)

the target echo to the discrimination of a single masker click from a pair of clicks (masker plus echo). In the backward masking condition, the 75% correct response threshold occurred when the masker click followed the echo by approximately 200 μs.

Moore et al. (1984) performed a particularly elegant backward masking experiment in which noise bursts could be timed to arrive in coincident with the target echo or be delayed up to 700 μs after the target echo. A *Tursiops truncatus* was trained to use its sonar to detect a water-filled aluminum cylinder of 9 cm O.D., 21 cm long, located at a range of 46 m. A raft supported the target and two transducers, one to detect the incident sonar signal and the other to project the 1-ms-long broadband noise burst, as shown in Figure 8.15. A digital delay device was used to delay the noise burst with respect to the target echo by 100, 300, 500, and 700 μs. The different delays were presented via a modified method of constants, with the delays randomly assigned to 10-trial blocks of equal numbers of target-present and -absent trials. A go/no-go response paradigm was used to test the dolphin's detection sensitivity.

The resultant psychometric function obtained by Moore et al. (1984) is shown in Figure 8.16. The dolphin's performance decreased monotonically with a decrease in the noise delay time. At noise delays of 500 and 700 μs, the dolphin's

detection performance was high, 95% and 98% correct, respectively. The dolphin's performance dropped to 50% correct at a delay of 100 μs, and the 70% correct response threshold was extrapolated to be 265 μs. Moore et al. (1984) related the backward masking threshold of 265 μs to the "critical interval" of Vel'min and Dubrovskiy (1976), stating that the critical interval represents a temporal resolving interval in dolphin echolocation. The backward masking threshold and critical interval will be discussed further in Section 8.4.1.

8.3.5 Target Detection Experiments with Other Species

Some earlier studies of the target detection of echolocating dolphins were performed by Busnel and Dziedzic (1967) with a blindfolded *Phocoena phocoena* and by Penner and Murchison (1970) with *Inia geoffrensis*. Busnel and Dziedzic (1967) used wires with different diameters (4 mm to 0.2 mm) and changed wires from session to session. The wires were stretched vertically spaced 1 m apart, to form a maze of 30 obstacles. The dolphin swam in the tank wearing rubber eye cups, and the number of contacts with the wires was counted. A threshold diameter of 0.35 mm corresponding to an avoidance performance of 79% was determined. Penner and Murchison (1970) also used wire targets and presented them

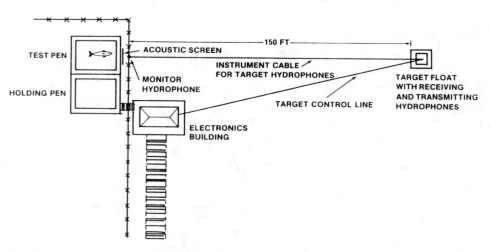

Figure 8.15. Experimental configuration for target detection in the presence of backward masking noise. (From Moore et al. 1984.)

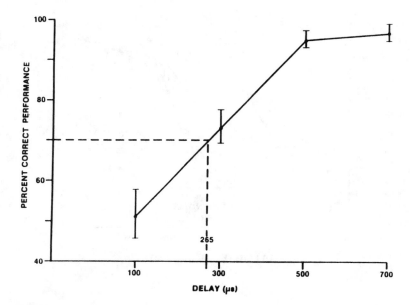

Figure 8.16. Psychometric function for backward masking as a function of masker delay. The threshold for 70% correct performance is shown. (From Moore et al. 1984.)

behind an acoustically transparent but visually opaque screen in a tank. The diameter of the wire at the Inia's detection threshold was 1.1 mm. Good source level recordings were not made in either study, and the target strength of the wires was not measured. Therefore, it is not possible to estimate the signal-to-noise ratio at the detection thresholds.

The target detection capabilities of *Delphinapterus leucas* were directly compared with those of *Tursiops truncatus* in a noise masking experiment reported by Turl et al. (1987). A similar experimental geometry as shown in Figure 8.7 was used, except that the noise source was 5 m from the animal's hoop station. A 7.62-cm sphere was used at ranges of 16.5 and 40 m and a 22.86-cm sphere at a range of 80 m. The noise source was 5 m from the hoop station, on a line between the target and the hoop. The experimental procedure was similar to that of Au and Penner (1981) except that a smaller noise increment of 3 dB was used. The dolphin's performance was shown in Figure 8.10. The beluga's performance results are given in Figure 8.17. The comparison indicates that the beluga's threshold was approximately 8 to 13 dB more sensitive than the bottlenose dolphin's. The beluga's superior performance is somewhat puzzling. We will show in Section 8.5 that this level of performance suggests that the animal's sensitivity was better

than that of an optimal detector, which is impossible. The beluga's superior performance probably is the result of imperfectly maintained alignment between the beluga, noise transducer, and target. During an earlier phase of the noise masking experiment with the beluga, Penner et al. (1986) discovered that the whale was minimizing the effects of the noise source by using a surface-reflected transmission path; a pair of schematics showing this is given in Figure 8.18. If the whale had a narrow receiving beam, the amount of noise received from the transducer would be less if the whale directed its beam upward and away from the noise source (Fig. 8.18B). When a rubber mat lined on one side with corprene was used to block the surface path, the whale's performance was altered considerably. The experience of Penner et al. (1986) indicates that the location of the masking noise source in the Turl et al. (1987) study may have allowed the whale to spatially filter the masking noise.

Thomas and Turl (1990) measured the target detection range of a false killer whale (*Pseudorca crassidens*) on the same biosonar range that Au and Snyder (1980) used for *Tursiops*. They initially used five 7.62-cm water-filled stainless steel spheres spaced 20 m apart, with the closest target at 40 m, to bracket the animal's performance. After determining that the threshold was near 100 m, they reconfigured the target spacing to

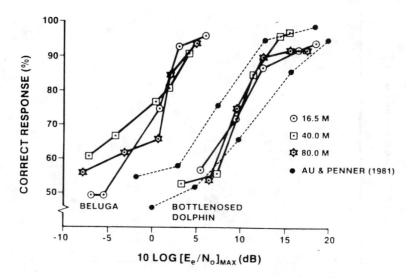

Figure 8.17. Beluga and bottlenose dolphin performance data at three distances as a function of the echo signal-to-noise ratio. (From Turl et al. 1987.)

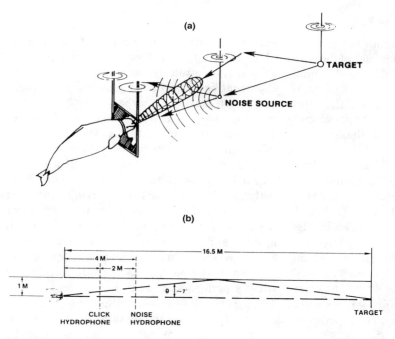

Figure 8.18 (*A*) Schematic showing how the beluga might achieve a higher signal-to-noise ratio by using a surface-reflected propagation path. (*B*) Side view of the beluga, the noise projector, and the test geometry showing a direct and a surface-reflected transmission path. (From Penner et al. 1986.)

5 m and collected data for three range configurations, starting at 80 m, then going to 90 m, and then 100 m. A two-alternative forced-choice response paradigm was used in which the animal was required to strike one of two paddles to indicate either "target present" or "target absent". Thomas and Turl conducted two series of measurements, one in which the target was presented from near to far in 10-trial blocks and one in which the target range was random in 10-trial blocks. The *Pseudorca* target detection performance results are shown in Figure 8.19. (The characteristics of the whale's sonar signals were discussed in Chapter 7).

The *Pseudorca*'s 50% correct detection threshold was approximately 113 m for the near-to-far

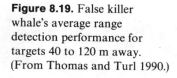

Figure 8.19. False killer whale's average range detection performance for targets 40 to 120 m away. (From Thomas and Turl 1990.)

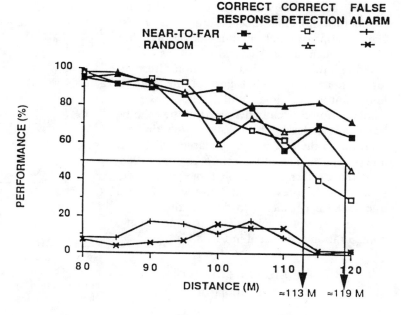

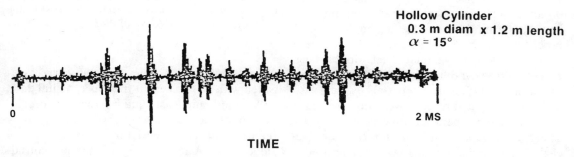

Figure 8.20. Example of a complex echo from a hollow cylinder. (From Au et al. 1988.)

8.4 Detection of Complex Echoes

series and 119 m when the target range was randomized in blocks of 10 trials. These data compare well with the data of Au and Snyder (1980) who measured a 50% correct detection threshold at approximately 113 m for *Tursiops*, with the target range being changed from near to far. The results of Thomas and Turl (1990) cannot be analyzed in terms of the echo signal-to-noise ratio since the receiving beam pattern and the directivity index of a *Pseudorca* have not been measured. An adult *Pseudorca* has a larger head than an adult *Tursiops*, so that its receiving beam should be narrower and its directivity index larger than that of a *Tursiops* by a few dB (see eq. 6-26).

We have seen in Section 8.1 that acoustic reflection can be a fairly complex process that can produce echoes much longer in duration than the transmitted signal and containing many resolvable highlights or echo components. An example of a complex echo from a simple water-filled aluminum cylinder having an outer diameter of 31 cm, wall thickness of 0.8 cm, and length of 1.2 m is shown in Figure 8.20. A simulated dolphin sonar signal projected at an incident angle of 15° from the broadside aspect was used to obtain the cylinder echo. Many resolvable highlights are part of the echo, and the echo duration is at least 2 ms, or approximately 40 times longer than that of the incident signal. Several questions can be asked regarding a dolphin's detection threshold for such an echo. What part of the echo is important to the animal? Will the energy over the whole duration or only over part of the echo be important to the dolphin? Will the dolphin concentrate on the area about the largest highlight and ignore the rest of the echo? Au et al. (1988) used the electronic phantom target system described in the previous section and performed a series of experiments to address the issue of detecting complex echoes. Their experimental configuration is depicted in Fig. 8.12.

8.4.1 Measurement of Integration Time

The auditory *integration time* of the dolphin for broadband transient signals was determined by first measuring the animal's threshold (staircase procedure), using a single-click echo for each click emitted by the animal. Then, double-click echoes with seven different separation times between clicks, varying from 50 to 600 μs, were used and the shifts in threshold were observed as a function of the separation time between clicks. The specific separation time for each double-click threshold measurement was randomized.

The results of the auditory integration time experiment are displayed in Figure 8.21. The ordinate is the E_e/N_0 at threshold based on an individual click. The abscissa is the separation time (ΔT) between double clicks. The threshold shifted approximately 3 dB as the stimulus changed from a single click to a double click. This shift is exactly what would be expected for an energy detector since there is 3 dB more energy in the double-click stimulus. As ΔT increased to 250 μs, the double-click threshold began to shift toward the single-click threshold, reaching it at a ΔT of about 300 μs and greater. Therefore, the presence of a second click with ΔT greater than 300 μs did not help the dolphin "detect" the phantom target. The solid curve in Figure 8.21 is the response of an ideal energy detector with an integration time of 264 μs. (The dolphin's performance data were fitted with several curves having

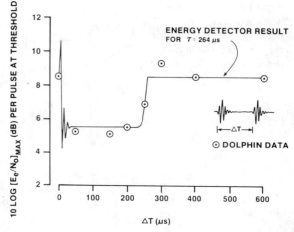

Figure 8.21. Results of integration time experiment, showing the average of the maximum E_e/N_0 per pulse at threshold as a function of the separation time between pulses. Each echo at 0 μs consisted of a single click, while each echo at the other separation times consisted of a double click. The solid curve is the response of an ideal energy detector that best fit the dolphin's data. (From Au et al. 1988.)

different integration times, and the curve associated with 264 μs best fit the experimental data, with a minimum least-square error.) The oscillations on the left side of the curve were caused by the first and second clicks overlapping at small values of ΔT. A 264-μs integration time is considerably shorter than the several milliseconds indicated by Johnson's (1968a) data for the pure-tone threshold as a function of duration (Fig. 3.4). The data of Au et al. (1988) suggest that dolphins process broadband transient signals differently than narrow-band pure-tone signals. Bullock and Ridgway (1972) found that different areas of a dolphin's brain are used to process broadband clicks and relatively longer narrow-band pure-tone signals. In Chapter 4, we discussed the contention of Vel'min and Dubrovskiy (1975) that dolphins may have two functionally "distinct subsystems for the perception of emotional and communicative sounds (0.001–30 kHz), or 'passive' hearing, and a subsystem for the perception of echo signals (20–170 kHz), or 'active' hearing."

The 264-μs auditory integration time corresponds to the backward masking thresholds

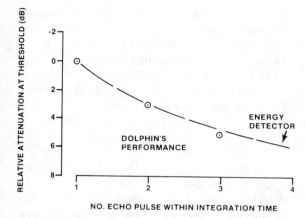

Figure 8.22. Dolphin's performance as a function of the number of clicks in the echo within the animal's integration time; the ordinate is the relative attenuation threshold. (From Au et al. 1988.)

reported by Vel'min and Dubrovskiy (1975) and by Moore et al. (1984) for echolocating *Tursiops truncatus*. Vel'min and Dubrovskiy used artificial echolocation pulses and Moore et al. used a burst of broadband noise to mask the target echoes. Moore et al. measured a backward masking threshold of 265 μs, essentially the same value as the integration time we measured. This suggests that backward masking may be related to integration time. The integration time also corresponds to the "critical interval" of approximately 260 μs measured for *Tursiops* by Vel'min and Dubrovskiy (1976), who defined this interval as a "critical time interval in which individual acoustic events merge into an acoustic whole."

However, our results suggest that Vel'min and Dubrovskiy's critical interval may merely be the analog of the integration time of an energy detector.

The dolphin as an energy detector was investigated further by considering how the detection threshold shifted as a function of the number of highlights within the integration time. Thresholds were determined for simulated echoes with one, two, and three highlights. The separation time for the two-click signal was 200 μs. The first separation time for the three-click signal was 50 μs, followed by a 150-μs separation time between the second and third clicks. The shift in the dolphin's detection threshold as a function of the number of clicks in the echo within the animal's integration time is displayed in Figure 8.22. The amplitude of the phantom echo depended on the value of the signal attenuator and the source level of the projected signal. The relative attenuation is an adjustment to the value of the signal attenuator at threshold that takes into account differences in the amplitudes of the emitted signals from trial to trial. The data indicate that the dolphin's performance followed the response of an energy detector.

8.4.2 Mixed Amplitude Highlights

The effects of mixed amplitude highlights on the dolphin's ability to detect echoes were tested with the three-click echoes displayed on the left side of Figure 8.23. The number above each highlight represents the relative amplitude of that high-

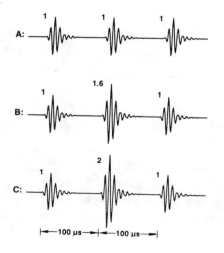

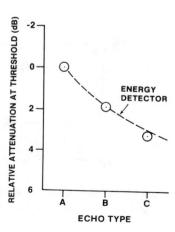

Figure 8.23. Mixed amplitude highlights experiment. (*Left*) Stimuli A, B, and C; (*right*) dolphin's relative threshold results. (From Au et al. 1988.)

light. All highlights occurred within the animal's integration time. Signal A consisted of equal amplitude highlights separated by 100 μs. The second highlights in signal B and signal C were 1.6 and 2 times greater, respectively, than the first and third highlights. The purpose of this test was to determine whether or not the first and third highlights of signals B and C would contribute to the detection process.

The results of this experiment are shown on the right side of Figure 8.23. The ordinate is the relative attenuation at threshold and the abscissa is the echo type. The dashed curve shows the response of an energy detector. The results indicate that the dolphin perceived the whole echo as one signal and processed the mixed-amplitude highlight signal as an energy detector would. Therefore, the first and third highlights did contribute to the detection process.

8.4.3 Multiple and Overlapping Integration Periods

The effects of multiple integration periods on the dolphin's detection performance were tested by using the signals shown on the left side of Figure 8.24. Signal A was the reference two-click echo,

with a separation time of 200 μs between clicks of equal amplitude. Signal B was a three-click echo having two distinct integration periods, with the energy in the first integration period being half (3 dB less) that of the second integration period. Signal C had a single highlight in the first integration period with twice (3 dB greater) the energy of the two clicks in the second integration period.

The dolphin's performance is shown in the right column of Figure 8.24. The threshold for signal B was essentially the same as for the reference signal. This suggests that the dolphin detected the signal using the second integration period, which had the same amount of energy as the reference signal and 3 dB more energy than the first integration period. By contrast, there was a 3.7-dB shift in relative threshold with signal C, suggesting that the dolphin used the energy in the first integration interval, which had 3 dB more energy than the second, to detect the signal. These results indicate that if an echo has highlights in multiple integration periods, the dolphin will respond to the integration period with the most energy.

The effects of overlapping integration periods on the dolphin's detection performance were

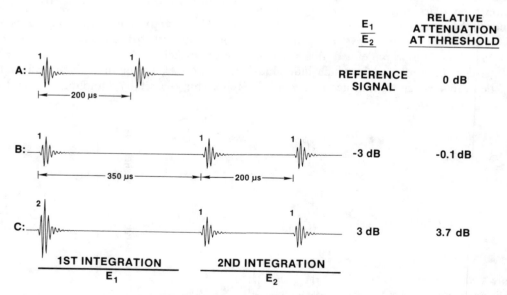

Figure 8.24. Multiple integration period experiment. (*Left*) Stimuli A, B, and C; (*right*) relative energy of both integration periods and dolphin's relative threshold results. (From Au et al. 1988.)

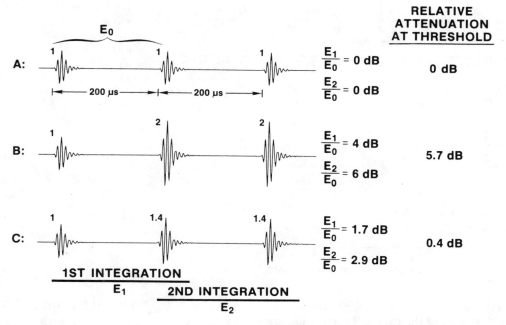

Figure 8.25. Overlapping integration period experiment. (*Left*) Stimuli A, B, and C; (*right*) relative energy of each integration period and dolphin's relative threshold results. (From Au et al. 1988.)

tested using the three-click echoes shown in Figure 8.25. Signal A was the reference signal, with a separation time of 200 μs between clicks of equal amplitudes. E_0 is the energy in the first two clicks of the reference signal. The second column from the right gives the relative amounts of energy in the first and second integration periods. The reference signal A had equal energy in both integration periods. Signal B had 4 and 6 dB more energy in the first and second integration periods, respectively, than the reference signal. Signal C had 1.7 and 2.9 dB more energy in the first and second integration periods, respectively, than the reference signal. The purpose of this experiment was to determine whether the dolphin would use the higher energy in the second integration period, which overlapped the first integration period, in detecting the echo signal.

The results of the overlapping integration period experiment are shown on the right side of Figure 8.25. There was a 5.7-dB shift in threshold with signal B and an insignificant threshold shift of 0.4 dB with signal C. This implies that in

analyzing signal B, the dolphin probably used the second integration period, which contained more energy than the first. The results with signal C suggest that the dolphin probably used the first integration period, because the amount of energy in the second period was not sufficiently greater than in the first.

8.4.4 Summation Effects

The effects of echoes with multiple integration periods on the signal detection performance of a dolphin were considered in the following experiment. The dolphin was first provided with a two-click echo in a single integration period and its detection threshold determined. Next, a four-click echo was played back to the dolphin in two successive integration periods and the threshold determined. The numbers were increased until a threshold had been determined for an eight-click echo played back to the dolphin in four successive integration periods. The eight-click echo stimulus is shown in the top part of Figure 8.26. The dolphin's detection performance as a func-

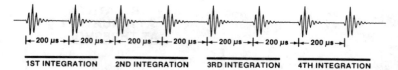

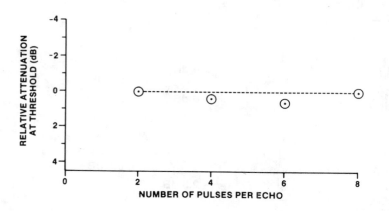

Figure 8.26. Summation effects experiment. (*Top*) Stimulus pattern; (*bottom*) dolphin's relative threshold results as a function of the number of clicks in the echo. (From Au et al. 1988.)

tion of the number of integration periods, relative to the threshold of the two-click, single-integration-period echo stimulus, is shown in the bottom part of the figure. The results indicate that the threshold for multiple integration periods did not shift from the threshold for the single integration period. This suggests that the dolphin did not sum the energy received in different integration periods but used the energy of a single integration period to detect the presence of the signal in noise.

8.5 Comparison Between a Dolphin and an Ideal Receiver

An ideal or optimal receiver is a receiver which theoretically yields the best possible performance in detecting a signal in white noise. No receiver can perform better than the ideal receiver, which will detect a signal in noise at the lowest signal-to-noise ratio. The ideal receiver is an important standard for comparing receivers. Such a comparison can be valuable in rating the effectiveness of receivers or detectors for the detection of signals in noise. In this section, the dolphin detection system will be compared with an ideal receiver. As a first step, we will formalize the concept of an ideal receiver and consider its performance in terms of ROC curves (see Section 1.5 on Signal Detection Theory).

8.5.1 The Ideal Receiver

In Chapter 1 we pointed out that Peterson, Birdsall, and Fox (1954) introduced a function called the *likelihood ratio*, which is defined as the ratio $f_{SN}(x)/f_N(x)$, where $f_{SN}(x)$ is the probability density function of the signal plus noise and $f_N(x)$ is the probability density function of the noise only. They concluded from their analysis that a receiver which calculates the likelihood ratio for each receiver input is the optimum, or ideal, receiver for detecting signals in noise. Whenever the likelihood ratio exceeds a threshold β, the decision is "yes," and whenever it is less than β, the decision is "no". Let us assume that the noise is bandlimited with bandwidth W and has a normal (Gaussian) distribution; then the sampled probability density function (pdf) for noise-only can be written as

$$f_N(x_i) = \frac{1}{\sqrt{2\pi\sigma^2}} e^{\frac{-x_i^2}{2\sigma^2}} \qquad (8\text{-}18)$$

The sampled pdf for signal plus noise will also be Gaussian, and can be expressed as

$$f_{SN}(x_i) = \frac{1}{\sqrt{2\pi\sigma^2}} e^{-\frac{(x_i-s)^2}{2\sigma^2}} \qquad (8\text{-}19)$$

where the noise variance, $\sigma^2 = N_0 W$, is equal to the rms noise level. The sampling theorem (Bracewell 1978) states that a bandlimited signal which is zero for frequencies greater than W can be fully reconstructed if it is sampled at a minimum interval of $\Delta t = \frac{1}{2W}$. The pdf samples are all statistically independent so that for a pdf sampled over a time period T the joint pdf of all $2WT$ samples is the product of the individual density functions,

$$f_N(x) = \prod_{i=1}^{2WT} f_N(x_i) \qquad (8\text{-}20)$$

and

$$f_{SN}(x) = \prod_{i=1}^{2WT} f_{SN}(x_i) \qquad (8\text{-}21)$$

The likelihood ratio can now be expressed as

$$l(x) = \frac{f_{SN}(x)}{f_N(x)} = \prod_{i=1}^{2WT} e^{-\frac{1}{2}\frac{(x_i - s_i)^2}{N_0 W}} e^{\frac{1}{2}\frac{x_i^2}{N_0 W}} \qquad (8\text{-}22)$$

Taking the natural logarithm of the likelihood ratio, and making use of the fact that the log of products equals the sum of the log, we arrive at

$$\ln l(x) = \frac{1}{N_0 W}\left(\sum_{i=1}^{2WT} x_i s_i - \frac{1}{2} \sum_{i=1}^{2WT} s_i^2 \right) \qquad (8\text{-}23)$$

For the situation where the signal is known exactly, $\sum s_i^2$ is a constant, and from the sampling theorem, $x(t)$ can be specified by $2WT$ samples, so that we can express the first summation as

$$\frac{1}{N_0}\sum_{i=1}^{2WT} x_i s_i = \frac{2W}{N_0}\int_0^T x(t)s(t)\,dt \qquad (8\text{-}24)$$

The integral is the cross-correlation function of the received signal $x(t)$ and the expected signal $s(t)$. Therefore, the optimal receiver for an exactly known signal is the cross-correlation between the received signal and the expected signal, a result derived by Peterson, Birdsall, and Fox (1954).

An alternative way of considering an ideal or optimum receiver involves the use of a *matched filter*. A matched filter has a response that is matched to the time-reversed form of the expected signal. If $x(t)$ is the input to a filter with a response $h(t)$, the output at time t is the convolution integral

$$y(t) = \int_0^t x(\tau)h(t - \tau)\,d\tau \qquad (8\text{-}25)$$

The response of a matched filter will be

$$h(t) = s(T - t) \qquad \text{for } 0 \le t \le T \quad (8\text{-}26)$$

From (8-25) the output of the matched filter for an input $x(t)$ will be

$$y(t) = \int_0^t x(\tau)s(T - t + \tau)\,d\tau \qquad (8\text{-}27)$$

The output of the matched filter will build up with time until a maximum value is reached at $t = T$, and then begin to drop. At $t = T$, the output is

$$y(T) = \int_0^T x(t)s(t)\,dt \qquad (8\text{-}28)$$

which is identical to the correlation integral of (8-24).

8.5.2 Dolphin Versus Ideal Receiver

Au and Pawloski (1989) reported on two experiments that compared the performance of a dolphin sonar in detecting a target in noise to the calculated performance of an ideal receiver. In both experiments, the electronic phantom target system depicted in Figure 8.14 was used. Experiment I was conducted to establish a realistic method of estimating E_e/N_0 at the dolphin's detection threshold. Two different types of echoes were used. The first echo type consisted of two clicks, separated by 200 μs, which were replicas of the respective transmitted click. This double-click echo was the standard used by Au et al. (1988) and thereby provided a way to check the dolphin's performance against past results. The amplitude of the echoes was directly proportional to the amplitude of the emitted clicks. With this echo type, the $(E_e/N_0)_{max}$ at threshold was determined. The second echo type consisted of a previously measured and digitized echolocation click from the animal, stored in an erasable programmable read-only memory (EPROM). The electronic target simulator was modified so that every time the dolphin emitted an echolocation signal, the EPROM was triggered to produce two pulses separated by 200 μs. The amplitude of the echo clicks was fixed for each trial independent of the dolphin's signal level, resulting in a fixed E_e/N_0. The difference between the threshold $(E_e/N_0)_{max}$ obtained with the normal variable-amplitude phantom echo and the E_e/N_0 obtained with the EPROM signal could then be determined.

The masked thresholds were determined by varying the signal attenuator in an up/down staircase procedure designed to produce a 50% correct detection threshold. After each target-present trial, the attenuator was increased (echo made weaker) if the animal's response was correct or decreased (echo made stronger) if the response was incorrect. During a session, the signal attenuator was therefore continuously increased and decreased, depending on the animal's responses on the target-present trials. Reversal points were the local maxima and minima of the signal attenuator during a session. At the start of each session, after the warm-up trials, the attenuator was increased (echo made weaker) in 2-dB steps until the first incorrect target-present response was made; thereafter 1-dB steps were used. The first 10 trials of each session were designated as warm-up trials, and the signal attenuator was fixed so that E_e/N_0 was approximately 10 dB above the animal's previously defined masked threshold. The last 10 trials of each session were "cool-off" trials, for which the attenuator was adjusted to the same level as in the warm-up trials.

After each session, the masked threshold was computed by averaging the values of the signal attenuator settings at the reversal points. A threshold estimate was considered complete when at least 20 reversals had been obtained over a minimum of two consecutive sessions and the average reversal values were within 2 dB of each other.

Four sessions resulting in 40 reversals were conducted for each of the two echo types. The average and standard deviation of $(E_e/N_0)_{max}$ at the reversal points for the normal variable-amplitude phantom echo was

$$(E_e/N_0)_{max} = 10.4 \pm 1.1 \text{ dB} \qquad (8\text{-}29)$$

This E_e/N_0 is approximately the same as the 75% correct response threshold obtained by Au et al. (1988) using a modified method of constants, and 1.8 dB greater than the threshold obtained with the staircase method.

The average and standard deviation of E_e/N_0 at the reversal points for the fixed EPROM echo were

$$E_e/N_0 = 7.5 \pm 1.5 \text{ dB} \qquad (8\text{-}30)$$

This number can be considered the most accurate estimate of E_e/N_0 since the echo strength for each trial was fixed and did not depend on the level of the dolphin's transmitted signal. There is a 2.9-dB difference between E_e/N_0 and $(E_e/N_0)_{max}$, indicating that an accurate estimate of the E_e/N_0 at the dolphin's threshold may be obtained by subtracting 2.9 dB from $(E_e/N_0)_{max}$, as was done in (8-19). This adjustment is strictly valid only for the dolphin used in this study; however, it can be used as an estimate with other dolphins.

Experiment II was conducted to obtain data that could be presented in an ROC format with the probability of detection, $P(Y/sn)$, plotted against the probability of false alarm, $P(Y/n)$. ROC curves for an ideal receiver could then be fitted to the dolphin data to estimate the d' or E_e/N_0 required by an ideal receiver to achieve a performance similar to that of the dolphin. Each session consisted of 50 trials divided into a 10-trial warm-up, a 20-trial easy detection task with a strong echo, and a 20-trial difficult detection task with a weak echo. During warm-up trials, the signal attenuator was set approximately 10 dB above a previously determined threshold value. The easy and the difficult detection trials were conducted with the signal attenuator set at approximately 6 and 2 dB, respectively, above the typical threshold value. Target-present and -absent conditions were randomized and balanced for the warm-ups and the two 20-trial blocks. The energy flux density associated with every click emitted by the dolphin during the target detection task was measured and stored on floppy disk.

The dolphin's response bias was manipulated by varying the payoff matrix (number of pieces of fish reinforcement for correct responses). Schusterman et al. (1975) found that the response bias of marine mammals could be manipulated in this way without significantly changing the animals' detection sensitivity. The animals were manipulated to be conservative in reporting the presence of a stimulus by rewarding them with four pieces of fish for each correct rejection versus one piece for each correct detection. Conversely, the animals were manipulated to be liberal by rewarding them with four pieces of fish for each correct detection versus one piece for each correct rejection. The payoff matrix was varied in terms of the ratio of correct detections to correct rejections in the following manner: 1:1, 1:4, 1:1, 4:1, 1:1, 8:1. Six consecutive sessions were conducted at each payoff setting, with the 1:1 payoff being the baseline. Each fish reward (Columbia River smelt) was cut in half to prevent satiation of the dolphin, especially during 8:1 sessions.

The changes in the dolphin's target detection performance as its response bias was manipulated are plotted in an ROC format in Figure 8.27. The detection sensitivity d' represents the minimum value of E_e/N_0 necessary to lead to the performance of an ideal receiver. Each data point represents 120 trials each for the 1:4, 4:1, and 8:1 payoff ratios. At the 1:1 payoff condition, 360 trials were performed. The ideal isosensitivity curves associated with d' values of 2.2 and 1.6 for strong and weak echoes, respectively, are included in Figure 8.27. These isosensitivity curves best matched the dolphin's performance in a least-square error manner.

Changes in the animal's performance with changes in the payoff matrix were relatively systematic and predictable. As the payoff ratio for correct detections increased from 1:4 to 1:1 and 4:1, the dolphin became progressively more liberal in reporting on the presence of the target, coupled with an increase in the false alarm rates and a decrease in β. However, even as the

Table 8.1. The average and standard deviation of $(E_e/N_0)_{max}$ from every target-present trial and for different payoff matrix conditions (from Au and Pawloski 1989)

Payoff Matrix	Strong Echo $(E_e/N_0)_{max}$	Weak Echo $(E_e/N_0)_{max}$
1:4	13.9 ± 1.3 dB	9.1 ± 1.0 dB
1:1	15.3 ± 1.4	10.6 ± 0.9
4:1	16.0 ± 1.3	11.2 ± 1.0
1:8	15.0 ± 1.1	10.2 ± 0.8
Ave	15.1 ± 1.3	10.3 ± 0.9

dolphin's response bias varied, its detection sensitivity remained relatively constant. As the payoff ratio for correct detection increased to 8:1, the dolphin did not become more liberal than at the 4:1 payoff for either the strong or the weak echo cases.

The average and standard deviation of $(E_e/N_0)_{max}$ for each trial associated with $\overline{SE_{max}}$ at different payoff conditions are given in Table 8.1. The composite average of $(E_e/N_0)_{max}$ across the various payoff matrix for the strong and weak echoes is also included. The difference in average $(E_e/N_0)_{max}$ between strong and weak echoes was 4.8 dB, which agreed closely with the 4.0-dB difference in the signal attenuator settings for both echoes.

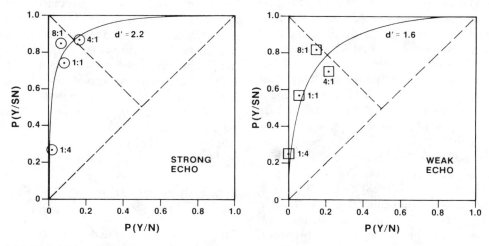

Figure 8.27. Dolphin performance results plotted in an ROC format, along with isosensitivity curves that best matched the results. (From Au and Pawloski 1989.)

An optimal receiver having a d' of 2.2 for the strong echo and of 1.6 for the weak echo will approximate the dolphin's behavioral performance. The echo energy-to-noise ratio $(E/N_0)_{op}$ for this optimal receiver can be calculated from the definition of d' given by (1-34) as

$$d' = \sqrt{2\left(\frac{E}{N_0}\right)_{op}} \qquad (8\text{-}31)$$

The echo energy-to-noise ratio in dB is

$$(E_e/N_0)_{op} = 10\log\left(\frac{d'^2}{2}\right) \qquad (8\text{-}32)$$

Therefore, for an optimal receiver approximating an echolocating dolphin

$$\left(\frac{E_e}{N_0}\right)_{op} = \begin{cases} 3.8 \text{ dB} & \text{(strong echo)} \\ 1.1 \text{ dB} & \text{(weak echo)} \end{cases} \qquad (8\text{-}33)$$

An estimate of the echo energy-to-noise ratio for the dolphin at the two different echo conditions can be obtained by subtracting 2.9 dB from the average value of $(E_e/N_0)_{max}$ shown in Table 8.1. The difference in E_e/N_0 between an optimal receiver and the dolphin can be expressed as

$$\left(\frac{E_e}{N_0}\right)_{op} - \left(\frac{E_e}{N_0}\right)_{dol} = \begin{cases} 8.4 \text{ dB} & \text{(strong echo)} \\ 6.3 \text{ dB} & \text{(weak echo)} \end{cases}$$
$$(8\text{-}34)$$

Averaging the differences for strong and weak echoes, we conclude that an optimal receiver would outperform the dolphin by approximately 7.4 dB. The differences between an optimal receiver and the dolphin should be the same for both types of echoes. The 2.1-dB discrepancy in the experimental results is well within the range of error of biosonar experiments.

The beluga results presented by Turl et al. (1987), discussed in Section 8.3.5, are puzzling in view of the difference of approximately 7.4 dB between the dolphin and an optimal receiver. Turl et al. (1987) found that the beluga was approximately 8 to 13 dB better than the dolphin in detecting a target in noise. The *Tursiops* used in this study (Au et al. 1988) and by Turl et al. (1987) had virtually the same detection threshold; $(E_e/N_0)_{max}$ at the 75% correct response threshold was within 1 dB for both animals. This would indicate that the beluga is as good or better (a physical impossibility) than an optimal detector.

Perhaps the beluga discovered some way to reduce the received noise by spatially filtering the target echo from the masking noise, as it did using a surface-reflected path in an earlier experiment (Penner et al. 1986).

8.6 Target Detection in Reverberation

The target detection capability of any echolocation system, man-made or biological, is limited by both interfering noise and *reverberation*. Reverberation differs from noise in several aspects. It is caused by the sonar itself and is the total contribution of unwanted echoes scattered back from objects and inhomogeneities in the medium and on its boundaries. The spectral characteristics of reverberation are similar to those of the projected signal and its intensity is directly proportional to the intensity of the projected signal. Therefore, in a reverberation-limited situation, target detection cannot be improved by increasing the intensity of the projected signal. Target detection becomes dependent on the ability of the system to discriminate between the target of interest and false targets and clutter that contribute to the reverberation.

Reverberation is caused by scatterers in the sea and is usually separated into two categories. The first type occurs in the volume, or body, of the sea and is called *volume reverberation*. It is caused by large numbers of tiny (with respect to wavelength) scatterers such as marine life (deep scattering layer, schools of fish) and inanimate matter (bubbles, algae bloom, suspended particulate matter) and by the inhomogeneous structure of the sea itself (Urick 1983). The second type of reverberation is *surface reverberation*, caused by scatterers on or near a surface such as the ocean bottom or the sea surface. Surface reverberation is usually subdivided further into two categories: bottom reverberation and sea surface reverberation.

8.6.1 Reverberation-Limited Form of the Sonar Equation

The reverberation-limited form of the sonar equation can be used to describe the target detection performance of a sonar limited by

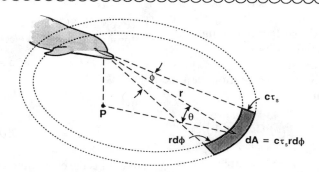

Figure 8.28. Geometry of a dolphin ensonifying an area dA of the bottom. θ is the angle between the beam axis and the horizon, τ_s the duration of the bottom reverberation.

reverberation. The usual form of the sonar equation equates the detection threshold with the ratio of the echo and reverberation intensities. However, for broadband transient signals the generalized form of the sonar equation should be used (Urick 1962, 1983), as was done in Section 8.2 for the noise-limited case. The generalized form of the sonar equation equates the detection threshold DT_R with the ratio of echo and reverberation energy flux densities (Urick 1962, 1983) and can be written as

$$DT_R = EE - RE \qquad (8\text{-}35)$$

where EE = energy flux density of the echo
RE = energy flux density of the reverberation

The energy flux density of the target echo can be expressed as

$$EE = SE - 2TL + TS_E \qquad (8\text{-}36)$$

SE is defined in Eq. (8-8), TL is the one-way transmission loss, and TS_E is defined in (8-11). The received reverberation is associated with a fundamental ratio called the scattering strength. It is analogous to target strength and is the ratio in dB of the energy of the sound scattered by a unit area referenced to a distance of 1 m, to the energy of the incident plane wave. Let E_s be the energy of sound scattered by 1-m^2 area or a 1-m^3 volume referenced to 1 m from the scatterers and let E_i be the energy of the incident plane wave; then the scattering strength, $S_{(s,v)E}$, is defined as

$$S_{(s,v)E} = 10 \log \frac{E_s}{E_i} \qquad (8\text{-}37)$$

The subscript s refers to surface reverberation and the subscript v to volume reverberation. The scattered energy is calculated by integrating the instantaneous intensity over the integration time of the dolphin, τ_{int}. The analytical expression for RE will be different for surface and volume reverberation. For the case of surface reverberation, consider the geometry in Figure 8.28, showing a dolphin ensonifying a small surface (dA) of scatterers representing the ocean bottom. The duration of the backscattered signal from the bottom is denoted as τ_s. However, the period of reverberation that is meaningful to the dolphin is its integration time so that τ_s should be replaced by τ_{int}. Following Urick (1983), we denote the transmission beam pattern as $b_t(\theta, \phi)$ so that the axial energy flux density 1 m from the dolphin is $E_0 b_t(\theta, \phi)$. The incident energy flux density at the small area dA located at range r can be expressed as

$$incident\ E = \frac{E_0 b_t(\theta, \phi)}{r^2} \qquad (8\text{-}38)$$

If we define s_s as the linear form of the surface scattering strength ($S_s = 10 \log s_s$), the energy of the sound scattered back at a point 1 m from dA will be

$$Reflected\ E = \frac{E_0 b_t(\theta, \phi) s_{sE} dA}{r^2} \qquad (8\text{-}39)$$

The backscattered energy arriving at the dolphin will be the reflected energy in (8-39) divided by another r^2 term to account for the transmission loss of the reflected signal traveling back to the receiver, multiplied by the receiving beam pattern $b_r(\theta, \phi)$, or $(E_0/r^4)b_t(\theta, \phi)b_r(\theta, \phi)s_{sE}\,dA$. The total contribution of the surface of homogeneous scatters can be summed up by integrating over the elemental area dA, so that the energy flux density of the received reverberation can be expressed in log form as

$$RE_s = 10\log\left(\frac{E_0}{r^4}s_s\int b_t(\theta, \phi)b_r(\theta, \phi)\,dA\right) \quad (8\text{-}40)$$

The integral in (8-40) is often a complex function without a simple analytical form and must be evaluated numerically. Some useful graphs to evaluate the integral are provided by Urick and Hoover (1956). In many situations the axis of the beam is only slightly inclined toward the scattering surface, i.e., the scattering surface corresponds to $\theta \approx 0$, so that only the horizontal beam patterns, $b_t(0, \phi)$ and $b_r(0, \phi)$, need be used in (8-40). Let dA be taken as a portion of a circular annulus, as shown in Figure 8.28, at a horizontal range of r_h; then it can be expressed as

$$dA = c\tau_{\text{int}}r_h\,d\phi \quad (8\text{-}41)$$

The received reverberation level can now be expressed as

$$RE_s = 10\log\left(\frac{E_0}{r^4}s_{sE}c\tau_{\text{int}}r\int_0^{2\pi} b_t(\theta, \phi)b_r(\theta, \phi)\,d\Phi\right) \quad (8\text{-}42)$$

Let

$$\Phi = \int_0^{2\pi} b_t(\theta, \phi)b_r(\theta, \phi)\,d\Phi \quad (8\text{-}43)$$

Then the reverberation energy flux density can be written as

$$RE_s = SE - 2TL + S_{sE} + 10\log A \quad (8\text{-}44)$$

where

$$A = c\tau_{\text{int}}r_h\Phi \quad (8\text{-}45)$$

For low values of θ, the slant range r and the horizontal range r_h will be approximately the same. Equation (8-45) is the same as the equation for short transients derived by Urick (1962) with $c\tau_e/2$ replaced by $c\tau_{\text{int}}$. Substituting (8-36) for EE and (8-44) for RE_E into (8-35), we arrive at the reverberation-limited form of the sonar equation for surface reverberation:

$$DT_{RS} = TS_E - S_{sE} - 10\log(\Phi r_h\tau_{\text{int}}c) \quad (8\text{-}46)$$

An expression similar to (8-44) can be derived for volume reverberation (see Urick 1983):

$$RE_v = SE - 2TL + S_{vE} + 10\log(\Psi r^2 c\tau_{\text{int}}) \quad (8\text{-}47)$$

where

$$\Psi = \int_0^{4\pi} b_t(\theta, \phi)b_r(\theta, \phi)\,d\Omega \quad (8\text{-}48)$$

Here $d\Omega$ is the elemental solid angle subtended by dV at the source. The the volume-reverberation-limited form of the sonar equation can be expressed as

$$DT_{RV} = TS_E - S_{vE} - 10\log(\Psi r^2\tau_{\text{int}}c) \quad (8\text{-}49)$$

8.6.2 Biosonar Target Detection in Reverberation

Titov (1972), as reported by Ayrapet'yants and Konstantinov (1974), investigated the capability of *Tursiops* to detect targets in the presence of interfering objects. Smooth rocks, varying in size from 5 to 30 mm in 5-mm increments, were used to form two 40-cm circles on the bottom of a tank; each circle contained 300 rocks, 50 of each size. A target was first lowered from the surface to the middle of a circle, to the bottom amidst the rocks; then it was raised until the animal could detect it. The dolphin could detect a 50-mm lead sphere lying on the bottom 75% of the time at a distance of 5 m. A 33-mm solid steel sphere had to be raised 1.7 cm above the largest rock before the animal could detect it 75% of the time at a distance of 5 m. The dolphin approached the target swimming close to the bottom rather than at the surface, presumably to minimize the reverberation. Unfortunately, Titov (1972) did not characterize the reverberation and did not report on the target strength of the target. Therefore, it is not possible to determine the echo-to-reverberation ratio (E/R) at the dolphin's performance threshold.

Murchison (1980a,b) studied the effects of bottom reverberation on the target detection capabilities of two *Tursiops* in Kaneohe Bay, using the experimental configuration shown in Figure 8.4. A 6.35-cm diam. solid steel sphere was used at depths varying from 1.2 to 6.3 m. At a depth of 6.3 m, the target was on the bottom. The animals' 50% correct detection threshold ranges for the different target depths are plotted in Figure 8.29. As the target depth increased, the animals' detection ranges decreased, showing the effects of bottom reverberation. The bottom at the location of Murchison's experiment consisted of sandy silt composed of 19% clay, 48% silt, and 33% sand, with a mean grain size of 0.015 mm. Murchison (1980a,b) did not measure the bottom reverberation so that the echo-to-reverberation ratio (E/R) at the animals' performance threshold could not be determined.

Au (1992) used a monostatic echoranging system emitting simulated dolphin sonar signals to measure the scattering strength of the bottom where Murchison's experiment was conducted. The transducer was tilted below the horizontal plane by 4.7° to correspond to the grazing angle of a dolphin at a depth of 1 m searching for a target lying on the bottom at a range of 70 m. The horizontal product beam $b_t(\phi)b_r(\phi)$ for the measuring transducer is shown in Figure 8.30A. The integral of (8-43) for the product beam was found to be equal to 0.14. Using (8-46), the scattering strength of the bottom was found to

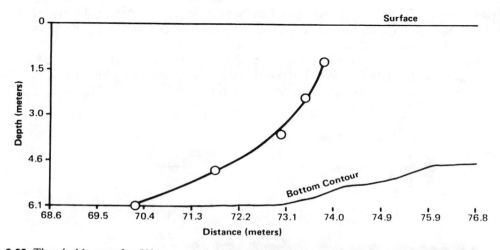

Figure 8.29. Threshold range for 50% correct detection of two *Tursiops truncatus* as a function of the depth of the 6.35-cm solid spherical target. (From Murchison 1980a,b.)

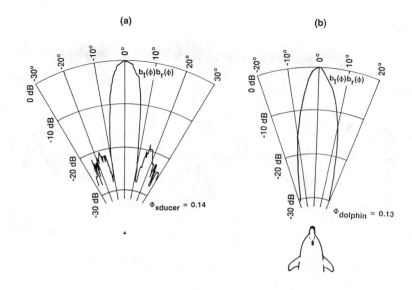

Figure 8.30. (*A*) Horizontal product beam for the transducer used to measure the scattering strength of the bottom at the location of Murchison's (1980a,b) experiment; (*B*) horizontal product beam for the dolphin transmission and reception beam shown in Fig. 6.9.

be -44.4 dB. The product beam for the dolphin is shown in Figure 8.30B. It was obtained by multiplying the horizontal transmission and reception beams displayed in Figure 6.9. The integral of (8-43) was found to be 0.13. Therefore, for a target range of 70 m, $10 \log c\tau_{int} r_h \phi_{dol} = 5.6$ dB. The target strength of the 6.35-cm diam. solid steel sphere was measured with a simulated dolphin sonar signal and found to be approximately -33.4 dB. Inserting the appropriate values into (8-46), Au (1992) obtained for the detection threshold

$$DT_{RS} = -34.8 + 44.4 - 5.6 = 4.0 \text{ dB} \quad (8\text{-}50)$$

The target echo and the bottom reverberation corresponding to a DT_{RS} of 4 dB are shown in Figure 8.31. Also included in the figure are the sum of the target echo and the bottom reverberation (simulating what the dolphins would receive), and the spectra of the summed echo denoted by

a solid line and the target echo denoted by a dashed line. The spectrum of the summed echo was obtained by first multiplying the summed echo with a cosine tapered window (Otnes and Enochson 1978) having a half-power width of 264 μs. The line above the summed echo indicates the width and position of the window. Although the spectrum extends to 200 kHz, the upper frequency limit of hearing for *Tursiops truncatus* is aproximately 150 kHz. The figure shows the effects of the bottom reverberation in masking the signal in both the time and frequency domains. The echo-plus-reverberation waveform does not contain the distinct highlight structure of the target echo measured in the free field. Only the two largest highlights of the target can be clearly seen in the echo. In the spectrum of the summed echo, the distinctive peaks and nulls of the target echo are no longer distinguishable for frequencies below 120 kHz.

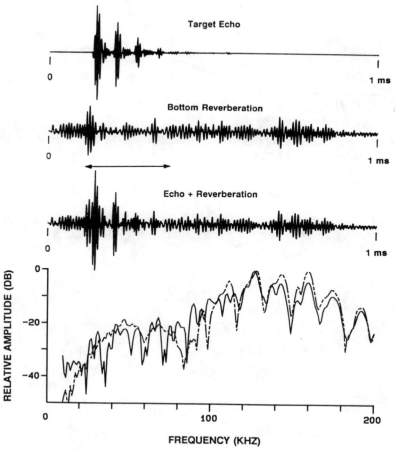

Figure 8.31. Echo from the 6.35-cm solid sphere target used by Murchison (1980a,b) and the bottom reverberation measured by Au (1992). The relative amplitudes of the echoes were scaled for a DT_{RS} of 4 dB.

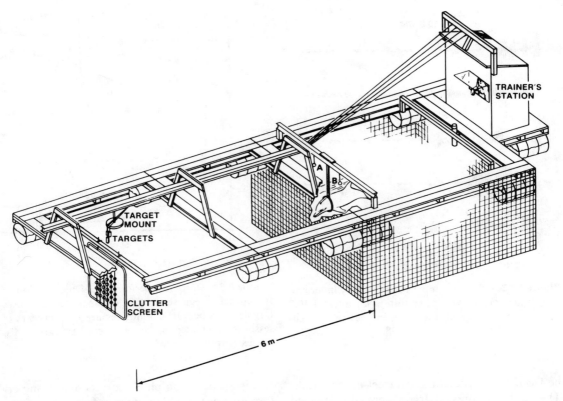

Figure 8.32. Experimental configuration showing dolphin in a hoop station with targets located 6 m from the hoop and a clutter screen directly behind the targets. (From Au and Turl 1983.)

Au and Turl (1983) investigated the capability of an echolocating *Tursiops* to detect targets placed near a clutter screen (Fig. 8.32). The targets were hollow aluminum cylinders (3.81-cm diam., 0.32-cm wall thickness) of 10.0, 14.0, and 17.8 cm length. By using cylinders of the same diameter and wall thickness but of different length, the amount of energy in the echo returning to the animal could be varied without a change in echo structure, so that the echoes from each cylinder "sounded" the same to the animal. A constant target range of 6 m from a hoop station and a target depth of 1.2 m were used throughout the experiment. The clutter screen consisted of forty-eight 5.1-cm diam. cork balls with their centers spaced 15.2 cm apart in a 6 × 8 rectangular array. The cork balls were tied eight to a line and attached to a 1.9 × 1.9-m frame made of 3.2-cm diam. water-filled PVC pipes. The clutter screen was located behind the target at a range of $6 + \Delta R$ m from the hoop, where ΔR

is the range from the target to the screen. The center of the screen coincided with the depth of the target.

Backscatter measurements of the targets and the clutter screen were made using a simulated dolphin echolocation signal. The measurement of the clutter screen was made with the transducer positioned so that its 3-dB beamwidth covered approximately the same area of the screen as the dolphin's beam would at a range of 6 m. Echoes from five consecutive pings with the smallest (10.0-cm) cylinder placed 10.2 cm in front of the clutter screen are shown in Figure 8.33. The echoes from the cylinder were relatively similar from ping to ping, whereas the backscatter from the clutter screen varied slightly from ping to ping. The echoes from the clutter screen were the result of a series of complex destructive and constructive interferences of scattered signals from individual balls and the frame.

A session consisted of 66 trials, divided equally

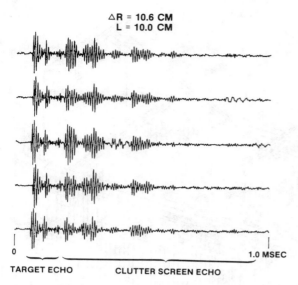

ΔR = 10.6 CM
L = 10.0 CM

0   1.0 MSEC

TARGET ECHO   CLUTTER SCREEN ECHO

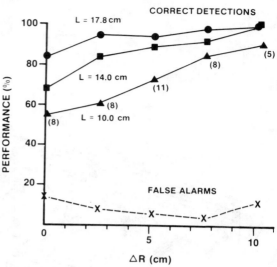

Figure 8.33. Echoes from the 10-cm long cylinder and the clutter screen, with a separation distance of 10.2 cm between target and clutter screen, for five consecutive pings. (From Au and Turl 1983.)

Figure 8.34. Dolphin's performance as a function of the clutter screen separation distance ΔR, for three targets. Numbers in parentheses are the numbers of sessions conducted at the different separation distances. (From Au and Turl 1983.)

between target-present and target-absent trials. The 33 target-present trials were also divided equally among the three targets, so that each target was presented to the animal 11 times during each session. The dolphin was required to swim into a hoop and "echolocate upon command," a technique developed by Schusterman et al. (1980). If the dolphin echolocated before the activation of the tone cue, the trial was aborted. A hydrophone click detector which triggered a speaker was used by the experimenter to monitor the dolphin's pulse emission. Performance data were collected as a function of the clutter screen distance from the target. Five separation distances (ΔR) of 0 to 10.2 cm in 2.54-cm increments were used. The specific ΔR was randomly chosen before each session, with the constraint that no more than two consecutive sessions could be at the same separation distance. For conditions other than ΔR = 0, the experiment was similar to a backward masking experiment except that the masker consisted of echoes from the clutter screen rather than broadband noise. The experiment could also be considered a measure of the dolphin's sonar range resolution capability when ΔR ≠ 0.

Results of the dolphin's correct detection performance as a function of the separation distance (ΔR) between the targets and the clutter screen are shown in Figure 8.34. Only the target-present trials were used to generate the correct detection curves. The numbers in parentheses are the numbers of sessions at the various separation distances. The animal's accuracy decreased both as the separation distance decreased and as the targets got smaller. The false alarm rate was low (below 14.5%), comparable to those obtained by Au and Snyder (1980) for a target detection task as a function of range. This suggests that the dolphin was biased in favor of responding "target absent" and used a conservative criterion when reporting on the presence of the target.

The echo-to-reverberation ratio for the different targets was determined for the case in which the targets were within the plane of the clutter screen (ΔR = 0 cm). Au and Turl (1983) used an integration window of 1 ms to compute the reverberation, which is not appropriate in light of the results of Au et al. (1988) showing an integration time of 264 μs for *Tursiops*. Integrating the reverberation over a 264-μs window would reduce the reverberation strength by 2 dB.

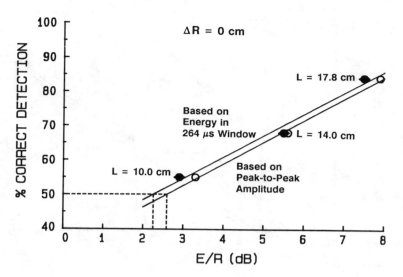

Figure 8.35. Dolphin's target detection performance as a function of the echo-to-reverberation ratio for a zero separation distance (targets were within the plane of the clutter screen). The echo-to-reverberation ratios based on the energy in a 264-μs window are indicated by closed circles and those based on peak-to-peak amplitudes are indicated by open circles. *L* is the target length. (Modified from data presented by Au and Turl 1983.)

The dolphin's performance as a function of the E/R ratio based on the energy in a 264-μs window and on the peak-to-peak values of the target echoes and the reverberation is shown in Figure 8.35. The linear least-square lines fitted to the data indicate that the 50% detection threshold corresponds to an E/R_E value of 2.3 dB and an E/R_{pp} value of 2.5 dB. Therefore, the energy in the target echo must be at least 2.3 dB above the reverberation echoes before the animal can detect the targets. The straight lines indicate that the animal's target detection sensitivity is directly proportional to the E/R. The 2.3-dB E/R value is relatively similar to the 4.0 dB determined by Au (1992) for the case of bottom reverberation. However, there is a subtle difference between the detection of a target lying on the bottom and a target that is coplanar with an array of cork balls. In the clutter screen situation, the target and clutter echoes arrived simultaneously at the dolphin, corresponding to a simultaneous masking experiment. In the bottom target case, the dolphin probably received reverberation from the bottom before receiving the target echo, so that forward masking as well as simultaneous masking occurred. Target detection is probably slightly more difficult in the latter situation than in the former.

Turl et al. (1991) conducted a similar experiment as Au and Turl (1983) in San Diego Bay to measure the capability of a beluga to detect targets in the presence of reverberation. A clutter

screen consisting of 300 cork balls 5.1 cm in diameter with their centers spaced 15.2 cm apart was used at a range of 8 m from the animal's hoop station. Five hollow stainless steel cylinders, of 3.2 cm O.D. and 0.38 cm wall thickness, with lengths of 14, 10, 7, 5, and 3 cm, served as targets. Therefore, the targets had similar echo structures but different target strengths. The target strength of the steel cylinders and the scatter strength of the clutter screen were measured in the same manner as in the study of Au and Turl (1983). The transducer was located at an appropriate range from the clutter screen so that its 3-dB beamwidth covered approximately the same area of the clutter screen as the beluga's beam would at a range of 8 m.

The five targets were located at five separation distances (ΔR) of 0, 2.5, 5.1, 7, and 10 cm from the front of the screen. The position of each target during a session was randomly chosen, with the constraint that at the end of the experiment each target had been at each ΔR for a total of 100 trials. A session consisted of 50 trials divided into 10-trial blocks, with one of the five preselected separation distances randomly assigned to each block. Each block also consisted of equal numbers of target-present and target-absent trials, randomly determined. The results of Turl et al. (1991) are shown in Figure 8.36, which also includes the revised results of Au and Turl (1983) for *Tursiops*. The 50% detection threshold corresponded to an E/R_E value of -2.7 dB and an

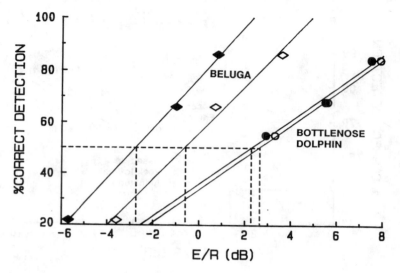

Figure 8.36. Beluga and bottlenose dolphin's target detection performance as a function of the echo-to-reverberation ratio for a zero separation distance (targets were within the plane of the clutter screen). The echo-to-reverberation ratios based on the energy in a 264-μs window are indicated by closed symbols and those based on peak-to-peak amplitudes are indicated by open symbols. The solid lines are linear curves fitted to the data. (Modified from data presented by Turl et al. 1991.)

E/R_{pp} value of -0.6 dB. A 264-μs window was used to calculate the reverberation energy R_E, without any evidence, however, that a 264-μs window is appropriate for a beluga. Very little is known on how belugas process broadband sonar signals. It is extremely interesting that the target echo could be smaller on both a peak-to-peak and an energy basis than the clutter screen echo, and yet the beluga was able to detect the target. The results indicate that the beluga was approximately 3.2 to 5.0 dB more sensitive than the bottlenose dolphin in detecting a target in reverberation.

In an earlier section of this chapter, we questioned the interpretation of the results of Turl et al. (1987) that posited that a beluga was more sensitive by 8 to 13 dB than a bottlenose dolphin in detecting a target in noise. The comparison between a dolphin and an ideal detector showed that the dolphin was approximately 7.6 dB less sensitive than an ideal detector, indicating that it would be impossible for a beluga to be more sensitive than a dolphin by 13 dB. The beluga target detection in reverberation results obtained by Turl et al. (1991) also support the contention that the whale's target detection sensitivity in noise was overestimated by Turl et al. (1987), probably because of inaccurate alignment between the noise transducer, target, and animal hoop station. Differences in relative detection sensitivity between a bottlenose dolphin and a beluga should be similar for noise and reverberation masking.

8.7 Summary

The capabilities of the bottlenose dolphin and beluga whale to detect targets in noise and reverberation with their sonar have been determined. The false killer whale target detection capabilities have only been studied for the noise-limited case. The noise-limited and reverberation-limited forms of the generalized sonar equation presented in this chapter have been applied only to *Tursiops*, since there is insufficient knowledge of some important parameters associated with the beluga and false killer whales. Bottlenose dolphins appear to process broadband echo signals in a similar manner as an energy detector having an integration time of 264 μs. The integration time is probably related to the "critical interval" of Vel'min and Dubrovskiy (1976) and the forward-masking threshold measured by Moore et al. (1984). Averaging the results of five experiments, the echo energy-to-noise ratio at the 75% correct response threshold is approximately 6 dB, which is comparable to humans detecting signals in noise. An ideal receiver is approximately 7.4 dB better than *Tursiops* at detecting a target masked by white noise. The echo-to-reverberation ratio corresponding to a 50% correct detection threshold is approximately 2.3

to 4.0 dB. Target detection experiments in noise and reverberation indicate that the beluga may have a keener target detection capability than the bottlenose dolphin. However, little is known about the beluga active sonar system in comparison with *Tursiops*, so that an accurate comparison of detection capabilities in terms of echo energy-to-noise ratio or echo-to-reverberation ratio cannot be made.

References

Au, W.W.L. (1988). Sonar target detection and recognition by odontocetes. In: P.E. Nachtigall and P.W.B. Moore, eds., *Animal Sonar: Processes and Performance*. New York: Plenum Press, pp. 451–465.

Au, W.W.L. (1992). Application of the reverberation-limited form of the sonar equation to dolphin echolocation. J. Acoust. Soc. Am. (In press).

Au, W.W.L., Moore, P.W.B., and Martin, S.W. (1987). Phantom electronic target for dolphin sonar research. J. Acoust. Soc. Am. 82: 711–713.

Au, W.W.L., and Pawloski, D.A. (1989). A comparison of signal detection between an echolocating dolphin and an optimal receiver. J. Comp. Physiol A, 164: 451–458.

Au, W.W.L., and Penner, R.H. (1981). Target detection in noise by echolocating Atlantic bottlenose dolphins. J. Acoust. Soc. Am. 70: 687–693.

Au, W.W.L., and Snyder, K.J. (1980). Long-range target detection in open waters by an echolocating Atlantic bottlenose dolphin (*Tursiops truncatus*). J. Acoust. Soc. Am. 68: 1077–1084.

Au, W.W.L., and Turl, C.W. (1983). Target detection in reverberation by an echolocating Atlantic bottlenose dolphin (*Tursiops truncatus*). J. Acoust. Soc. Am. 73: 1676–1681.

Au, W.W.L., Floyd, R.W., Penner, R.H., and Murchison, A.E. (1974). Measurement of echolocation signals of the Atlantic bottlenose dolphin, *Tursiops truncatus* Montagu, in open waters. J. Acoust. Soc. Am. 56: 1280–1290.

Au, W.W.L., Moore, P.W.B., and Pawloski, D.A. (1988). Detection of complex echoes in noise by an echolocating dolphin. J. Acoust. Soc. Am. 83: 662–668.

Ayrapet'yants, E. Sh., and Konstantinov, A.J. (1974). *Echolocation in Nature*, Part 2. Arlington, Va.: Joint Publication Research Service.

Barnard, G.R., and McKinney, C.M. (1961). Scattering of acoustic energy by solid and air-filled cylinders in water. J. Acoust. Soc. Am. 33: 226–238.

Bel'kovich, V.M., and Dubrovskiy, N.A. (1976). *Sensory Basis of Cetacean Orientation*. Leningrad: Nauka.

Bracewell, R.N. (1978). *The Fourier Transform and Its Applications*. New York: McGraw-Hill.

Bullock, T.H., and Ridgway, S.H. (1972). Evoked potentials in the central auditory systems of alert porpoises to their own and artificial sounds. J. Neurobiol. 3: 79–99.

Busnel, R.G., and Dziedzic, A. (1967). Resultats metrologiques experimentaux de l'echolocating delphinid. In: R.G. Busnel, ed., *Animal Sonar Systems: Biology and Bionics*. Laboratoire de Physiologie Acoustique, Jouy-en-Josas, France, pp. 307–335.

Green, D.M., Birdsall, T.G., and Tanner, W.P., Jr. (1957). Signal detection as a function of signal duration. J. Acoust. Soc. Am 29: 523–531.

Jeffress, L. A. (1968). Mathematical and electrical models of auditory detection. J. Acoust. Soc. Am. 44: 187–203.

Johnson, C.S. (1968a). Relation between absolute threshold and duration of tone pulses in the bottlenosed porpoise. J. Acoust. Soc. Am. 43: 757–763.

Johnson, S.C. (1968b). Masked tonal thresholds in the bottlenosed porpoise. J. Acoust. Soc. Am. 44: 965–967.

McFadden, D. (1966). Masking-level differences with continuous and with burst masking noise. J. Acoust. Soc. Am. 40: 1414–1419.

Moore, P.W.B., Hall, R.W., Friedl, W.A., and Nachtigall, P.E. (1984). The critical interval in dolphin echolocation: what is it? J. Acoust. Soc. Am. 76: 314–317.

Murchison, A.E. (1980a). Maximum detection range and range resolution in echolocating bottlenose porpoise (*Tursiops truncatus*). In: R.G. Busnel and J.F. Fish, eds, *Animal Sonar Systems*. New York: Plenum Press, pp. 43–70.

Murchison, A.E. (1980b). Maximum detection range and range resolution in echolocating bottlenose porpoises, *Tursiops truncatus* (Montagu). Ph.D. dissertation, University of California, Santa Cruz.

Otnes, R.K., and Enochson, L. (1978). *Applied Time Series Analysis*. New York: John Wiley and Sons.

Neubauer, W.G. (1986). *Acoustic Reflection from Surfaces and Shapes*. Washington, D.C.: Naval Research Lab.

Penner, R., and Murchison, A.E. (1970). Experimentally demonstrated echolocation in the Amazon River porpoise, *Inia geoffrensis* (Blainville). T. Poulter, ed., *Proc. 7th Annual Conf. Biol. Sonar and Diving Mammals*, Menlo Park, Cal.: Stanford Research Institute, pp. 17–38.

Penner, R.H., Turl, C.W., and Au, W.W.L. (1986).

Target detection by the beluga using a surface-reflected path. J. Acoust. Soc. Am. 80: 1842–1843.

Peterson, W.W., Birdsall, T.G., and Fox, W.C. (1954). The theory of signal detectability. Inst. Radio Engr. PGIT 4, 171–212.

Schusterman, R.J., Barrett, R., and Moore, P.W.B. (1975). Detection of underwater signals by a California sea lion and a bottlenose porpoise: variation in the payoff matrix. J. Acoust. Soc. Am. 57: 1526–1532.

Schusterman, R.J., Kersting, D.A., and Au, W.W.L. (1980). Stimulus control of echolocation pulses in *Tursiops truncatus*. In: R.G. Busnel and J.F. Fish, eds., *Animal Sonar Systems*. New York: Plenum Press, pp. 981–982.

Shirley, D.J., and Diercks, K.J. (1970). Analysis of the frequency response of simple geometric targets. J. Acoust. Soc. Am. 48: 1275–1282.

Thomas, J.A., and Turl, C.W. (1990). Echolocation characteristics and range detection by a false killer whale (*Pseudorca crassidens*). In: J.A. Thomas and R.A. Kastelein, eds., *Sensory Abilities of Cetaceans: Laboratory and Field Evidence*. New York: Plenum Press, pp. 321–334.

Titov, A.A. (1972). Investigation of sonic activity and

phenomenological characteristics of the echolocation analyzer of Black Sea delphinids. Candidatorial dissertation, Karadag.

Turl, C.W., Penner, R.H., and Au, W.W.L. (1987). Comparison of target detection capabilities of the beluga and bottlenose dolphin. J. Acoust. Soc. Am. 82: 1487–1491.

Turl, C.W., Skaar, D.J., and Au, W.W.L. (1991). The echolocation ability of the beluga (*Delphinapterus leucas*) to detect target in clutter. J. Acoust. Soc. Am. 89: 896–901.

Urick, R.J. (1962). Generalized form of the sonar equations. J. Acoust. Soc. Am. 34: 547–550.

Urick, R.J. (1983). *Principles of Underwater Sound*. 3rd ed. New York: McGraw-Hill.

Urick, R.J., and Hoover, R.M. (1956). Backscattering of sound from the sea surface: its measurement, causes, and application to the prediction of reverberation levels. J. Acoust. Soc. Am. 28: 1038–1042.

Vel'min, V.A., and Dubrovskiy, N.A. (1975). On the analysis of pulsed sounds by dolphins. Dokl. Akad. Nauk. SSSR 225: 470–473.

Vel'min, V.A., and Dubrovskiy, N.A. (1976). The critical interval of active hearing in dolphins. Sov. Phys. Acoust. 2: 351–352.

9

Biosonar Discrimination, Recognition, and Classification

One of the outstanding characteristics of the dolphin sonar system which distinguishes it from any man-made sonar is the ability to make fine distinctions in the features or properties of targets. This ability has amazed and sparked the interest of many involved in the development and use of active sonar systems. The abilities to perform fine target discrimination, recognition, and classification are often considered synonymous; however, there are subtle differences between these functions. Target discrimination means the ability to discern from their echoes that two targets are different. Target recognition means the ability to recognize features of the echoes from specific targets compared with echoes from any other targets; it involves a discrimination capability, an ability to recall from memory the echo features of specific targets, and the ability to compare present sonar echoes with those stored in memory. Target classification means the ability to separate targets into different classes according to some arbitrary criteria such as metal versus nonmetal, organic versus inorganic, edible versus inedible, smooth surface versus rough surface, etc. Most of the experiments discussed in this chapter involve target discrimination; a few involve target recognition. Target classification experiments involving many different classes of targets are generally difficult to construct and implement with animals.

Many dolphin sonar experiments have been conducted to investigate the capability of dolphins to discriminate between objects differing in size, structure, shape, and material composition. Most of these experiments have been reviewed by Nachtigall (1980). The majority of references to Soviet research are found in Ayrapet'yants and Konstantinov (1974) and Bel'kovitch and Dubrovskiy (1976). The primary emphasis in this chapter will be on possible cues available to dolphins in performing different sonar discrimination tasks, in order to demystify the target discrimination capability of dolphins.

Dolphin sonar discrimination experiments are generally designed to require the animal to choose between at least two targets, presented either simultaneously or successively. The targets usually differ along a single physical dimension, and the animal's discrimination threshold is determined by making the difference progressively smaller. Although differences in targets may exist in a single physical dimension (diameter, wall thickness, length, etc.), several acoustic features may be affected as this single dimension is varied. Therefore, an important objective in sonar discrimination experiments is to determine what acoustic features are being tested and how these features change as the physical characteristics of the targets change. Unfortunately, this is easier said than done since the backscattering process is often quite complex, even with simple geometrically shaped targets, as was discussed in Chapter 8.

9.1 Mathematical Tools for the Analysis of Target Echoes

Before proceeding to a discussion of various discrimination experiments, it will be helpful to consider some mathematical tools that are useful in the analysis of target echoes. One of the simplest ways to analyze the properties of a target echo is to transform the receive time domain echo into the frequency domain using a Fast Fourier Transform (FFT) computer algorithm. The echo is digitized and fed into a computer which then computes the FFT of the echo transforming it into the frequency domain. The echoes shown in Figure 8.1 are examples of echoes represented in both the time and frequency domains. Another useful way of analyzing an echo signal is to examine the envelope of the echo (ignoring the higher frequency oscillations) and determine the relative spacing and amplitude of the echo highlights.

The *envelope* of a signal can be computed by taking the "real" time domain signal and making it complex, or "analytic," using the *Hilbert transform*. Any actual time domain signal is real (does not have an imaginary part). Its frequency domain representation obtained through the Fourier Transform (eq. 1-20) is complex (has both a real and an imaginary part). A complex representation of a real signal can be obtained by adding an imaginary part (with special properties) to the real signal, creating a new signal that is referred to as an analytic signal. The *analytic signal* is derived by suppressing the negative frequency portion of the real-signal spectrum and doubling positive frequency components. Let $s(t)$ be the real signal and $\hat{s}(t)$ the imaginary part so that

$$s_a(t) = s(t) + j\hat{s}(t) \qquad (9\text{-}1)$$

The requirement that the analytic signal have zero negative frequencies and that its positive frequency components be double those of the real signal implies that

$$\Im[\hat{s}(t)] = \begin{cases} -jS(f) & \text{for } f > 0 \\ jS(f) & \text{for } f < 0 \end{cases} \qquad (9\text{-}2)$$

where $S(f)$ is the Fourier transform of $s(t)$. Equation (9-1) can also be expressed as

$$s_a(t) = |s_a(t)|e^{j\phi(t)} \qquad (9\text{-}3)$$

where

$$|s_a(t)| = \sqrt{s^2(t) + \hat{s}^2(t)} \qquad (9\text{-}4)$$

and

$$\phi(t) = \arctan\frac{\hat{s}(t)}{s(t)} \qquad (9\text{-}5)$$

The magnitude of the analytic signal in (9-4) is the envelope of the real signal $s(t)$. Therefore, the envelope of a signal can be obtained by simply adding a complex portion having the properties expressed in (9-2) to the real signal and then determining the absolute value of the analytic signal using (9-4). The result of performing this mathematical operation is depicted in Figure 9.1, where the envelope calculated for the echo from a 7.62-cm water-filled sphere (Fig. 8.1) is shown. The functions $s(t)$ and $\hat{s}(t)$ are also related through the Hilbert transform, where $\hat{s}(t)$ is the Hilbert transform of $s(t)$, and $s(t)$ is the inverse Hilbert transform of $\hat{s}(t)$. For a more comprehensive and thorough discussion of the analytic signal, the reader should consult the two excellent books on radar signal processing by Rihaczek (1969) and Burdic (1968).

Echoes contaminated with noise can be analyzed by first processing the echoes with a filter whose response is matched to the transmitted signal. From Section 8.1, we saw that echo highlights tend to resemble the transmitted signal, so that a matched filter would be a good way to reduce the effects of white Gaussian noise. Therefore, if $s(t)$ is the outgoing signal, then the response of the matched filter represented by (8-26) will be $s(T - t)$. Let $X(f)$ and $S(f)$ be the Fourier transforms of the echo $x(t)$ and the transmitted signal $s(t)$, respectively; then according to the Correlation Theorem in Fourier analysis (Brigham 1988) the Fourier transform of the output can be written as

$$\Im[y(t)] = Y(f) = X(f)S^*(f) \qquad (9\text{-}6)$$

where $S^*(f)$ is the complex conjugate of $S(f)$. The output of the matched filter in the time domain is the inverse Fourier Transform of (9-6),

$$y(t) = \Im^{-1}[X(f)S^*(f)] \qquad (9\text{-}7)$$

The envelope of $y(t)$ can then be calculated to

Figure 9.1. Echo signal from a 7.62-cm water-filled sphere and its envelope (dashed contour).

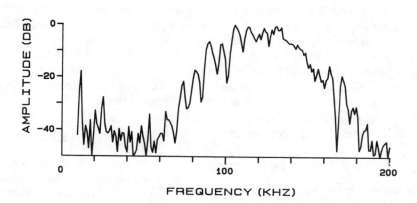

determine the spacing and relative amplitude of the highlights in the echo.

9.2 Target Size Discrimination

9.2.1 Cylinder Length and Diameter

Ayrapet'yants et al. (1969) performed a discrimination experiment with Black Sea bottlenose dolphins (*Tursiops truncatus*) in which the length of steel cylinders was varied. The animals were required to respond by pulling a ring whenever the 110-mm × 25-mm (diam. × length) standard was presented and not respond when a nonstandard-length cylinder was presented. They found that the dolphins could discriminate a 30-mm-long cylinder from the 25-mm-long standard at a 70% correct response level. Zaslavskiy et al. (1969) also performed a cylinder length discrimination experiment, using the harbor porpoise (*Phocoena phocoena*) and a simultaneous target presentation procedure. From a range of 2.5 m, the porpoise could discriminate between cylinders that were 75 mm versus 95 mm in length 80% of the time.

Cylinder size discrimination experiments were also performed at the Naval Ocean Systems Center (NOSC) in San Diego with an Atlantic bottlenose dolphin (*Tursiops truncatus*) and an Amazon River dolphin (*Inia geoffrensis*) (Evans 1973). Solid chloroprene cylinders were presented simultaneously and the blindfolded dolphins were required to recognize the standard from the nonstandard cylinder. The diameter of the nonstandard cylinders was varied in increments corresponding to target strength increments of 1 dB. The results indicated that both species could discriminate based on target strength differences of 1 dB at a 70% correct performance level.

The cylinder length discrimination experiments of Ayrapet'yants et al. (1969) and Zaslavskiy et al. (1969) were actually target strength discrimination experiments. Highlights were probably present in the echoes from the sonar signal penetrating and propagating along different acoustic paths within the targets and from circumferential waves (Neubauer 1986). However, for an incident signal that is normal to the longitudinal axis of a cylinder, the structure of the echo will only be affected by the diameter and material composition and not the length. Since in these studies the diameter and material composition were fixed and the length varied, only the amplitude of the target echoes was affected. The target strength of an acoustically rigid or a soft cylinder of finite length can be expressed as (Urick 1983)

$$TS = 10\log(aL^2/2\lambda) \qquad (9-8)$$

where a is the radius, L is the length of the cylinder, and λ is the wavelength of the signal. The differences in target strength were approximately 1.6 and 2.1 dB for the targets used by Ayrapet'yants et al. (1969) and Zaslavskiy et al. (1969), respectively. These values compare well with the 1-dB difference observed in the NOSC experiments (Evans 1973). However, since the diameter of the cylinders was varied in the NOSC experiments, additional cues from circumferential waves (Barnard and McKinney 1961; Diercks et al. 1963) may have been present.

9.2.2 Sphere Diameter

Numerous sphere size discrimination experiments have been performed with metallic targets using *Tursiops*. A summary of these experiments is given in Table 9.1. The fourth column (T.S. Diff.) gives the difference in calculated target strength between the standard and comparison sphere at the animal's threshold. The target strength of a large, rigid or soft sphere can be expressed as (Urick 1983)

$$TS = 20\log\left(\frac{a}{2}\right) \qquad (9-9)$$

where a is the radius of the sphere.

Two cues associated with sphere size discrimination are the differences in target strength and the highlight structure of the echoes. The incident signal penetrating and propagating along different paths within a sphere will result in the presence of many highlights or echo components (Shirley and Diercks 1970). Circumferential wave components will also contribute to the echo structure (Wille 1965; Uberall et al. 1966). Examples of the echo structure and frequency spectrum of echoes from a 2.54-cm solid steel sphere and a 7.62-cm water-filled steel sphere were shown in Figure 8.1. The highlight structure (e.g., position and amplitude of the highlights) is determined by the diameter and material composition of the sphere.

9.2.3 Planar Targets

Barta (1969) conducted a size discrimination experiment using circular aluminum disks covered with neoprene. A *Tursiops* was trained to choose the smaller of two simultaneously presented targets. A divider between the targets restricted the minimum range between the dolphin and targets to 0.7 m. The dolphin discriminated a 16.1-cm from a 15.2-cm diam. disk at a 75% correct threshold. Bel'kovich et al. (1969) used plastic foam square targets and trained a common porpoise (*Delphinus delphis*) to choose the larger of two simultaneously presented targets. The dolphin discriminated between a 100-cm² and a 90.25-cm² target at a 77% correct response level.

Table 9.1. Results of biosonar size discrimination experiments with spherical targets. Stand. Diam. is the diameter of the standard sphere. Increm. Diam. is the diameter of the comparison sphere at the discrimination threshold. T.S. Diff. is the difference in target strength between the standard and comparison target (calculated)

Stand. Diam. (cm)	Material	Increm. Diam. (cm)	T.S. Diff. (dB)	Range (m)	Reference
5.71	Ni–steel	0.64	0.9	>0.5	Turner and Norris (1966)
10.40	steel	3.90	2.8	2–6	†Dubrovskiy et al. (1971)
57.10	steel	6.40	0.9	2–6	"
5.00	lead	0.50	0.8	8	†Dubrovskiy (1972)
1.02	lead	0.15	1.2	3	*Fadeyeva (1973)
1.40	lead	0.20	1.2	4.8	*Dubrovskiy and Krasnov (1971)
10.20	lead	1.50	1.2	2–6	Ayrapet'yants and Konstantinov (1974)

† cited in Ayrapet'yants and Konstantinov (1974)
* cited in Bel'kovich and Dubrovskiy (1976)

The main cue available in the planar target size discrimination was differences in target strength. The target strength of a rigid or soft planar target at normal incidence of the signal can be expressed as (Urick 1983)

$$TS = 20\log\left(\frac{A}{\lambda}\right) \qquad (9\text{-}10)$$

where A is the area of the target and is the wavelength of the signal in water. The target strength differences between the standard and comparison targets at threshold were 1 dB and 0.9 dB for the targets used by Barta (1969) and Bel'kovich et al. (1969), respectively. Backscatter measurements with Barta's targets indicated that threshold size discrimination was performed with a 1-dB differential in target strength.

9.3 Target Structure Discrimination

9.3.1 Thickness and Material Composition of Plates

Evans and Powell (1967) were the first to demonstrate that a blindfolded, echolocating *Tursiops* could discriminate between metallic plates of different thickness and material composition. The dolphin was trained to recognize a 30-cm diam. circular copper disc of 0.22-cm thickness among comparison targets of the same size. The

standard copper disc and a comparison target were presented side by side, separated by a center-to-center distance of 50 cm. The dolphin was required to station at the far end of a 9-m tank facing away from the targets. Upon command from the experimenter, the dolphin turned and swam toward the target, echolocating along the way. A typical trajectory and scanning motion of the dolphin in 1-second intervals during the last 2 meters of its swim pattern is shown in Figure 9.2. Also included in the figure is a list of the comparison targets. The dolphin's performance along with the performance of another *Tursiops* and a *Lagenorhynchus* reported later by Evans (1973) are shown in Figure 9.3. The three dolphins performed similarly, being able to differentiate the standard from all comparison targets at a performance level of 75% or better except for the 0.32-cm copper and 0.32-cm brass targets. *Tursiops* #1 recognized the 0.32-cm copper plate and *Tursiops* #2 discriminated the 0.32-cm brass plate approximately 75% of the time. However, *Tursiops* #1 had difficulties distinguishing the standard from the 0.32-cm brass and *Tursiops* #2 had difficulties with the 0.32-cm copper plate. The *Lagenorhynchus* had difficulties distinguishing the standard from both the 0.32-cm copper and 0.32-cm brass plate.

Johnson (1967) argued that since the reflectivity of the metal plates used by Evans and Powell (1967) was the same, the only possible cue that would allow the dolphin to discriminate between

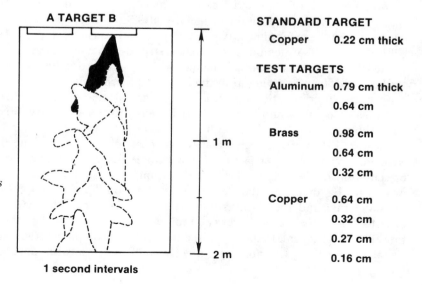

Figure 9.2. Typical swim pattern of a *Tursiops truncatus* during the final 2 m of the plate discrimination and recognition experiment of Evans and Powell (1967). The dolphin's position is given in 1-sec intervals.

A TARGET B

1 second intervals

STANDARD TARGET

Copper	0.22 cm thick

TEST TARGETS

Aluminum	0.79 cm thick
	0.64 cm
Brass	0.98 cm
	0.64 cm
	0.32 cm
Copper	0.64 cm
	0.32 cm
	0.27 cm
	0.16 cm

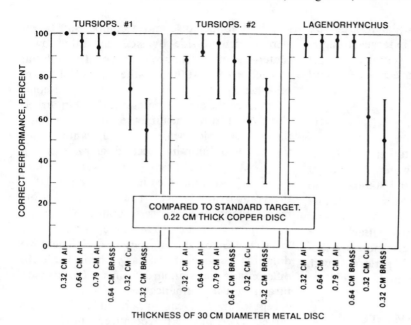

Figure 9.3. Comparative performance of two *Tursiops* and one *Lagenorhynchus* in the plate discrimination and recognition experiment, showing the mean and range of performance scores. (From Evans 1973.)

the metallic plates was the difference in phase of the reflected signals. MacKay (1967) originally suggested the metallic plate discrimination experiment to test whether the dolphin might use phase difference information. Weisser et al. (1967) examined the metal plates used by Evans and Powell by insonifying them at normal incident angle using simulated dolphin echolocation signals. There were no indications from the results that phase difference played a role in making the echoes discriminable since no obvious differences between echo waveforms were observed. How the dolphins were able to discriminate the thickness and material composition of metal plates remained a mystery for almost two decades.

Fish et al. (1976) demonstrated that instrumented human divers wearing a helmet with a sending and two receiving transducers could perform the discrimination task using the same metal plates as Evans and Powell (1967). Broadband simulated dolphin clicks with energy centered at 60 kHz were projected and the echoes were time-stretched by a factor of 128 before presentation to the divers. The time stretching was done by digitizing the echoes at a fast sampling rate and then converting the signal back to analog form at a slower ($\frac{1}{128}$) sampling rate. The 60-kHz peak frequency of the incident signal was effectively transformed to 469 Hz, well

within the human audio range. A photograph of a diver and the apparatus and a schematic of the diver's orientation are shown in Figure 9.4. The average performance of three divers along with that of dolphin #1 from the Evans and Powell (1967) experiment are shown in Figure 9.5. The dolphin data include some cases not reported in Evans (1973). The results indicate that the divers could perform the metallic plate discrimination as well or better than the dolphin, however, no explanations or discussion of the cues used by the divers or the dolphin were presented.

Au and Martin (1988) examined the same metal plates used in the experiments of both Evans and Powell (1967) and Fish et al. (1976) with a monostatic system which projected simulated dolphin sonar signals. They obtained echoes from the metal plates at both normal incident and at 14° from normal (approximating the divers' geometry). Backscatter results at normal incident for the standard copper and three other plates of the same 0.32 cm wall thickness are shown in Figure 9.6. The echo waveforms are shown individually while the spectra are plotted on the same graph. As with the measurement of Weisser et al. (1967), very little difference can be observed in the echoes that would allow either the dolphin or the instrumented diver to discriminate between the plates. Echoes from the same

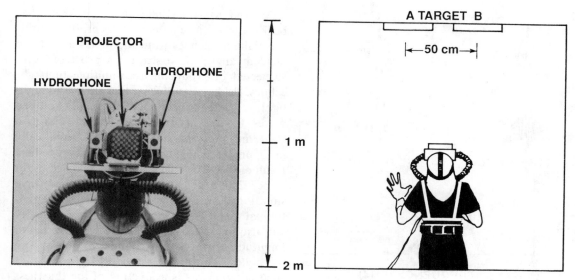

Figure 9.4. (*Left*) Apparatus used by instrumented divers (from Fish et al. 1976); (*right*) geometry of the metal plate discrimination experiment.

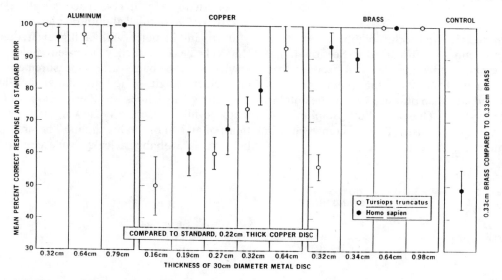

Figure 9.5. Comparative performance of the dolphin in the Evans and Powell (1967) experiment and human divers. (From Fish et al. 1976.)

four plates at the 14° incident angle are shown in Figure 9.7. The echo waveforms are shown on the left and the envelopes of the matched-filter response on the right. The reference signal for the matched filter was the transmitted signal. The matched-filter response was calculated using equation (9-7) and the envelope determined using

(9-1) to (9-4). Differences in the echoes are obvious from the figure. Obvious differences also existed between the standard copper and the other plates, including the other copper plates of differing thickness. The numbers above the matched-filter envelopes are the times of arrival of different echo highlights in microseconds,

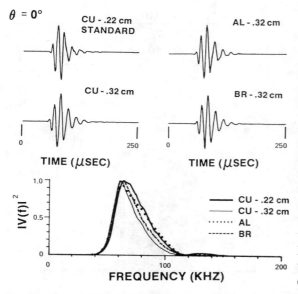

Figure 9.6. Echoes from four of the plates used by Evans and Powell (1967) at normal incident. (From Au and Martin 1988.)

Au and Martin (1988) also examined what backscattering mechanism would produce multiple highlight echoes from the plates at oblique incident angles. Two scattering processes were suspected of producing the multiple highlight echoes: "leaky" Lamb waves for the initial highlight and edge reflection of internally trapped waves for the secondary highlight. The two scattering processes are described schematically in Figure 9.8. The initial highlight may also be a result of specular reflection of the bounded beam rather than of a leaky wave. The trapped wave situation shown in Figure 9.8 refers to the longitudinal wave. Transverse waves of lower velocity will also be excited in the plates and converted to longitudinal waves at a boundary upon exiting the plate. The time of arrival of the secondary echo components is a function of the thickness and material composition (velocity of sound in the material) of the plates. In order to confirm their conjectures about the backscattering processes, Au and Martin (1988) acoustically examined two 0.79-cm thick aluminum plates 30.5 and 61.0 cm in diameter with a higher frequency (120-kHz peak frequency) transient pulse. The results of the backscatter measurement for an incident angle of 15° are shown in Figure 9.9. Below the echoes are the envelopes of the matched-filter responses showing the relative times of arrival of the highlights. The first highlight, A, is probably a leaky wave component

relative to the first echo highlight. Multiple highlight echoes (as discussed in Section 4.2) of the type shown in Figure 9.7 will produce time separation pitch (TSP) in the human auditory system and may also produce TSP in the dolphin auditory system. Therefore, both dolphins and humans may have used TSP difference cues in discriminating the metal plates.

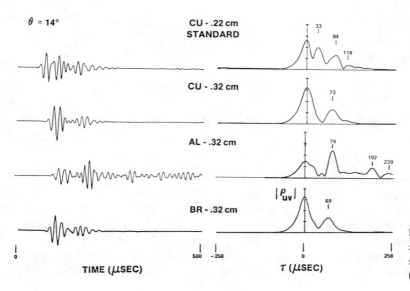

Figure 9.7. Backscatter results at 14° incident angle for the same four plates as in Fig. 9.6. (From Au and Martin 1988.)

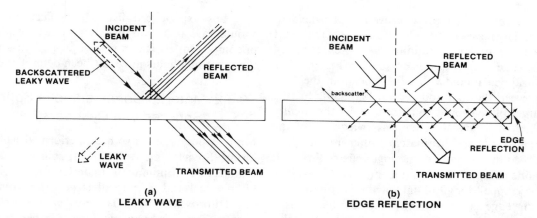

Figure 9.8. Schematic representation of (*A*) leaky Lamb wave backscatter and (*B*) trapped longitudinal wave with an edge reflection. (From Au and Martin 1988.)

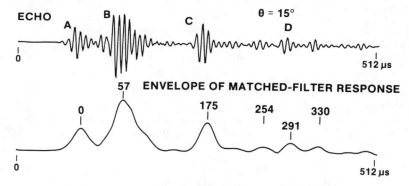

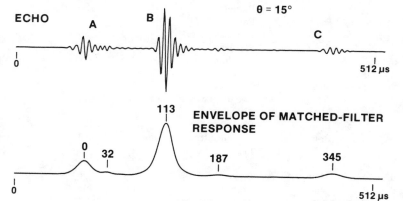

Figure 9.9. Backscatter results for the 30.5-cm and 61.0-cm diam. plates at an incident angle of 15° (From Au and Martin 1988.)

since it is always present, even as the incident angle increases to 40°. Component B is the result of the trapped longitudinal wave reflecting off the edge of the plate as depicted in Fig. 9.9. For a trapped mode with edge reflection, the time between the first and second highlights for the larger plate should be twice that for the smaller plate, since its diameter was twice that of the smaller plate. The matched-filter results confirmed the presence of the edge-reflected trapped mode component. Component C is the result of the trapped longitudinal wave reflecting off the same edge as component B, then off the opposite edge, and back off the original edge. For this mode, travel time should be three times the travel time for component B. Again, the matched-filter results confirmed the presence of the triple-reflected trapped component. Component D for the 30.5-cm plate is a trapped mode component

that has experienced five edge reflections. It should have an arrival time that is five times that of component B. The modes of propagation for the other highlights have not been determined.

9.3.2 Material Composition of Solid Spheres and Cylinders

Many experiments have been performed in the former Soviet Union on the capability of delphinids to discriminate the material composition of spherical and cylindrical targets (Bel'kovich and Dubrovskiy 1976). Three different dolphin species, *Tursiops truncatus*, *Phocoena phocoena*, and *Delphinus delphis*, have been used with targets constructed from a host of different materials. A summary of these material composition discrimination experiments with echolocating dolphins is provided in Table 9.2. The results

Table 9.2. Summary data of (Soviet) biosonar material composition discrimination experiments (after Bel'kovich and Dubrovskiy 1976)

Target/ (cm)	Stand. Target	Comp. Target	% Corr.	Range (m)	References/Species
Cyl. d = 7.5 l = 11.5	Steel Wood	Wood, Plastic, Glass Plastics	75	4–6	Zaslavskiy et al. 1969 (*Phocoena*)
Spheres d = 5.0	Steel	Duralumin.	46–67	3–11	Dubrovskiy et al. 1970 (*Tursiops*)
Spheres d = 5.0	Brass	Alum. Steel, Texolite Ebonite, Fluoroplastic	>91 100	— —	Abramov et al. 1971 (*Tursiops*)
Sphere d = 5.0	Lead	Steel, Duralumin., Wax Rubber, Paraffin, Plexigl.	>97 >96	5–11 "	Babkin et al. 1971 (*Tursiops*)
	Steel	Duralumin Wax, Rub., Paraf. Plexigl.	62 >93	" "	
	Duralumin	Wax, Rub., Paraf. Plexigl.	>92	"	
	Wax	Rubber Paraffin Plexiglas	61 72 100	" " "	
	Rubber	Paraffin Plexiglas	81 86	" "	
	Paraf.	Plexiglas	93	"	
Spheres d = 5.0	Steel Brass	Duralumin. Duralumin. Ebonite, Steel	58–65	3–10	Titov 1972 (*Tursiops*)
Spheres d = 5.0	Steel	Duralumin. Ebonite, Lead, Plexiglas	70 >92	5–11 "	Titov 1972 (*Delphinus*)
	Ebon.	Plexiglas	78	"	
	Lead	Plexiglas	100	"	
	Brass	Ebonite, Steel, Duralumin. Plexiglas, Lead	>93 >90	" "	
Spheres d = 7.0 d = 1.0	Alum. Alum.	Brass Brass	96 46	— —	Yershova et al. 1973 and Golubkov et al. 1973 (*Tursiops*)

indicate that dolphins could discriminate rather easily among most of the materials tested except for steel versus duralumin and wax versus rubber. It is interesting to note that three different experiments (Dubrovskiy et al. 1971; Babkin et al. 1971 and Titov 1972) comparing steel versus duraluminum resulted in relatively poor performance by the animals.

Dubrovskiy et al. (1970, 1971) used the analytical findings of Hickling (1962) to explain how dolphins discriminate among spherical targets based on differences in material composition and diameter. Hickling (1962) showed that the steady-state echo from a sphere can be expressed as

$$p_e = \frac{P_0}{2r} f_\infty(ka) e^{ika(r-\tau)} \qquad (9\text{-}11)$$

where p_0 is the pressure at the source, r is the distance between the source and the sphere, τ is the delay time for the acoustic signal to reach the target, k is the wavenumber and is equal to $2\pi/\lambda$, and $f_\infty(ka)$ is a dimensionless form factor describing the frequency response of the scattering process. An example of how f_∞ varies with frequency, for aluminum spheres in this case, is shown in Figure 9.10. Note how f_∞ oscillates with frequency, exhibiting many maxima and minima. The echo from a 12.7-cm diam. solid aluminum sphere measured by the author using a simulated dolphin sonar signal is shown in Figure 9.11. The ripples in f_∞ should resemble ripples in the spectrum of the echo. Hickling

(1962) also found that the average frequency interval (Δ) between the maxima, or minima, in the form function is directly related to the shear velocity of the material of the sphere, as can be seen in the graph presented by Hickling (Fig. 9.12). The oscillations in the form function are almost but not exactly periodic, so an average interval between maxima or minima was taken. Dubrovskiy et al. (1970, 1971) postulated that dolphins recognize differences in the material composition and diameter of spherical targets by detecting differences in $\delta\Delta$, where $\delta\Delta = \Delta_1 - \Delta_2$. Since $k = 2\pi/\lambda$, the difference in average oscillation frequency for two solid spheres of diameters a_1 and a_2 can be expressed as

$$\delta f = \frac{c}{2\pi}\left(\frac{\Delta_1}{a_1} - \frac{\Delta_2}{a_2}\right) \qquad (9\text{-}12)$$

Dubrovskiy et al. (1971) and Dubrovskiy and Krasnov (1971) experimentally found that the dolphin discrimination threshold was approximately

$$|\delta\Delta| \approx 2 \text{ to } 3 \text{ kHz}$$

Golubkov et al. (1973), Dubrovskiy and Fadeyeva (1973), and Yershova et al. (1973) considered the time domain characteristics of echoes from spheres and related the oscillations in the frequency domain to the separation between the primary echo and secondary echo (Δt) which is equal to $1/\Delta$. The difference between the Δt for one sphere and the Δt for another sphere can be expressed as

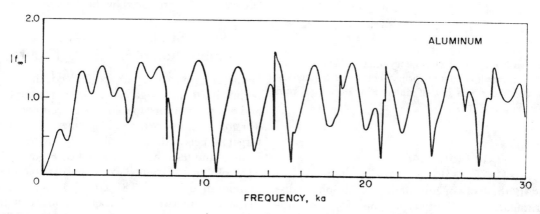

Figure 9.10. Variations in the form factor, f_∞, as a function of frequency for solid aluminum spheres. (From Hickling 1962.)

Figure 9.11. Echo from a 12.7-cm solid aluminum sphere using a simulated dolphin sonar signal.

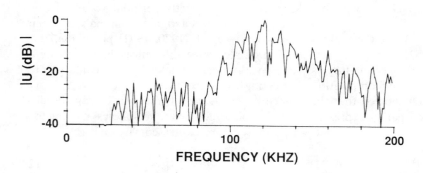

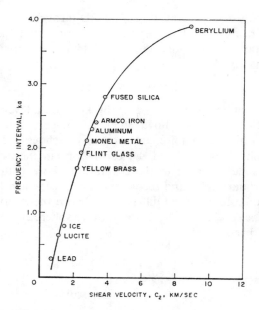

Figure 9.12. Average frequency interval between minima or maxima in the form function as a function of the shear velocity of different materials. (From Hickling 1962.)

$$\delta t = \Delta t_1 - \Delta t_2 = \frac{\Delta_2 - \Delta_1}{\Delta_1 \Delta_2} \qquad (9\text{-}13)$$

Another way of considering the dolphin's discrimination capability is in terms of time separation pitch (TSP). Arrival time differences in the highlights may be perceived as a time separation

pitch (TSP), especially if the echo components are highly correlated. Humans presented with a correlated pair of sound pulses perceive a pitch that is equal to $1/T$, where T is the separation time between pulses (Small and McClellan 1963; McClellan and Small 1965). We discussed in Chapter 4 that signals with ripples in the frequency domain or highlights in the time domain can produce TSP in the human auditory system. The presence of multiple highlights in the echo waveform produces ripples in the form function f_∞ and the frequency spectrum and may also produce TSP in the dolphin's auditory system. Therefore, the predominant cue for dolphins in discriminating among solid spheres based on material composition and diameter may be differences in TSP produced by the different spheres.

9.3.3 Structure and Material Composition of Hollow Cylinders

Hammer and Au (1980) performed three experiments to investigate the target recognition and discrimination capability of an echolocating *Tursiops*. Their experimental geometry, with two target funnels at distances of 6 and 16 m from the front of the test pen, is depicted in Figure 9.13. The dolphin was trained to recognize two hollow aluminum cylinders 3.81 cm and 7.62 cm in diameter and two coral rock cylinders of the same diameters (Fig. 9.14). All of the cylinders were 17.8 cm in length. The coral rock targets

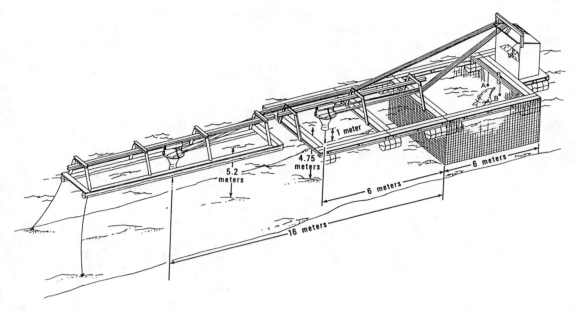

Figure 9.13. Experimental geometry for the target recognition experiment. (From Hammer and Au 1980.)

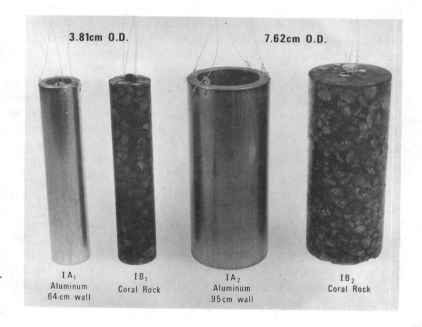

Figure 9.14. Standard aluminum and coral targets used in the target recognition and discrimination experiments of Hammer and Au (1980).

were constructed of coral pebbles encapsulated in degassed epoxy. The targets were presented 6 m and 16 m from the animal's pen. During a trial only one target was presented and the dolphin was required to echolocate the target and respond to paddle A if it was one of the aluminum standards or paddle B if it was one of the coral rock standards. After baseline performance exceeded 95% correct with the standard targets, probe sessions were conducted to investigate the dolphin's ability to discriminate probe targets, which were also cylinders 17.8 cm in length, but

Table 9.3. Cylindrical probe targets used in Hammer and Au experiments. Abbreviations for the material compositions are AL, aluminum; CPN, corprene; CRK, coral rock; PVC, polyvinylchloride

General Discrimination							
Probe No.	1	2	3	4	5	6	7	8
Composition	AL	CPN	AL	CPN	AL	CRK	AL	PVC
Wall (cm)	0.48	solid	solid	solid	0.64	solid	solid	0.79
O.D. (cm)	6.35	6.35	3.81	4.06	11.43	11.43	7.62	7.62

Wall Thickness Discrimination							
	3.81-cm O.D. aluminum				7.62-cm O.D. aluminum			
Probe No.	1	2	3	4	5	6	7	8
Wall (cm)	0.32	0.48	0.79	0.95	0.32	0.69	1.29	7.62

Material Composition Discrimination					
	3.81-cm O.D./0.32-cm wall			7.62-cm O.D./0.40-cm wall		
Probe No.	1	2	3	4	5	6
Composition	bronze	glass	steel	bronze	glass	steel

of varying structure and composition. Two probe targets were used in each of these sessions; only 8 of 64 trials in each session were probe trials, 4 for each probe target. The various probe targets used in all three experiments are described in Table 9.3.

The first experiment of Hammer and Au (1980) involved a general discrimination task in which two solid aluminum cylinders with the same diameters as the two standard (hollow) aluminum cylinders, two hollow aluminum cylinders of different diameter and wall thickness than the aluminum standards, one coral rock cylinder with a larger diameter than either coral rock standard, one PVC tube, and two corprene (cork-neoprene) cylinders were used as probe targets. A photograph of the probe targets used in this experiment is shown in Figure 9.15. Initially the dolphin classified three of the aluminum probes with the standard aluminum targets, but eventually, after the target range was moved from 6 to 16 m and then back to 6 m, it classified all of the probes with the coral rock standards. The dolphin began to focus on the salient features of the two aluminum standards and rejected all other targets as being "not A." In other words, the dolphin became a null detector and searched for the presence or absence of the aluminum standards.

In the second experiment, Hammer and Au (1980) investigated the dolphin's ability to discriminate by wall thickness. Hollow aluminum probe targets with the same outer diameters but different wall thickness compared to the aluminum standards were used. The results showed that the dolphin could reliably discriminate wall thickness differences of 0.16 cm for the 3.81-cm O.D. cylinders and of 0.32 cm for the 7.62-cm O.D. cylinders. However a thickness differential threshold was not determined.

In the third experiment, Hammer and Au (1980) investigated the dolphin's ability to discriminate by material composition using bronze, glass, and stainless steel probe cylinders that had the same dimensions as the aluminum standards. The results of the final probe trial sequence are displayed in Figure 9.16 in terms of the percentage of time the dolphin reported the probe target as either of the aluminum standards. The dolphin mistook the small bronze and steel cylinders for an aluminum standard 25% and 12.5% of the time, respectively. The glass cylinders were mistaken for the aluminum standards 100% and 87.5% of the time for the small and large cylinders, respectively.

All of the targets used by Hammer and Au (1980) were examined acoustically in a test pool using simulated dolphin sonar signals. Echoes

Figure 9.15. Probe targets used in the general discrimination experiment of Hammer and Au (1980).

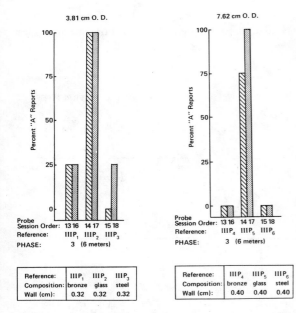

Figure 9.16. Probe target results for the material discrimination test of Hammer and Au (1980).

the comparison or probe targets are shown in (b–e). Similarities or dissimilarities between the echoes can be more readily determined by studying the envelopes of the matched-filter response rather than the echo waveforms or their frequency spectra. The echo waveforms and their corresponding frequency spectra exhibit many minor variations that tend to hinder comparison of the target echoes. The envelopes of the matched-filter responses, on the other hand, are fairly simple yet accentuate details such as time of arrival of highlights, correlation between echo components and transmitted signal, and the relative strength of the various highlights. The matched-filter responses of the steel and bronze cylinders are readily discernable from the aluminum standard. The response for the glass is fairly similar to that for the aluminum cylinder. The matched-filter results are useful to determine the time of arrival of the various highlights in the echo. From visual inspection of the echo structures we can see that all of the targets have different arrival times for the secondary echo components and therefore, different echo structures.

Hammer and Au (1980) suggested that the predominant cue used by the dolphin in discriminating among the various cylinders was probably TSP, generated mainly by the presence of the first and second echo components. In Figure 9.19, the

from the cylinders used in the material composition experiment are shown in Figures 9.17 and 9.18. The echo structure is shown on the left, the frequency spectrum is in the middle, and the envelope of the matched-filter response is on the right. The aluminum standard is shown in (a) and

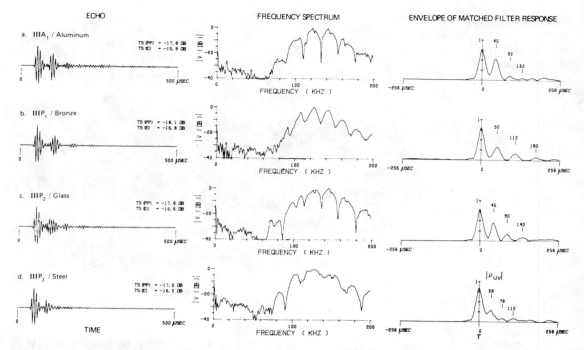

Figure 9.17. Echoes from the 3.81-cm O.D. cylinders used in the material composition discrimination experiment. (From Hammer and Au 1980.)

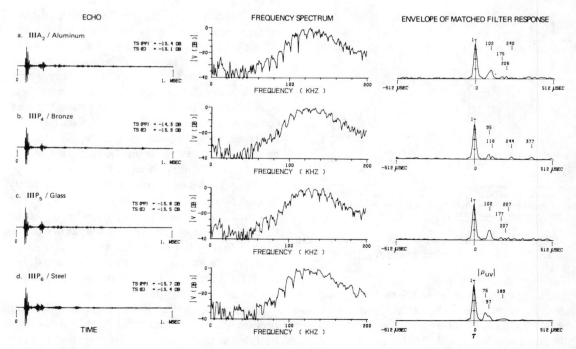

Figure 9.18. Echoes from the 7.62-cm O.D. cylinders used in the material composition discrimination experiment. (From Hammer and Au 1980.)

Figure 9.19. Echo from a 3.82-cm O.D. aluminum cylinder showing the frequency spectrum for the total signal and for a truncated signal (dashed curve) with only the first two echo components (portion of time domain spectrum between vertical bars) present. (From Hammer and Au 1980.)

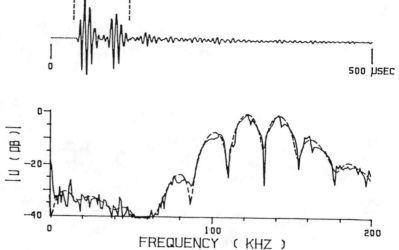

frequency spectrum of the first and second echo component for one of the aluminum targets used by Hammer and Au (1980) is overlaid on the total echo spectrum. Note how well the total spectrum is described by the rippled spectrum for the first two echo components. As discussed in Chapter 4, such a rippled spectrum is perceived as TSP by humans (Bilsen 1966).

The glass and aluminum cylinders described in Table 9.3 were next used by Schusterman et al. (1980) in a two-alternative forced-choice discrimination experiment involving the same *Tursiops* used in the Hammer and Au (1980) study. Targets were located at a range of 6 m from the front of the experimental pen and were submerged in vertical orientation at a depth of 1 m. A trial consisted of having either a glass or aluminum cylinder presented; the animal was required to perform a sonar scan and respond to paddle A if it thought that the target was either one of the two aluminum cylinders, or respond to paddle B for the glass cylinders. After 30 sessions, the dolphin could perfectly discriminate between the small (3.61-cm O.D.) aluminum and glass cylinders. However, the animal was never able to discriminate between the large (7.62-cm O.D.) aluminum and glass cylinders.

Au and Turl (1991) performed a material composition experiment with cylinders to test if there existed aspect-independent cues that the dolphin could utilize to discriminate between cylinders. Their experimental configuration, a

dolphin in a hoop station with a target rotor 6 m from the hoop, is depicted in Figure 9.20. An aluminum cylinder with a 7.62-cm O.D., 0.4-cm wall thickness, and a 17.1-cm length was used as the standard target. Two comparison targets, a hollow stainless steel cylinder having the same dimensions as the standard and a cylinder of coral rock encapsulated in degassed epoxy, were used. The standard and a comparison target were supported on the rotor bar with their longitudinal axis parallel to the bar and the horizon. A trial consisted of the dolphin swimming into the hoop and initiating a sonar search when it received a command from the experimenter—the procedure we have referred to earlier as "echolocation on command" (Schusterman et al. 1980). The dolphin then backed out of the hoop and touched paddle A to report the presence of the aluminum standard, or paddle B if the comparison target was present.

The dolphin was trained to discriminate between two targets at baseline aspects of 0°, 45°, and 90°, where 0° was the broadside aspect (longitudinal axis perpendicular to the animal). After the dolphin performed the discrimination at a level of 90% correct for the baseline aspects, probe trials at novel aspect angles were introduced and the animal's response was recorded. In the first discrimination task, the dolphin was trained with the large aluminum and steel cylinders used by Hammer and Au (1980). The dolphin was able to achieve the 90% correct re-

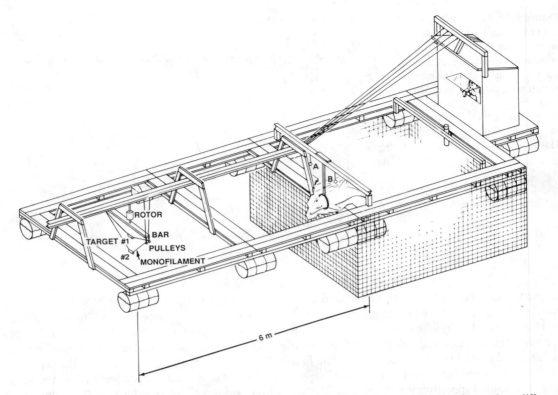

Figure 9.20. Experimental configuration for the cylinder material composition discrimination task at different target aspect angles. (From Au and Turl 1991.)

sponse criterion for the 0° aspect but was unable to reach the criterion for the 45° and 90° aspect. Therefore, probe sessions were not conducted with this set of targets, but additional sessions were conducted to obtain the animal's performance at 10° and 80° aspect angles. The dolphin's discrimination performance as a function of target aspect angle is shown in Figure 9.21. The results indicate that the dolphin could easily discriminate between the aluminum and steel cylinders at different aspect angles. The worst performance occurred at an aspect angle of 45°, but even at this angle the dolphin's overall performance, which is the average of the performances for the individual targets, was above 80° correct. A minimum of 300 trials per angle were conducted for the baseline aspect angles.

An easier discrimination task, involving the small aluminum cylinder and the small coral rock cylinder, was attempted next. The dolphin was able to reach a modified criterion with this discrimination task, as can be seen in Table 9.4.

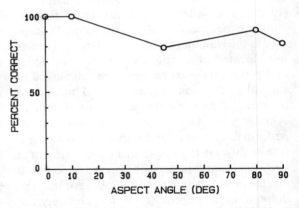

Figure 9.21. Dolphin's aluminum versus steel discrimination performance as a function of the target aspect angle. (From Au and Turl 1991.)

The results of novel aspect trials are also shown in the table. The baseline results represent a minimum of 220 trials per angle. Each novel aspect angle was used in a minimum of 25 probe trials except for 60°, where only 10 probe trials were

Table 9.4. Dolphin discrimination results as a function of aspect angle for the aluminum and coral rock targets (from Marrtin and Au 1986)

		Baseline Performance				
	0°		45°		90°	
	AL	Rock	AL	Rock	AL	Rock
	100%	94%	91%	89%	100%	96%

						Probe Sessions							
0°		15°		30°		45°		60°		75°		90°	
AL	Rock	AL	Rock	AL	Rock	AL	Rock	AL	Rock	AL	Rock	AL	Rock
100%	100%	98%	97%	96%	100%	100%	100%	100%	100%	100%	100%	100%	100%

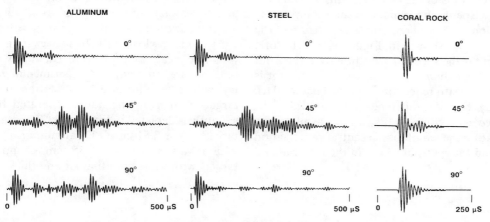

Figure 9.22. Echo waveforms from the aluminum, steel, and coral rock cylinders at different aspect angles. (From Au and Turl 1991.)

conducted. These results indicate that for simple discrimination there are aspect-independent cues that a dolphin can utilize in distinguishing by material composition.

The three cylinders used in the variable aspect experiment were acoustically examined using simulated dolphin sonar signals and echoes for three aspect angles, 0°, 45°, and 90°; the waveforms are shown in Figure 9.22. The echoes for the aluminum and steel cylinders at 0° should be the same as those in Figure 9.16. Difference between the aluminum and steel echoes can be clearly seen. At the 0° aspect the highlights occurred at different times. The echoes for the aluminum cylinder also did not decay as rapidly as for the steel cylinder. At the 45° and 90° aspect

angles, the times of occurrence of the different highlights were similar for the aluminum and steel cylinders, but there were differences in the amplitude and shape of the highlights. The difference between the echoes from the aluminum and coral rock cylinders is obvious. Secondary highlights tended to attenuate quickly for the coral rock target so that the echo structure of this target was not very complex.

9.3.4 Wall Thickness of Cylinders

The capability of an echolocating *Tursiops* to discriminate differences in the wall thickness of hollow steel cylinders was studied by Titov (1972). Titov's wall thickness experiment was briefly described by both Ayrapet'yants and Konstantinov (1974) and Bel'kovich and Dubrovskiy (1976). Presumably, a two-alternative forced-choice procedure with simultaneous target presentation was used. Both the outer diameter and length of the cylinders were 50 mm and wall thickness varied between 0.1 and 2 mm. The dolphin was trained to choose the thinner of two cylinders presented simultaneously at a range of 5 m. The animal was able to react to a wall thickness difference of 0.2 mm at the 75% correct response level.

The predominant cue for the dolphin in Titov's wall thickness experiment was probably the difference in arrival time of the echoes from the front and back walls of the cylinder. This arrival time difference between the first two echo components has a dominant influence on the spectrum of the echo, as was shown in Figure 9.18. The first highlight is the specular reflection from the front wall of the cylinder and the second highlight is the central path reflection from the back wall (cf. Fig. 8.2). The time difference in the arrival of the first two echo components can be calculated by calculating the length of the central propagation path and the speed of sound in the material and in water

$$\tau = \frac{2th}{c_1} + \frac{2(\text{O.D.} - 2th)}{c_0} \qquad (9\text{-}14)$$

where c_1 is the longitudinal sound velocity of the material making up the cylinder, c_0 is the sound velocity in water, th is the wall thickness cylinder and O.D. is the outer diameter of the cylinder. The difference in arrival time for one cylinder versus another cylinder can be expressed as

$$\Delta t = \left| 2\frac{\Delta th}{c_1} - 4\frac{\Delta th}{c_0} \right| \qquad (9\text{-}15)$$

where $\Delta th = th_1 - th_2$. Inserting the longitudinal sound velocity for steel and seawater along with the dolphin's wall thickness discrimination threshold of 0.2 mm into equation (9-15), we obtain an estimate of the dolphin's time difference discrimination threshold, which is $\Delta t \approx 0.5 \ \mu s$.

Au and Pawloski (1992) also examined the wall thickness discrimination capability of an echolocating Tursiops. The experimental geometry, with a dolphin in a hoop station and two targets 8 m from the hoop, separated by 22° azimuth, is depicted in Figure 9.23. The standard target was a 3.81-cm O.D. aluminum cylinder with a wall thickness of 6.35 mm. Comparison targets with walls both thinner and thicker than

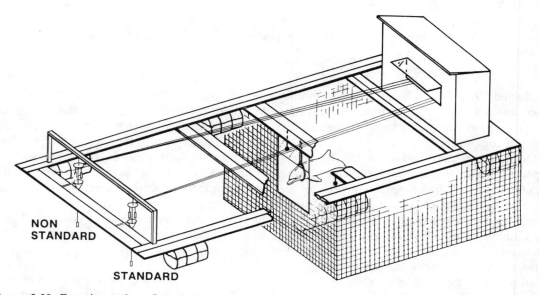

NON STANDARD

STANDARD

Figure 9.23. Experimental configuration of the wall thickness discrimination experiment. The targets were located 8 m from the hoop station. At each target location, a funnel was used to control the position of the targets. (From Au and Pawloski 1992.)

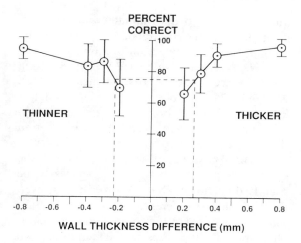

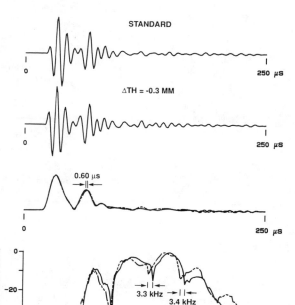

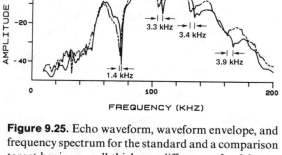

Figure 9.24. Dolphin wall thickness discrimination performance as a function of wall thickness difference. (From Au and Pawloski 1992.)

Figure 9.25. Echo waveform, waveform envelope, and frequency spectrum for the standard and a comparison target having a wall thickness difference of −0.3 mm. The dashed curves for the envelope and the frequency spectrum are for the comparison target. (From Au and Pawloski 1990.)

the standard were used; the incremental differences in wall thickness, compared to the standard, were ±0.2, ±0.3, ±0.4, and ±0.8 mm. During a trial, the dolphin was required to echolocate and to respond to the paddle that was on the same side of the center line as the standard target. The dolphin's discrimination and target recognition capabilities were tested in the free field first with the thinner comparison targets, then with the thicker comparison targets. The dolphin's performance as a function of wall thickness difference is shown in Figure 9.24. The 75% correct response threshold corresponded to a wall thickness difference of −0.23 mm for thinner targets and +0.27 mm for thicker targets. These results compare well with the results of Titov (1972) who used a different paradigm and targets of different sizes and material composition. From equation (9-15), a Δth of 0.23 and 0.27 mm corresponds to a time difference threshold of $\Delta t \approx 0.5$ to 0.6 μs, respectively, which compares well with the Δt of approximately 0.5 μs obtained from Titov's (1972) results.

The targets were acoustically examined using a simulated dolphin sonar signal. The echo waveform, envelope of the waveform, and the frequency spectrum for the standard and for comparison targets with wall thickness differences of −0.3 mm and −0.2 mm are displayed in Figures 9.25 and 9.26, respectively. The dolphin was able

to perform above threshold for the targets associated with Fig. 9.25, but not for the targets associated with Figure 9.26. The echo signals were digitized at a 1-mHz sample rate so that the time interval between sampled points was 1 μs. In order to determine the relative time between the first two peaks in the envelope of the echo waveform, values between the sampled points were interpolated at 0.1-μs intervals using the Sampling Theorem, which can be found in most textbooks on digital signal analysis (e.g., Brigham 1988). Calculations using equation (9-15) indicated that $\Delta t \approx 0.67$ μs and 0.45 μs the standard and comparison targets with wall thickness differences of −0.3 mm and −0.2 mm, respectively. The measured values in Figures 9.25 and 9.26 are in reasonable agreement with the calculated values. The envelope curves in these two figures

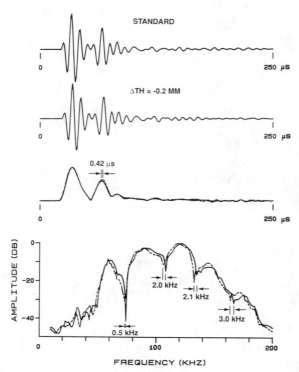

Figure 9.26. Echo waveform, waveform envelope, and frequency spectrum for the standard and a comparison target having a wall thickness difference of −0.2 mm. The dashed curves for the envelope and the frequency spectrum are for the comparison target. (From Au and Pawloski 1990.)

suggest that if the dolphin used time domain cues, it must have been able to perceive incremental time differences of approximately 500 ns between highlight intervals.

Figures 9.25 and 9.26 also show differences in the frequency spectra of the echoes from the standard and comparison targets. The frequency spectra for the thinner comparison targets resemble the spectrum of the standard target, but are shifted slightly towards lower frequencies. The spectra for the thicker comparison targets were shifted towards higher frequencies. The average frequency differences, Δf's were 3.2 and 2.2 kHz for a wall thickness difference of −0.3 mm and −0.2 mm, respectively. The frequencies of the nulls in an echo spectrum were interpolated by padding the 1024 sampled points with zero's to create a signal consisting of 8192 points at 1 μs intervals. If the dolphin used this shift in fre-

quency spectra to discriminate wall thickness difference, then the spectral data suggest that the dolphin could perceive a shift of approximately 3.3 kHz, but not a shift of 2.1 kHz. Splitting the difference in the two cases depicted in Figs. 9.25 and 9.26, the dolphin's capability to detect frequency shifts in broadband spectra is approximately 2.7 kHz for broadband echolocation signals with peak frequencies in the vicinity of 120 kHz. The spectral shifts also caused the spectral amplitudes at some frequencies to be slightly different by about 3 to 5 dB. However, these amplitude differences were probably not the main cues used by the dolphin since comparable amplitude differences can be seen in the two cases depicted in Figs. 9.25 and 9.26, and if the dolphin used the amplitude difference cue, it would have been able to discriminate the targets associated with Fig. 9.26.

Instead of detecting the time interval difference between the first and second highlights in the echoes from the standard and comparison targets or detecting a frequency shift in the spectra, the dolphin may have relied on TSP cues. The first two highlight should generate a TSP of approximately 28.30 kHz for the standard, 27.94 kHz for the −.2 mm and 27.77 kHz for the −0.3 mm comparison targets. If the dolphin used differences in TSP to discriminate the targets, we can infer that the animal could discern a TSP difference of 530 Hz between the standard and the −0.3-mm comparison target, but could not discern the 360-Hz difference between the standard and the −0.2-mm comparison target. Therefore, the dolphin's TSP discrimination threshold should be approximately 450 Hz. Similar results would have been obtained by considering the thicker comparison targets.

The dolphin's sonar signals were monitored with a computer system which measured the interval, amplitude, and energy flux density of each emitted signal. Although the dolphin was trained to examine both of the targets that were presented simultaneously, the animal became a null detector very early on and thus behaved like the animal in the Hammer and Au (1980) experiment. The dolphin typically examined the target on the right side and made its response without examining the target on the left. In other words, the dolphin remembered the characteristics of

Figure 9.27. Dolphin wall thickness discrimination performance in masking noise. (From Au and Pawloski 1992.)

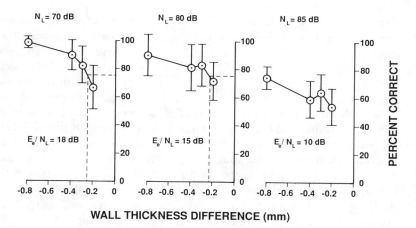

WALL THICKNESS DIFFERENCE (mm)

the standard target and responded either "A" or "not A," in a manner similar to the dolphin used by Hammer and Au.

The dolphin's capability to make fine wall thickness discriminations was also tested in the presence of white masking noise. The noise was projected by two spherical transducers located 2 m from the hoop, in line with the center of the hoop and each target location. Three different noise levels were used—70, 80, and 85 dB re 1 $\mu Pa^2/Hz$—and the dolphin was tested with the thinner comparison targets. The dolphin's performance in the presence of masking noise is shown in Figure 9.27. Also included with the performance results is the average of the maximum echo energy-to-noise ratio $(E_e/N_L)_{max}$ at the three noise levels. The dolphin's performance was relatively constant as the noise level increased from the ambient to 70 and 80 dB. At a noise level of 85 dB, $(\overline{E_e/N_L})_{max}$ was approximately 14 dB and the dolphin was not able to reach the 75% correct response criterion. Therefore, the dolphin required an echo energy-to-noise ratio between 14 and 19 dB to perform the discrimination task. A good estimate of the threshold signal-to-noise for this discrimination task would be 16.5 dB, the average difference between 14 and 19 dB. If the average energy flux density per trial were used to compute the energy-to-noise ratio, each ratio shown in Figure 9.27 would be reduced by 4 dB, since the average SE per trial was approximately 4 dB lower than the maximum SE.

An example of an echo from the standard target in noise for a signal-to-noise ratio of 19 dB

is shown in Figure 9.28. The masking noise was windowed by a cosine taper (Otnes and Enochson 1978) with a width (between the half-power points) of 264 μs, corresponding to the integration time of *Tursiops*. The dashed curve is the spectrum of the echo from the standard target in a free field. All but the first two highlights in the time domain presentation are masked by noise. Most of the characteristics of the spectrum for the noise-free condition are lost with the addition of noise. The results shown in Figure 9.28 seem to suggest that both time domain or

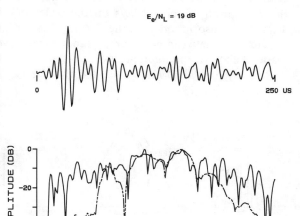

Figure 9.28. Example of echo from standard target in broadband noise with an energy-to-noise ratio of 19 dB. The dashed curve is the noise-free spectrum.

TSP cues may be more salient than frequency domain cues in discriminating the targets in noise.

9.3.5 Pyramid Steps

Bel'kovich et al. (1969) trained two *Delphinus delphis* to discriminate between a standard three-stepped pyramid and various comparison targets. The geometry of the standard and some of the comparison targets is shown in Figure 9.29. The standard pyramid was constructed of foam plastic squares steps, each 12 mm thick. The base step had an area of 100 cm^2, the middle, 49 cm^2, and the top step, 9 cm^2. The dolphins easily discriminated between the standard target and the single-layered targets and the two-stepped pyramid. When the area of the top step of a three-stepped comparison target was reduced to 8.4 cm^2 and then to 6.25 cm^2, the dolphins' performance fell to near 70% for both cases. The dolphins easily identified the comparison three-stepped pyramid when the thickness of the top two steps was reduced from 12 mm to 6 mm. When only the thickness of the middle step was reduced to 6 mm, the dolphins' discrimination performance dropped to 86.7%. By varying the thickness of the second and third steps, Bel'kovich et al. (1969) obtained a differential threshold in the perception of echo delays on the order of 4 μs. They also found that the dolphins could discriminate between equal-area planar targets made of ebonite and foam plastic.

Bel'kovich et al. (1969) concluded that "it is reasonable to assume that the dolphins distinguished between the figures (step thickness differences) by using the change in spectral composition of the reflected echo signals, including change in the relationship between the time of their return and the elements constituting the truncated stepped pyramid." The echoes from stepped pyramids should contain highlights associated with each step. The highlights should be highly correlated, and could produce TSP in the auditory system of the dolphins. Varying the step size will affect the time of arrival, and varying the area will affect the relative amplitude of the highlights.

In a follow-up experiment utilizing one of the common dolphins of the previous study, Bel'kovich and Borisov (1971) measured the ability of the animal to differentiate between a standard 10-cm square plate and 10-cm square comparison plates with different-sized square holes in them. The plates consisted of 3-mm thick Plexiglas covered with a 3.5-mm layer of resin. The dolphin could differentiate between the plates at a 10-m range for holes that were at least 6.2% of the plate area. The cues provided in this experiment were probably differences in target strength and the presence of secondary highlights caused by reflections off the edges of the circular holes. Bel'kovich and Borisov concluded however, that differences in reflectivity and in the frequency spectrum of the echo were the primary cues. The presence of secondary highlights for the squares with holes may have produced TSP and also caused the frequency spectrum of the echoes

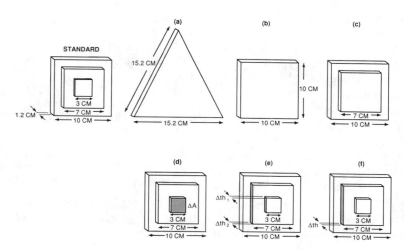

Figure 9.29. Geometry of the standard and comparison pyramid targets used by Bel'kovich et al. (1969).

to be rippled compared with the relatively smooth spectrum for the uncut squares, which would not have any TSP cues.

9.4 Target Shape Discrimination

9.4.1 Planar Targets

Barta (1969) trained a blindfolded dolphin to choose between circular squares and triangular aluminum disks covered with neoprene as targets, using the same two-alternative forced-choice paradigm as in the size discrimination experiment reported on earlier. The animal reliably distinguished circles from squares and triangles of the same cross-sectional area. Bagdonas et al. (1970) used targets made from ebonite (10 mm thick) and trained a *Delphinus delphis* to discriminate between a 100-cm² square and a 50 cm² triangle.

The dolphins in the experiments of Barta (1969) and Bagdonas et al. (1970) probably perceived noticeable changes in echo amplitude as they scanned across different-shaped targets. Such amplitude fluctuations could have provided the basis for the discrimination. Polar plots of the relative intensity of sound reflected from the different targets used by Barta (1969) are shown in Figure 9.30. It is clear that the different-shaped targets had different angular variations in the echoes. Differences in target strength also could have provided an additional cue in the targets

used by Bagdonas et al. (1970). There should have been about 6 dB difference between the 100-cm² square and the 50-cm² triangle.

9.4.2 Spheres and Cylinders

Au et al. (1980) conducted an experiment to determine if an echolocating dolphin could discriminate between foam spheres and cylinders located 6 m from a hoop station. Three spheres and five cylinders of varying size but overlapping target strength were used so that target strength differences would not be a cue. The dimensions of the spheres and cylinders along with their corresponding target strengths are listed in Table 9.5. Two spheres and two cylinders were used in

Table 9.5. Dimensions of the spheres and cylinders used in the shape discrimination experiment of Au et al. (1980) and their corresponding target strength

Target	Diameter	Length	Target Strength
Spheres			
S_1	10.2 cm	—	−32.1 dB
S_2	12.7 cm	—	−31.2 dB
S_3	15.2 cm	—	−28.7 dB
Cylinders			
C_1	1.9 cm	4.9 cm	−31.4 dB
C_2	2.5 cm	3.8 cm	−32.3 dB
C_3	2.5 cm	5.1 cm	−28.7 dB
C_4	3.8 cm	3.8 cm	−30.1 dB
C_5	3.8 cm	5.1 cm	−27.6 dB

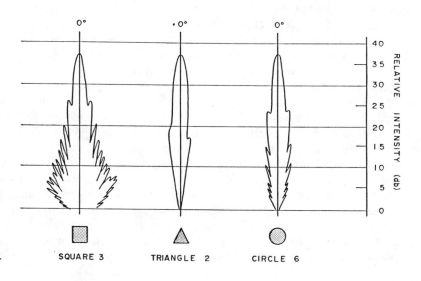

Figure 9.30. Polar plots of echo radiation patterns from the targets used by Barta (1969). (From Fish et al., 1976).

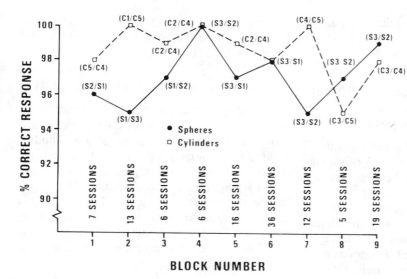

Figure 9.31. Summary of dolphin sphere–cylinder discrimination performance. (From Au et al. 1980.)

each 64-trial session, with one target present per trial. The dolphin was required to swim into a hoop station and "echolocate on command" when presented with a specific acoustic cue (Schusterman et al. 1980). Two response paddles, one associated with spheres and the other with cylinders, were used. The dolphin's performance is summarized in Figure 9.31 for different combinations of spheres and cylinders. Each block consisted of at least 5 and at most 36 sessions. The task was not difficult for the dolphin since it could perform the various discriminations at a correct response level of at least 94%.

We found that echoes returned to the animal by a direct path and an air-water surface-reflected path. Since the surface-reflected component was larger for the spheres than for the cylinders, we concluded that this reflection mechanism provided the major cue to the dolphin. However, when a sound-absorbing mat was introduced in a session to eliminate the surface-reflected component of the target echo, the dolphin still performed perfectly. Therefore, the dolphin probably performed the task based on differences in the characteristics of the echoes returning via the direct path. Examples of echoes from a foam sphere and a cylinder are shown in Figure 9.32. A possible cue involved the presence of circumferential waves following the specular reflection off the front surface of the targets. Such circumferential wave component can be seen in the echo

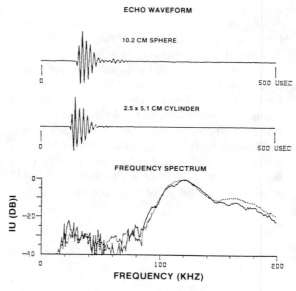

Figure 9.32. Results of backscatter measurements of a foam sphere and a foam cylinder used in the sphere–cylinder discrimination experiment of Au et al. (1980). The dotted frequency spectrum is for the cylinder echo.

from the sphere, but is not present in the echo from the cylinder. The cylinders had diameters that were much smaller than the diameters of the spheres; thus any circumferential wave component of the echoes would arrive shortly after the specular reflection and would not be separable from the specular reflection. Another possible cue involved the generation of the circumferential wave component in the targets. Circumferential waves are best generated when the longitudinal axis of the cylinder is exactly perpendicular to the incident signal. In a bay environment, wind and wave actions will constantly vary the alignment of the cylinders when submerged under water, producing conditions not very favorable for exciting circumferential waves. Finally, target strength for a finite-length foam cylinder increases logarithmically with frequency according to (9-8) and is constant with frequency for a sphere according to (9-9). However, this effect is probably not very significant since from Figure 9.32 the echo of the cylinder is greater than that of the sphere only for frequencies greater than approximately 150 kHz, the upper frequency limit of hearing for *Tursiops truncatus*.

9.4.3 Cylinders and Cubes

Nachtigall et al. (1980) trained a blindfolded echolocating *Tursiops* to discriminate between foam cylinders and cubes. The animal was trained to station in a circular tank and echolocate down a water-filled trough, as shown in Figure 9.33. Two targets (a cylinder and a cube) were presented simultaneously 2 m from the entrance to the trough; the animal was required to touch the response paddle on the side of the cylinder. Targets were placed between four 30-lb test monofilament lines attached with stretchable elastic cord to a wood frame, as shown in the diagram. Three different-sized cylinders were repeatedly paired with each of three different-sized cubes. Once the ability of the animal to distinguish cylinders from cubes was well established with the targets in an upright position, a probe technique was used to examine the effects of changing the target aspect. Baseline performance trials were conducted on 56 of the 63 trials per session, but on the other seven trials one of the targets was presented either on its side or with the flat face toward the dolphin, as shown in Figure 9.34. The experimental results shown in Figure 9.34 indicate that the dolphin could correctly discriminate between upright cylinders and cubes in 91% of the trials, and that two of the probe orientations did not affect the animal's ability to discriminate between targets. However, the results also show that the animal could not discriminate between targets when the probes were in the "flat-face-forward" orientation.

After the experiment, the targets were examined acoustically with a monostatic sonar measurement system using simulated dolphin sonar signals. The backscatter measurements indicated high variability in the echo amplitudes for the cubes and the cylinders in flat-face-forward position and low variability for the cylinders in upright position. The standard deviations of 15 target strength measurements for the cubes with flat faces forward were 2.7, 5.9, and 6.2 dB, compared with 1.2, 1.0, and 1.0 dB for the cubes standing upright. The targets were removed and placed back on the holder after each measurement to simulate the experimental situation.

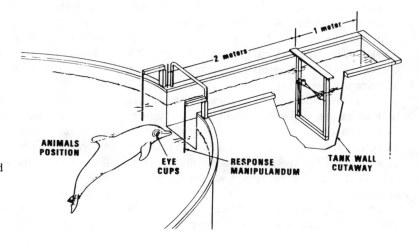

Figure 9.33. Dolphin tank and apparatus in the cylinder and cube discrimination experiment. (From Nachtigall et al. 1980.)

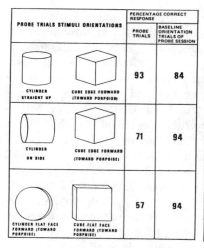

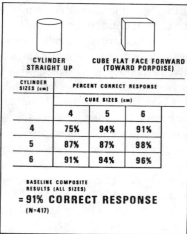

Figure 9.34. Results of cylinder–cube discrimination experiment, in terms of percent correct choice of the cylinder. (From Nachtigall et al. 1980.)

Measurements from cylinders with a flat face forward yielded similar results as measurements from cubes with a flat face forward. Standard deviations of 2.6, 4.9, and 5.9 dB were obtained. These differences in variability paralleled the performance of the dolphin. The animal most likely received echoes varying in amplitude when scanning across the flat surfaces of the cubes or the tops of the cylinders and received relatively uniform-amplitude echoes when scanning across the curved portion of the cylinders.

9.5 Delayed Matching to Sample

Roitblat et al. (1990) performed a *matching-to-sample* experiment with an echolocating *Tursiops* in which the dolphin was required to match targets from a sample array with targets from comparison arrays. A target set consisted of a large hollow PVC tube (10 cm O.D., 25 cm length, 0.3 cm wall), a small hollow PVC tube (7.5 cm O.D., 15 cm length, 0.3 cm wall), a solid aluminum cone (10 cm base, 10 cm length), and a 5-cm water-filled sphere. Since the targets differed in material composition, shape, dimensions, and structure, this experiment should be classified as a multidimensional discrimination experiment. The experimental configuration (Fig. 9.35) included a sample array and three comparison arrays, with an identical target set each. The sample array was located 4.8 m directly ahead of the animal's station and the comparison arrays

were suspended from a bar located 3.9 m from the observing aperture. A trial consisted of the blindfolded dolphin inserting its head into the observation aperture while a sample target from the sample array was gently lowered into the water. The dolphin began echolocating as soon as the acoustic shutter was lowered. The sample presentation ended when the dolphin ceased to emit signals for approximately five seconds. The acoustic screen was then raised and the sample target removed. Next, a target from each comparison array was lowered into the water; one of the targets matched the sample and two were different. The acoustic screen was again lowered and the dolphin echo-located on the comparison targets to determine which was the matching target. The animal responded by touching one of three response paddles to indicate the location of the matching target. One of the purposes of this delayed matching-to-sample experiment was to study the dolphin's decision-making process.

The dolphin's choice accuracy during the final 48 sessions of the experiment averaged 94.5% correct (Roitblat et al. 1990). Error choices were more frequent when the sphere was the sample. Errors were also more common when the matching target was from the right comparison array. When errors were committed, the dolphin more likely chose the middle array. The dolphin's typical search pattern consisted of examining the left target first, the middle one second, and finally the right target. Generally more clicks were emitted during the first scan when the matching

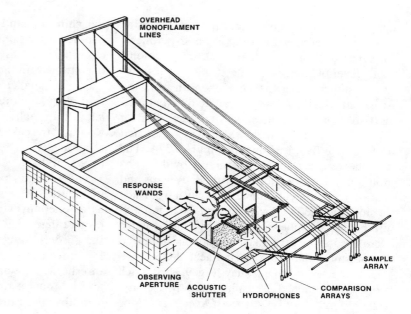

Figure 9.35. Diagram of the matching-to-sample apparatus. (From Roitblat et al. 1990 © Am. Psychol. Assoc. Reprinted by permission.)

stimulus was being observed then when a non-matching target was being scanned, suggesting that it was easier to reject a nonmatching then to identify a matching target. Roitblat et al. (1990) were able to develop a sequential sample model using Bayesian decision rules that responded to the matching-to-sample task in a similar manner as the dolphin.

The target echoes, collected again using simulated dolphin sonar signals, are shown in Figure 9.36. The discrimination task was not difficult for the dolphin since the target echoes were very discriminable in both the time and frequency domains, as can be seen from the figure. The use of easily discriminable targets was deliberate, since the purpose of the original experiment was to demonstrate that a dolphin could grasp the concepts of same–difference and matching-to-sample in a sonar modality (Nachtigall and Patterson 1981).

9.6 Target Range Difference Discrimination

Murchison (1980) conducted a study to determine the capability of an echolocating dolphin to indicate which of two targets was at closer range. Although Murchison (1980) considered his experiment to be a *range resolution* determination

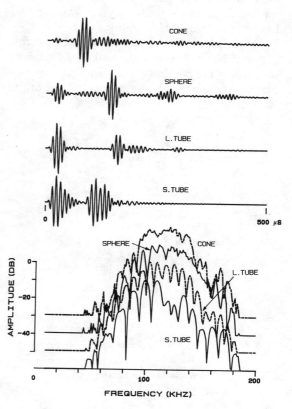

Figure 9.36. Echo waveforms and frequency spectra for the targets used in the delayed matching-to-sample experiment. The frequency spectra are offset along the vertical axis for easier viewing.

test, it should be more accurately classified as a range difference discrimination experiment. In sonar and radar terminology, range resolution generally refers to the minimum separation distance along the same azimuth from the sensor at which two targets can be differentiated as two separate entities. The configuration of Murchison's experiment, which was conducted in Kaneohe Bay, is shown in Figure 9.37A. The targets were separated in azimuth by 40°. The dolphin was trained to wear rubber eyecups and station in a chin cup that could swivel from side to side (Fig. 9.37B). Two identical 7.62-cm polyurethane foam spheres with internal lead weights were used as targets. Each target was stabilized by four monofilament lines connected to a cross that was attached to a trolley (Fig. 9.37 B). Each trolley was attached to an aluminum traveler and the amount of travel on the traveler could be adjusted by mechanical stops. The movement of the trolley was controlled by lines and pulley from the experimenter's station.

The dolphin was trained to station in the chin cup and begin its sonar scan when an acoustic screen was lowered out of the way. Upon completing its sonar scan, the animal backed out of the chin cup and responded by touching the paddle on the same side of the center line as the closer target. Sessions were conducted in blocks of 10 trials with the relative range difference, ΔR, becoming smaller after each block. Within a 10-trial block, a modified random schedule (right–left, 5 trials each) was used to specify which target would be the closer target. The dolphin's relative range acuity was tested for absolute target ranges of 1, 3, and 7 m.

The dolphin's performance results are shown in Figure 9.38, with percentage correct plotted as a function of ΔR for the different absolute target ranges. The results clearly indicate that as the absolute range increased, the dolphin's performance decreased. The 75% correct response thresholds were at ΔRs of 0.9, 1.5, and 3 cm for absolute target ranges of 1, 3, and 7 m. At the 1-m absolute range, the dolphin's performance approached the theoretical performance of a matched filter. The theoretical performance of a matched filter was determined by using one of the measured signals emitted by the dolphin and a technique used by Simmons (1969). The matched-filter response was determined using equation (9-7); the results are shown in Figure 9.39. The

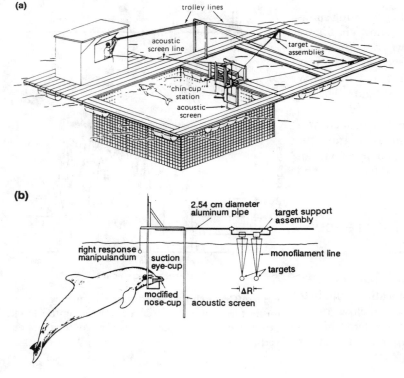

Figure 9.37. (*A*) Experimental configuration of the relative range determination experiment; (*B*) close-up view of chin cup station and target assembly. (From Murchison 1980.)

Figure 9.38. Target range difference discrimination results. (From Murchison 1980.)

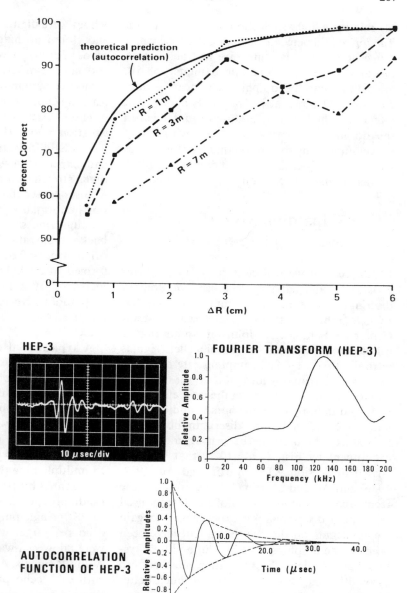

Figure 9.39. Echolocation pulse produced by the dolphin during the relative range discrimination experiment, along with its Fourier transform and autocorrelation function. (From Murchison 1980.)

echo from a foam sphere should resemble an inverted version of the outgoing signal, so that the matched-filter operation is equivalent to determining the autocorrelation function of the outgoing signal. Let $y(\tau)$ be the autocorrelation function defined by (9-7), and the theoretical performance of a matched filter can be expressed as

$$\mathrm{PC}(\tau) = 0.5 + (1 - Env|y(\tau)|) \quad (9\text{-}17)$$

where $\mathrm{PC}(\tau)$ is the probability of a correct response as a function of the delay time, and $Env|y(\tau)|$ is the envelope of the absolute value of the autocorrelation function. The theoretical performance curve is merely an inverted version of the envelope of the matched-filter response with two end points matched to 50% and 100% correct. The theoretical curve should apply to all of the absolute target ranges used by Murchison (1980) since the signal-to-noise ratio was probably relatively high. In Section 8.3, we saw that a *Tursiops* could detect a 7.62-cm O.D. sphere at a range of 113 m. The difference in transmission loss at 113 m and at 7 m is approximately 58 dB, indicating that the dolphin echolocating at a 7.62-cm O.D. target at 7 m probably operated at a relatively high signal-to-noise ratio. However, the dolphin's performance varied with target range, supporting the notion that a dolphin does not process echoes like an ideal or matched filter, as was discussed in Section 8.5.

9.7 Insights from Human Listening Experiments

In this section we will consider a different approach to analyzing targets used in dolphin discrimination experiments that was pioneered by C.S. Johnson at the Naval Ocean Systems Center in San Diego. Johnson found that if echoes obtained with ultrasonic click signals were digitized at a fast sampling rate and then reconverted to analog form at a slower sampling rate (essentially compressing the frequency spectrum and shifting it into the human audio range), humans could make fine discriminations. This procedure was used in the instrumented-diver experiment by Fish et al. (1976) that was discussed in Section 9.3.1. Martin and Au (1982, 1986) and Au and Martin (1989) used this same technique to examine some of the targets used in dolphin discrimination experiments, to gain some insights on the cue that might have been used by the dolphins. The human auditory system has excellent discrimination and pattern recognition capabilities and is still much better at analyzing complex sounds than any instrument or computer software presently available. Furthermore, as discussed in Chapter 3, various psychoacoustic experiments with *Tursiops truncatus* suggest that the inner ear of dolphins functions similarly to the human inner ear, but in a higher frequency range.

Target echoes for Au and Martin's experiments were collected using a computer-controlled monostatic echo measurement system which transmitted a broadband, dolphin-like echolocation signal. The incident signal had a duration of approximately 50 μs, a peak frequency of 122 kHz, and a 3-dB bandwidth of 39 kHz. Target echoes were digitized at a sample rate of 1 mHz and stored on magnetic tape. Ten consecutive echoes per target were normally stored on tape and later transferred to disk using a PDP-11 computer system which controlled the human listening experiments.

The subjects listened to signals in a sound isolation booth (Industrial Acoustics Co.) via Koss ESP-9B electrostatic headphones. Preliminary experiments with non-test echoes indicated that a stretch factor of 50 and a repetition rate of 4 pulses per second provided the best discrimination performance. The stretch factor is defined as the digitizing sample rate divided by the playback sample rate. With a stretch factor of 50, the original peak frequency of 122 kHz was transformed to 2.4 kHz, and the echo duration was increased by a factor of 50. The signal peak amplitudes were adjusted to be the same so that target strength would not be a discrimination cue.

A typical trial consisted of a subject being presented with prerecorded echoes from either one of two or one of four targets. Subjects were required to assign targets to one of two categories by pressing push-button switches labeled A and B. The stimulus was present for 15 seconds or until the subject responded. Correct response feedback was provided by lights labeled A and B. In multiple-ping (MP) sessions, each target was represented by 10 echoes but only one of them, randomly chosen, was used in a given trial. A few SP (SP) single-ping sessions using only a single echo per target were also conducted. Subjects were allowed a warm-up period to listen to the A and B signals. A session consisted of 64 trials, with each echo presented an equal number of times in random order.

9.7.1 Cylinder Discrimination

The first set of echoes was taken from the cylinders used in the Hammer and Au (1980) experiment. Only one subject was tested for the probe targets used in the general discrimination task shown in Figure 9.15 and described in Table 9.3, because the task was found to be trivial and no prior training was required (Martin and Au 1982). Multiple cues were available, since the echoes were dissimilar in both the frequency and time domain. A second test was conducted in which subjects had to discriminate between solid and hollow aluminum cylinders and between the large and small standard hollow aluminum cylinders. Four subjects without any prior learning had an average performance of 98% correct for the hollow versus solid aluminum cylinders. Martin and Au (1982) reported that the predominant cue was the longer duration of the hollow cylinder echoes. This difference was described as "click and hiss" for the hollow cylinder and "click only" for the solid cylinder. Time separation pitch cues were also reported although they were not as obvious as the duration cues. A 92% response accuracy was reported for discrimination between the small and large aluminum standards. Two dominant cues were reported: the longer duration of the large-cylinder echo and the higher TSP associated with the small-cylinder echo.

The next test involved the targets used in the material discrimination portion of the Hammer and Au (1980) and Schusterman et al. (1980) experiments discussed in Section 9.3.3. Targets were aluminum, steel, bronze, and glass cylinders with an outer diameter of either 3.81 or 7.62 cm and a length of 17.8 cm. The aluminum target echoes were always used as the reference echoes. The echo waveforms, frequency spectra, and matched-filter responses for the targets are displayed in Figures 9.17 and 9.18. The aluminum versus bronze discrimination was performed with two pairs of targets per material, each pair consisting of a 3.81-cm and a 7.62-cm O.D. cylinder. Single-ping data were used in such away that one of four echoes occurred on each trial and each echo was used in 16 trials per session, randomly distributed. The aluminum versus steel discrimination was performed in the same manner. Five subjects without any prior training participated in two sessions, or 128 trials, for each discrimination task. The aluminum versus glass discrimination task was performed under three different conditions: (a) single ping with the 3.81-cm O.D. targets followed by single ping with the 7.62-cm O.D. targets; (b) single ping with one of four targets; (c) multiple pings with the 3.81-cm O.D. targets followed by multiple pings with the 7.62-cm O.D. targets.

The average performance of three subjects in the aluminum versus bronze and the aluminum versus steel discrimination was 98% and 95%

correct, respectively. The subjects all reported that they first determined whether an echo originated from a large or small cylinder based on a duration cue. Echoes from a large cylinder had longer durations. Subjects reported that discrimination between the small aluminum and bronze cylinders was based on the presence of a lower TSP in the bronze than in the aluminum. From the envelope of the matched-filter response in Figure 9.16 we can see that the time separation between the first and second echo component was 52 μs for the small bronze cylinder and 45 μs for the small aluminum cylinder. After stretching of the signals by a factor of 50, the resulting TSP should be 385 Hz for the bronze and 444 Hz for the aluminum. The subjects reported that discrimination between the large aluminum and bronze cylinders was based on the presence of TSP with the aluminum cylinder and the absence of TSP with the bronze cylinder. Figure 9.16 shows interference between the second and third echo component for the bronze target, which may have affected the perception of TSP.

The subjects reported that the aluminum–steel discrimination was made on the basis of clearly perceptible TSP with the echoes of both the small and large aluminum cylinder. The presence of TSP was not as definite for the steel cylinders. The envelopes of the matched-filter responses in Figures 9.16 and 9.17 suggest that the aluminum targets should produce clearer TSPs since the first and second highlights were more highly corrected.

The results of the aluminum versus glass discrimination task are shown in Table 9.6. These results represent data obtained after the subjects' performance stabilized. The data indicate that all of the subjects could discriminate between aluminum and glass with performance accuracy varying between 72.3 and 97.9% correct. The subjects indicated that the echoes from the aluminum and glass targets sounded very similar and that the introduction of variances due to multiple pings made the task more difficult.

The reported discrimination cue was the difference in echo duration between the aluminum and glass echoes for both the small and large targets. From Figures 9.17 and 9.18 we can see that the echoes from the glass targets damped out sooner than the echoes from the aluminum targets.

Table 9.6. Results of the aluminum–glass discrimination task for different conditions (from Au and Martin 1989)

Subject	3.81-cm O.D. Cylinder		7.62-cm O.D. Cylinder	
	No. of Trials	% correct	No. of Trials	% correct
Single ping—one of two targets				
DM	192	94.3	256	94.5
KD	192	95.3	191	97.9
PT	318	87.7	256	93.4
DS	384	75.8	382	72.3
GP	384	74.7	382	74.9
Single ping—one of four targets				
DM	210	92.9	210	95.2
KD	139	96.2	125	97.0
PT	191	86.4	193	97.9
Multiple pings—one of two targets				
DM	384	85.2	384	94.3
KD	256	88.3	192	84.4
PT	384	74.0	384	78.4
GP	320	76.6	192	76.6

Visual inspection of the small-target echoes indicates that the glass echo damped out approximately 14 ms (0.28 ms before stretching) before the aluminum echo. For the larger targets, the glass echo damped out approximately 5 to 7 ms (0.10 to 0.14 ms before stretching) before the aluminum echo. It was mentioned previously that Schusterman et al. (1980) trained a dolphin to perform the small aluminum–glass discrimination, but could not train the animal to perform the large aluminum–glass discrimination. The dolphin probably did not perceive the duration difference if it integrated the echoes over a 264-μs window and ignored portions of the echoes beyond the integration period. It may also be possible that the animal could not detect duration cues because these cues are contained in the portions of the signals which are approximately 32 dB below the peak and may have been masked by the ambient noise of the bay. A third possibility for the dolphin is that the initial peaks in the echoes could have forward-masked later portions of the echo (Resnick and Feth 1975), since the total echo duration is approximately 0.5 to 1.0 ms.

A further examination of the aluminum versus glass discrimination was performed with echoes from the large targets systematically truncated between groups of echo highlights, as shown in Figure 9.40. The results of the experiment are plotted in Figure 9.41. Discrimination accuracy decreased as the signals were truncated, with the exception of one data point for subject PT. The results also demonstrate the importance of duration cues in the human listener experiment, since performance accuracy decreased when the signal durations were made the same upon the first truncation. Because the tail portion of the aluminum echo was approximately 32 dB below the level of the primary echo component, the subjects were probably using information over a 32-dB dynamic range before truncation.

Performance remained significantly above chance after the duration cue was eliminated upon the first truncation, and remained above chance with further truncations. The final truncation eliminated all but the first two echo components, yet the subjects were able to discriminate between signals above 70% correct. The time between the first and second echo components is virtually the same for both targets; thus, the discrimination probably was based on cues other than differences in TSP. The subjects indicated that the glass target had a slightly higher "click pitch" than the aluminum target when truncated signals were used. Click pitch is defined here as the pitch associated with the peak frequency of a broadband transient signal. This cue was difficult to extract and not always reliable. Examining the frequency spectra of Figure 9.40, we can see that

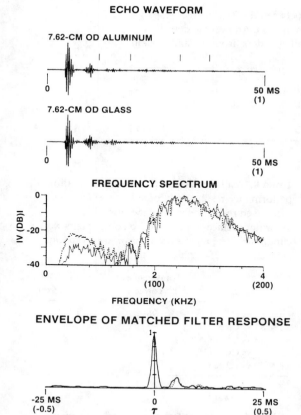

ECHO WAVEFORM

7.62-CM OD ALUMINUM

0   50 MS
                                                  (1)

7.62-CM OD GLASS

0   50 MS
                                                  (1)

FREQUENCY SPECTRUM

|V (DB)|

0

-20

-40

0   2   4
                     (100)   (200)

FREQUENCY (KHZ)

ENVELOPE OF MATCHED FILTER RESPONSE

-25 MS   0   25 MS
(-0.5)   T   (0.5)

Figure 9.40. Typical echo waveforms, frequency spectra, and matched-filter responses for the 7.62-cm aluminum and glass cylinders. The solid spectrum is for the aluminum cylinder, and the dotted spectrum is for the glass cylinder. The tic marks shown above the aluminum echo indicate where the signals were truncated. (From Au and Martin 1989.)

the minimum for frequencies above 1.8 kHz for the glass spectrum is approximately 67 Hz higher than that of the aluminum spectrum. The figure only shows the spectra of the total signals; however, the spectra for the first and second echo components were shown by Hammer and Au (1980) to be similar to the total echo spectrum.

Several of the material composition discrimination tasks were also performed in white noise. Performance of discrimination tasks in noise can be used to determine the difference in signal-to-noise (S/N) ratio between the point where echoes are just detectable and the point where they can be discriminated. This information is a direct measure of task difficulty and can give insight into the importance of particular discrimination cues. The differences between the discrimination and detection thresholds measured in the experiment with white noise are listed in Table 9.7. Simple tasks such as the aluminum–bronze and the 3.81-cm O.D. aluminum–glass discriminations required a S/N ratio of 7 to 11 dB above the detection threshold to obtain 75% correct. For the most difficult material discrimination, 7.62-cm O.D. aluminum versus glass cylinders, a S/N ratio of 21 to 30 dB above the detection threshold was required for 75% correct discrimination.

9.7.2 Sphere–Cylinder Discrimination

The foam spheres and cylinders used in the dolphin discrimination study of Au et al. (1980) were next used as targets with human listeners.

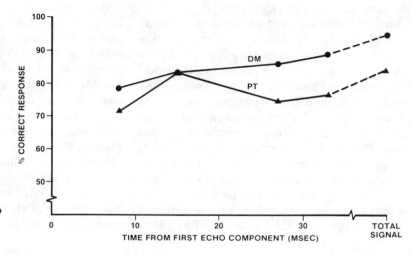

Figure 9.41. Discrimination performance results with the 7.62-cm aluminum and glass cylinders as a function of echo duration. (From Au and Martin 1989.)

100

90

80

70

60

50

% CORRECT RESPONSE

DM

PT

0   10   20   30   TOTAL
                                            SIGNAL

TIME FROM FIRST ECHO COMPONENT (MSEC)

Table 9.7. Difference in S/N ratio between the 75% correct response thresholds for detection and discrimination. An average detection threshold of 10.5 dB was used for all cylinders (from Au and Martin 1989)

Task	DM	PT	RB
Hollow aluminum vs glass; 7.62-cm O.D.	24 dB	30 dB	21 dB
Hollow aluminum vs glass: 3.82-cm O.D.	11 dB	—	—
Hollow aluminum vs bronze: 3.81-cm O.D.	7 dB	—	11 dB

Tests were conducted using both two-target (one sphere and one cylinder) and four-target (two of each) conditions. Discrimination experiments were also conducted with foam target echoes modified by applying a time window to the signals. This time window eliminated an air-water surface–reflected component from the echoes. Target sizes were chosen such that the target strengths of the two classes overlapped, eliminating target strength as a useful discrimination cue. An example of echoes from one of the foam spheres and one of the cylinders was shown in Figure 9.32.

Discrimination results pooled across the four subjects for the foam targets are shown in Table 9.8. The diameters of the foam spheres were given in Table 9.3. The average of the correct discrimination results varied between 84% and 96% correct for the unwindowed echoes and between 81% and 88% correct for the windowed echoes.

Subjects reported using two cues for these discriminations: a higher pitch for cylinder echoes and a larger low frequency reverberation for sphere echoes. The pitch difference probably occurred because the target strength of a finite cylinder at normal incident increases with frequency and is constant for a sphere (Urick 1983). The low frequency reverberation resulted from acoustic energy reflecting off the target toward the surface and bouncing off the air-water interface back toward the transducer, and from a circumferential echo component present for the spheres. It was previously suggested (Section 9.4.2) that a cylinder had to be aligned with its longitudinal axis perpendicular to the direction of the sonar signal, a condition that was difficult to achieve with the relatively lightweight cylindrical foam targets suspended by monofilament lines. For tests with echoes that had no surface-

Table 9.8. Sphere versus cylinder discrimination performance results with foam targets. The windowed results refer to echoes for which the air-water surface–reflected echo components were eliminated. The results are the average from four subjects, with 256 trials per subject

Task	Total Echoes (% correct)	Windowed Echoes (% correct)
S2 vs C4	96	88
S2 & S3 vs C3 & C4	93	85
S1 & S3 vs C1 & C5	88	81
S1 & S2 vs C4 & C5	84	—
S1 & S2 vs C2 & C4	91	83

reflected component, the subjects' discrimination performance dropped an average of 8% (windowed data of Table 9.8). However, performance exceeded 80% correct on all tasks considered.

9.7.3 Target Detection in Reverberation By Human Subjects

The clutter screen experiment of Au and Turl (1983) was simulated in the human listening experiment by taking echoes from the clutter screen and mixing them with echoes from the cylindrical targets. The human listener was required to specify whether a target was present or not during a trial. The target echo was aligned in time with the clutter screen echo and superimposed upon it to simulate a situation in which the target and clutter screen were in the same plane ($\Delta R = 0$). One of ten different clutter screen echoes chosen randomly for each trial and one of ten target echoes also chosen randomly were used in each target-present trial. The results of the detection in clutter experiment for two human subjects, along with the results of Au and

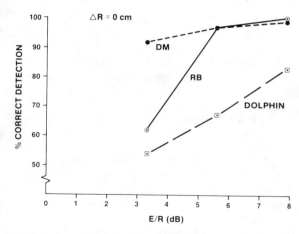

Figure 9.42. Performance results for two humans and a dolphin for target detection in clutter.

Turl (1983) for the dolphin, are shown in Figure 9.42. The cue used by the humans was the presence of a "click" sound from the target. The echoes from the clutter screen sounded diffuse whereas the echoes from the aluminum cylinders sounded like compact clicks with a definite TSP. Subjects seemed to have learned to integrate only over the duration of the target echo.

Martin and Au (1986) performed other discrimination experiments with human subjects that were peripherally but not directly related to specific dolphin experiments not discussed here. These human listening experiments have been very useful in providing insights into target discrimination which can be summarized as follows:

1. Time duration cues can be important even for signals approximately 30 dB below the peak amplitude.
2. TSP cues were present in almost all of the discrimination tasks except for the truncated aluminum versus glass cylinder and the sphere versus cylinder discriminations.
3. In cases in which TSP was not a factor, "click pitch" seemed to be a dominant cue.
4. Discrimination tests in noise showed that simple discrimination tasks required S/N ratios about 10 dB above the detection threshold for 75% correct discrimination. Difficult discrimination may require 30 dB more signal than noise at the detection threshold.

5. TSP cues can be useful in detecting a specific echo waveform amongst clutter.

9.8 Summary

Dolphins have a keen ability to perceive subtle differences in targets with their sonar. This capability is in part the result of being able to recognize differences on the order of 1 dB in the amplitude of echoes. The use of broadband short-duration transient-like sonar signals that can encode important target information also plays an important role in the dolphins' discrimination capabilities. Echoes from underwater targets usually have complex structures containing many highlights. Complex echo structures are the result of specular reflections from the front surface of a target combined with internal reflections that can propagate along different paths, reflections from different parts of a target, and contributions from circumferential waves traveling around a target. Dolphin sonar signals are usually short enough in duration so that these highlights are distinct and resolvable. Highlights convey important target information and are used by dolphins in discriminating between targets. The ability to scan across targets and also to ensonify targets from different angles is important in discriminating between targets of aspect-dependent reflectivity. Plausible accounts can be given of the strategies by which dolphins were able to perform the sonar discrimination tasks in most if not all experiments conducted. However, in situations where the major cues were derived mainly from the echo structure, we cannot weigh the relative importance of time domain, frequency domain, and TSP cues, assuming that dolphins can perceive TSP. Furthermore, the data are insufficient to quantify the properties of the basic auditory processes associated with discrimination between broadband click signals.

References

Abramov, A.P., Ayrapet'yants, E.S., Burdin, V.I., Golubkov, A.G., Yershova, I.V., Zhezherin, A.R., Krolev, V.I., Malyshev, Y.A., Ul'yanov, G.K., and Fradkin, V.B. (1971). Investigation of delphinid capacity to differentiate between three-dimensional objects according to linear size and material. In:

Summaries of Papers Delivered at the 7th All-Union Acoustics Conference, Leningrad, p. 3.

Au, W.W.L., and Hammer, C.E., Jr. (1980). Target recognition via echolocation by *Tursiops truncatus*. In: R.G. Busnel and J. F. Fish, eds., *Animal Sonar Systems*. New York: Plenum Press, pp. 855–858.

Au, W.W.L., and Martin, D.W. (1988). Sonar discrimination of metallic, plates. In: P.E. Nachtigall and P.W.B. Moore, eds., *Animal Sonar: Processes and Performance*. edited by P.E. New York: Plenum Press pp. 809–813.

Au, W.W.L., and Martin, D.W. (1989). Insights into dolphin sonar discrimination capabilities from human listening experiment. J. Acoust. Soc. Am. 86: 1662–1670.

Au, W.W.L., and Pawloski, D.A. (1992). Cylinder wall thickness discrimination by an echolocating dolphin. J. Comp. Physiol. A 172: 41–47.

Au, W.W.L., and Turl, C.W. (1991). Material composition discrimination of cylinders at different aspect angles by an echolocating dolphin. J. Acoust. Soc. Am. 89: 2448–2451.

Au, W.W.L., Schusterman, R.J., and Kersting, D.A. (1980). Sphere-cylinder discrimination via echolocation by *Tursiops truncatus*. In: R.G. Busnel and J.F. Fish, eds., *Animal Sonar Systems*. New York: Plenum Press, pp. 859–862.

Ayrapet'yants, E.S., and Konstantinov, A.I. (1974). *Echolocation in Nature*. Leningrad: Nauka.

Ayrapet'yants, E.S., Golubkov, A.G., Yershova, I.V., Zhezherin, A.R., Zvorkin, V.N., and Korolev, V.I. (1969). Echolocation differentiation and characteristics of radiated echolocation pulses in dolphins. Report of the Academy of Science of the USSR 188: 1197–1199.

Babkin, V.P., Dubrovskiy, N.A., Krasnov, P.S., and Titov, A.A. (1971). Discrimination of material of spherical targets by the bottlenose dolphin. In: Summaries of Papers Delivered at the 7th All-Union Acoustics Conference, Leningrad, p. 5.

Bagdonas, A.P., Bel'kovich, V.M., and Krushinskaya, N.L. (1970). Interaction between delphinid analyzers in discrimination. J. Higher Neural Act. 20: 1070–1074.

Barnard, G.R., and McKinney, C.M. (1961). Scattering of acoustic energy by solid and air-filled cylinders in water. J. Acoust. Soc. Am. 33: 226–238.

Barta, R.E. (1969). Acoustical pattern discrimination by an Atlantic bottle-nosed dolphin. Unpublished manuscript, Naval Undersea Center, San Diego, Cal.

Bel'kovich, V.M., and Borisov, V.I. (1971). Locational discrimination of complex configuration by dolphins. Tru. Akust. Inst. (Moscow) 17: 19–23.

Bel'kovich, V.M., and Dubrovskiy, N.A. (1976). *Sensory Bases of Cetacean Orientation*. Leningrad: Nauka.

Bel'kovich, V.M., Borisov, I.V., Gurevich, V.S., and Krushinskaya, N.L. (1969). Echolocating capabilities of the common dolphin (*Delphinus delphis*). Zool. Zhurn. 48: 876–883.

Bilsen, F.A. (1966). Repetition pitch: monaural interaction of a sound with the same but phase-shifted sound. Acustica 17: 295–300.

Brigham, E. O. (1988). *The Fast Fourier Transform and Its Applications*. Englewood Cliffs, N.J.: Prentice Hall.

Burdic, W.S. (1968). *Radar Signal Analysis*. Englewood Cliffs, N.J.: Prentice Hall.

Diercks, K.J., Goldsberry, T.G., and Horton, C.W. (1963). Circumferential waves in thin-walled air-filled cylinders in water. J. Acoust. Soc. Am. 35: 59–64.

Dubrovskiy, N.A. (1972). Discrimination of objects by dolphins using echolocation. Report of the 5th All-Union Conf. on Studies of Marine Mammals, Makhachkala, Part 2.

Dubrovskiy, N.A. (1990). On the two subsystems of auditory perception in *Tursiops truncatus*. In: J.A. Thomas and R. Kasterlein, eds., *Cetacean Sensory Systems: Field and Laboratory Evidences*. New York: Plenum Press, pp. 233–254.

Dubrovskiy, N.A., and Fadeyeva, L.M. (1973). Discrimination of spherical targets by delphinids. Tez. dokl. 4-y Vses. bion. konf., Moscow, pp. 29–34.

Dubrovskiy, N.A., and Krasnov, P.S. (1971). Discrimination of elastic spheres according to material and size by the bottlenose dolphin. Tr. Akust. Inst. (Moscow) 17: 9–18.

Dubrovskiy, N.A., Titov, A.A., Krasnov, P.S., Babkin, V.P., Lekomtsev, V.M., and Nikolenko, G.V. (1970). Investigation of resolution of echolocating system of the Black Sea bottlenose dolphin. In *Tr. Akust. Inst.* (Moscow) 10: 163–181.

Dubrovskiy, N.A., Krasnov, P.S., and Titov, A.A. (1971). Discrimination of solid elastic spheres by an echolocating porpoise, *Tursiops truncatus*. In: Proc. 7th Intern. Conf. Acoust., Budapest.

Evans, W.E. (1973). Echolocation by marine delphinids and one species of fresh-water dolphin. J. Acoust. Soc. Am. 54: 191–199.

Evans, W.E., and Powell, B.A. (1967). Discrimination of different metallic plates by an echolocating delphinid. In: R.G. Busnel, ed. *Animal Sonar Systems: Biology and Bionics*. Laboratoire de Physiologie Acoustique, Jouy-en-Josas, France, pp. 363–383.

Fadeyeva, L.M. (1973). Discrimination of spherical targets with different echo signal structure by the dolphin. Report of the 8th All-Union Acoust. Conf., Moscow.

Fish, J.F., Johnson, C.S., and Ljungblad, D.K. (1976). Sonar target discrimination by instrumented human divers. J. Acoust. Soc. Am. 59: 602–606.

Golubkov, A.G., Yershova, M.V., Zhezherin, A.R., and Fradkin, V.B. (1973). Experimental study of maximum range of discrimination of spheres by marine animals. Summary of Papers Delivered at the 4th All-Union Conference on Bionics, Moscow, Vol. 1, pp. 27–28.

Hammer, C.E., Jr. and Au, W.W.L. (1980). Porpoise echo-recognition: an analysis of controlling target characteristics. J. Acoust. Soc. Am. 68: 1285–1293.

Hickling, R. (1962). Analysis of echoes from a solid elastic sphere in water. J. Acoust. Soc. Am. 34: 1582–1592.

Johnson, C. S. (1967). Discussion. In: R.G. Busnel, ed., *Animal Sonar Systems: Biology and Bionics*. Laboratoire de Physiologie Acoustique, Jouy-en-Josas, France, pp. 384–398.

Mackay, R.S. (1967). Experiments to conduct in order to obtain comparative results. In: R.G. Busnel, ed., *Animal Sonar Systems: Biology and Bionics*. Laboratoire de Physiologie Acoustique, Jouy-en-Josas, France, pp. 1173–1196.

Martin, D.W., and Au, W.W.L. (1982). Aural discrimination of targets by human subjects using broadband sonar pulses. San Diego, Cal.: Naval Ocean Systems Center Techn. Rep. 847.

Martin, D.W., and Au, W.W.L. (1986). Broadband sonar classification cues. San Diego, Cal.: Naval Ocean Systems Center Techn. Rep. 1123.

McClellan, M.E., and Small, A.M. (1965). Time-Separation Pitch Associated with Correlated Noise Burst. J. Acoust. Soc. Am. 38: 142–143.

Murchison, A. E. (1980). Detection range and range resolution of porpoise. In: R.G. Busnel and J. F. Fish, eds., *Animal Sonar Systems*. New York: Plenum Press, pp. 43–70.

Nachtigall, P.E. (1980). Odontocete echolocation performance on object size, shape and material. In: R.G. Busnel and J.F. Fish, eds., *Animal Sonar Systems*. New York: Plenum Press, pp. 71–95.

Nachtigall, P. W., and Patterson, S.A. (1981). Echolocation and concept formation by an Atlantic bottlenosed dolphin: sameness-difference and matching-to-sample. (abstract). Fourth Biennial Conf. on the Biol. of Mar. Mamm., San Francisco, Cal.

Nachtigall, P.E., Murchison, A.E., and Au, W.W.L. (1980). Cylinder and cube discrimination by an echolocating blindfolded bottlenose dolphin. In: R.G. Busnel and J.F. Fish, eds., *Animal Sonar Systems*. New York: Plenum Press pp. 945–947.

Neubauer, W.G. (1986). *Acoustic Reflection from Surfaces and Shapes*. Washington D.C.: Naval Research Laboratory.

Otnes, R.K., and Enochson, L. (1978). *Applied Time Series Analysis*, Vol. 1. New York: John Wiley and Sons.

Resnick, S.B., and Feth, L.L. (1975). Discriminability of time-reversed click pairs: intensity effects. J. Acoust. Soc. Am. 57: 1493–1499.

Rihaczek, A.W. (1969). *Principles of High-Resolution Radar*. New York: McGraw-Hill.

Roitblat, H.L., Penner, R.H., and Nachtigall, P.E. (1990). Matching-to-sample by an echolocating dolphin. J. Exp. Psych.: Anim. Beh. Proc. 16: 85–95.

Schusterman, R.J., Kersting, D.A., and Au, W.W.L. (1980). Stimulus control of echolocation pulses in *Tursiops truncatus*. In: R.G. Busnel and J.F. Fish, eds., *Animal Sonar Systems*. New York: Plenum Press pp. 981–982.

Shirley, D.J., and Diercks, K.J. (1970). Analysis of the frequency response of simple geometric targets. J. Acoust. Soc. Am. 48: 1275–1282.

Simmons, J.A. (1969). "Depth Perception by Sonar in the Bat *Eptesicus fuscus*," Ph.D. Dissertation, Princeton University.

Small, A.M., and McClellan, M.E. (1963). Pitch associated with time delay between two pulse trains. J. Acoust. Soc. Am. 35: 1246–1255.

Titov, A.A. (1972). Investigation of sonic activity and phenomenological characteristics of the echolocation analyzer of Black Sea delphinids. Canditorial dissertation Karadag.

Turner, R.N., and Norris, K.S. (1966). Discriminative echolocation in a porpoise. J. Exp. Anal. Behav. 9: 535–544.

Uberall, H., Doolittle, R.D., and McNicholas, J.V. (1966). Use of sound pulses for a study of circumferential waves. J. Acoust. Soc. Am. 39: 564–578.

Urick, R.J. (1983). *Principles of Underwater Sound*. New York: McGraw-Hill.

Weisser, F.L., Diercks, K.J., and Evans, W.E. (1967). Analysis of short pulse echoes from copper plates. J. Acoust. Soc. Am.42: 1211 (A).

Wille, P. (1965). Experimentelle Untersuchungen zur Schallstreuung an schallweichen objekten. Acustica 15: 11–25.

Yershova, I.V., Zhezherin, A.R., and Ignat'yeva, V.A. (1973). Differences in echosignals used by delphinids to discriminate elastic spheres. In: Ref. Dokl. Vol. 8, Akust. Konf., Moscow, pp. 19–23.

Zaslavskiy, G.L., Titov, A.A., and Lekomtsev, V.M. (1969). Investigation of sonar capabilities of the common porpoise. In: Tr. Akust. Inst. (Moscow) 8: 134–138.

10

Signal Processing and Signal Processing Models

In this chapter we will examine some of the characteristics of the dolphin sonar signal from a signal processing point of view. Several sections will also be devoted to the discussion of signal processing techniques and the construction of mathematical models and electronic analogs to the dolphin sonar signal processing system. These models and analogs fall into the category called *bionic sonar*. Although there are still many large gaps in our understanding of biosonar processes, sufficient knowledge is available to allow the construction of simple mathematical models of biosonar signal processing that can be implemented in computer software. These models will not explain how dolphins process acoustic information but will instead explain how acoustic signals may be processed in order to obtain hardware and software systems with equivalent capabilities as the dolphin sonar system. These models on the whole are relatively simplistic and may match a dolphin's capabilities only in a limited context under ideal conditions. After all, even with many years of research in human auditory processing, we still do not have many good mathematical models that can be used to accurately perform a seemingly simple task such as speech recognition. Therefore it would be presumptuous and naive to expect that our understanding of the dolphin auditory system is sufficiently advanced to allow us to create sophisticated models of the dolphin sonar. Nevertheless, in recent years great strides have been made,

and continue to be made, in the fields of signal processing and artificial intelligence. These advances increase our ability to model or simulate biological processes through mathematical constructs and on the computer. One area that has attracted considerable attention in recent years is the theoretical development of artificial neural networks, along with appropriate mathematical development, computer simulation, and solid state semiconductor implementation. More will be said about this subject toward the end of this chapter.

10.1 Analysis of Dolphin Sonar Signals

In this section, some concepts from radar signal processing will be used to analyze the properties of dolphin sonar signals. We will assume that echoes are the result of reflection from point targets so that their waveforms will be the same as those of the transmitted signals, but with lower amplitudes. We will also assume that echoes are received and processed by a matched-filter receiver. We have already established the fact that dolphins do not actually process echoes like a matched filter (Chapter 8); however, the comparison can be helpful in understanding some of the ideal capabilities of dolphin sonar signals. For a deeper study of radar signal processing, readers are referred to books by Burdic (1968) and Rihaczek (1969).

10.1.1 Center Frequency and RMS Bandwidth

The frequency characteristics of dolphin sonar signals were described in Chapter 7 in terms of peak frequency and 3-dB bandwidth. Both parameters were relatively easy to determine. The peak frequency was defined as the frequency at which the spectrum has its maximum value. The 3-dB bandwidth was defined as the frequency band between the lower and upper halfpower (3 dB down from the maximum value) points in the frequency spectrum. In sonar signal analysis, a broadband signal is often described by its *center*, or *centroid*, *frequency* instead of its peak frequency. Let $s(t)$ be the signal waveform and $S(f)$ its Fourier transform. The center frequency of the signal is defined as

$$f_0 = \langle f \rangle = \frac{\int_{-\infty}^{\infty} f |S(f)|^2 \, df}{\int_{-\infty}^{\infty} |S(f)|^2 \, df} \qquad (10\text{-}1)$$

The center frequency, f_0, divides the energy of the spectrum into two equal parts, so that the energy in the upper frequency portion of the spectrum is equal to the energy in the lower frequency portion. Equation (10-1) can also be expressed in terms of energy density since by the Rayleigh and Parseval Theorem in Fourier Analysis (Bracewell 1978)

$$E = \int_{-\infty}^{\infty} |S(t)|^2 \, dt = \int_{-\infty}^{\infty} |S(f)|^2 \, df \qquad (10\text{-}2)$$

However, any actual physical waveform $s(t)$ is real, and its energy spectrum is an even function of frequency and is fully determined by its values for $f \geq 0$ (Bracewell 1978). Therefore, we can rewrite (10-1) as

$$f_0 = \frac{\int_0^{\infty} f |S(f)|^2 \, df}{\int_0^{\infty} |S(f)|^2 \, df} = \frac{1}{E/2} \int_0^{\infty} f |S(f)|^2 \, df \qquad (10\text{-}3)$$

The bandwidth of a signal can be described by first defining the mean square value of the function $S(f)$ as (Bracewell 1978)

$$\langle f^2 \rangle = \frac{\int_{-\infty}^{\infty} f^2 |S(f)|^2 \, df}{\int_{-\infty}^{\infty} |S(f)|^2 \, df} \qquad (10\text{-}4)$$

The *mean square bandwidth*, β^2, about the centroid of $|S(f)|^2$ can now be expressed as

$$\beta^2 = \frac{\int_{-\infty}^{\infty} (f - f_0)^2 |S(f)|^2 \, df}{\int_{-\infty}^{\infty} |S(f)|^2 \, df} \qquad (10\text{-}5)$$

which reduces to

$$\beta^2 = \frac{\int_0^{\infty} f^2 |S(f)|^2 \, df}{\int_0^{\infty} |S(f)|^2 \, df} - f_0^2 = \langle f^2 \rangle - f_0^2 \qquad (10\text{-}6)$$

The peak and center frequency, the 3-dB and *rms bandwidth*, β, and the *time resolution constant* of the signals depicted in Figures 7.5 and 7.17 are given in Table 10.1. (The time resolution constant will be defined in the next subsection and is given by equation (10-22).) Except for Sven, the center frequency of the *Tursiops truncatus* signals in Figure 7.5 tends to be approximately 10 to 15 kHz lower than the peak, which is not surprising since each spectrum appears to drop off faster for frequencies greater than f_p than for frequencies less than f_p, causing the centroid to be shifted to the left (lower frequencies) of f_p. However, the reverse is true for the longer duration, narrow-band signals used by the smaller dolphins (signals

Table 10.1. Signal parameters of the dolphin sonar signals shown in Figure 7.5 and 7.17. f_p, peak frequency; f_0, center frequency; BW, 3-dB bandwidth; β, rms bandwidth; $\Delta\tau$, time resolution constant

Dolphin	f_p(kHz)	f_0(kHz)	BW(kHz)	β(kHz)	$\Delta\tau(\mu s)$
T.T. (Ekiku)	115.2	101.5	41.0	21.5	15.3
T.T. (Ekahi)	115.2	93.1	41.0	28.1	11.8
T.T. (Heptuna)	117.2	101.3	44.9	21.5	14.5
T.T. (Sven)	121.1	124.7	39.1	21.4	13.6
Phocoena P.	136.7	131.2	10.7	16.6	35.9
Phocoenoides D.	127.0	130.7	10.7	17.0	38.8
Cephalorhynchus C.	114.3	123.1	18.0	15.8	26.4
Cephalorhynchus H.	117.2	116.6	9.8	25.9	27.2

from Fig. 7.17). The rms bandwidth is also smaller than the respective 3-dB bandwidth for all the *Tursiops* signals, but not for the other, smaller dolphin species.

10.1.1 Accuracy in Target Range Determination

An important piece of information a sonar should obtain from the received signal is the target range. One approach to determining the target range is to detect the envelope of the received echo (Section 9.1 discussed a procedure to mathematically determine the envelope of a signal) and measure the time difference t_0 between two corresponding points on the transmitted and received envelope, as depicted in Figure 10.1A. In the absence of any noise, this process can be performed very accurately. However, the presence of noise will distort the received waveform, as shown in Figure 10.1B, so that corresponding points on the envelopes of the transmitted signal and of the received echo cannot be accurately identified. The time of arrival of the noisy echo can be determined by shifting a replica of the transmitted signal along the time axis and calculating the position for which the integrated squared difference between this shifted waveform and the received noisy echo is a minimum. This procedure is equivalent to finding the best fit solution in comparing the two waveforms. Let the transmitted signal and the received noise be analytic (see Section 9.1) so that they can be expressed as $s_a(t)$ and $n_a(t)$, respectively. The received echo plus noise can be expressed as

$$r(t) = As_a(t) + n_a(t) \tag{10-7}$$

where $A \ll 1$ because of transmission and reflection losses. The time of arrival of the echo can be determined by finding the minimum of the integrated squared difference between the received echo and the shifted transmitted waveform,

$$\varepsilon^2 = \int_0^T |s_a(t + \tau) - r(t)|^2 \, dt \tag{10-8}$$

The upper limit of integration, T, represents the maximum range of interest and is assumed to be large compared with the signal duration. Expanding (10-8) we obtain

$$\varepsilon^2 = K - 2Re\left[\int_0^T s_a^*(t + \tau)r(t) \, dt\right] \tag{10-9}$$

where $s_a^*(t + \tau)$ is the complex conjugate of $s_a(t + \tau)$ and K is a constant based on the energy in the transmitted and received signal. ε will be minimum at the value of τ for which the integral has a maximum positive value. If t_0 is the delay time between the transmitted and received signal and τ_1 is the estimate of the delay time based on minimizing ε, the standard deviation of the error in estimating t_0 can be expressed as (Woodward 1953)

$$\sigma_\tau = (\tau_1 - t_0)_{rms} = \frac{1}{2\pi\beta_0\sqrt{2\dfrac{E}{N_0}}} \tag{10-10}$$

where β_0 is the rms bandwidth of (10-5), E is

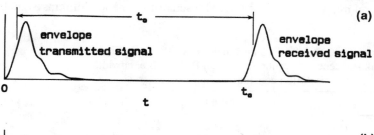

(a)

envelope transmitted signal

envelope received signal

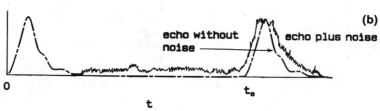

(b)

echo without noise

echo plus noise

Figure 10.1. (*A*) Relationship between the envelopes of the transmitted signal and received echo in the absence of noise; (*B*) envelope of received echo in the presence of noise.

the energy flux density and $N_0/2$ is the noise spectral density for the real noise waveform. Since $t_0 = 2R/c$, the standard deviation in the range estimation error can be expressed as

$$\sigma_R = \frac{1}{2\pi\beta_0 \sqrt{2\dfrac{E}{N_0}}} \frac{c}{2} \qquad (10\text{-}11)$$

The uncertainty in estimating range is inversely proportional to the product of the rms bandwidth and the square root of the signal-to-noise ratio.

In Section 9.5 we discussed the range determination experiment of Murchison (1980) involving a *Tursiops truncatus* named Heptuna. Typical values of the various parameters of Heptuna's sonar signals were shown in Table 10.1. Using Heptuna's signal shown in Figure 7.5, which has an rms bandwidth of 21.5 kHz, and processing the echo from a point target with a matched-filter receiver, the uncertainty in estimating the range from equation (10-11) is plotted in Figure 10.2 as a function of the signal-to-noise ratio. The results indicate that the dolphin sonar signal has a good range determination capability. For an E/N_0 of 10 dB, a matched-filter receiver would have an uncertainty of only 0.13 cm in determining range.

Let us now examine the relative range accuracy experiment of Murchison (1980) discussed in Section 9.5. The echo signal-to-noise ratio in dB can be calculated by using (8-12), which is rewritten here as

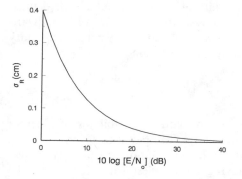

Figure 10.2. Range error estimate for a matched filter using Heptuna's sonar signals.

$$10\log(E_e/N_0) = SE - 2TL + TS - (NL - DI) \qquad (10\text{-}12)$$

Measurements of Heptuna's signal when the dolphin was performing the range accuracy task at 1 m indicated an average SL of 181 dB, which from (8-16) gives an SE of 123 dB. The target strength of a soft sphere can be calculated from the expression (Urick 1983)

$$TS = 20\log\left(\frac{d}{4}\right) \qquad (10\text{-}13)$$

where d is the diameter in meters. This expression results in a target strength of -34 dB for a 7.62-cm diameter sphere. From (8-4), the receiving directivity index for a peak frequency of 117 kHz is approximately 21 dB. The ambient noise level in Kaneohe Bay is found in Figure 1.5; at a peak frequency of 117 kHz it is equal to approximately 55 dB. Inserting the appropriate values into (10-12), we obtain $E_e/N_0 \approx 54$ dB. With such a high signal-to-noise ratio, a matched-filter receiver would be extremely accurate in determining target range, much more accurate than the dolphin.

10.1.3 Range Resolution

The *range resolution* capability of a sonar is another important parameter in evaluating its effectiveness for target identification and recognition. Range resolution involves resolving the echoes from two targets (or two highlights from an extended target) that are along the same azimuth with respect to the sonar. The range resolution capability of a signal is the minimum distance between two point targets at which they can be separated. At smaller distances, it is impossible to determine whether one or two targets (or highlights) are being detected by the receiver. Let $s_a(t)$ be the transmitted target and let the echo from two stationary point targets be $A \cdot s_a(t + t_0)$ and $A \cdot s_a(t + t_0 + \tau)$, where $t = 0$ is the time origin for transmission, t_0 is the round-trip delay for the first target, and τ is the difference in delay between the two echoes. We can use the integrated square magnitude of the difference between the two echoes as a measure of the distinctness of the two returns:

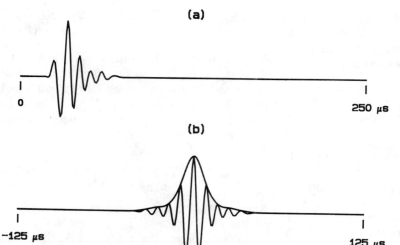

(a)

I
0

250 µs

(b)

I
−125 µs

I
125 µs

Figure 10.3. Example of (A) *Tursiops truncatus* (Heptuna) dolphin sonar signal; (B) auto-correlation function and its envelope.

$$\varepsilon^2 = A^2 \int_{-\infty}^{\infty} |s_a(t + t_0) - s_a(t + t_0 + \tau)|^2 \, dt$$

$$(10\text{-}14)$$

Expanding (10-14), we have

$$\varepsilon^2 = A^2 \int_{-\infty}^{\infty} (|s_a(t + t_0)|^2 + |s_a(t + t_0 + \tau)|^2) \, dt$$

$$- A^2 \int_{-\infty}^{\infty} [s_a(t + t_0)s_a^*(t + t_0 + \tau)$$

$$- s_a^*(t + t_0)s_a(t + t_0 + \tau)] \, dt$$

$$(10\text{-}15)$$

The first integral is the sum of the energy of both echoes and is equal to $2A^2$ times the energy in the transmitted signal. Therefore, (10-15) can be reduced to

$$\varepsilon^2 = 2A^2 \int_{-\infty}^{\infty} |s_a(t)|^2 \, dt$$

$$- 2A^2 \, Re \int_{-\infty}^{\infty} s_a(t)s_a^*(t + \tau) \, dt$$

$$(10\text{-}16)$$

The first integral is a constant and the second integral is the *autocorrelations function* of $s_a(t)$. The difference between the echoes will be at a minimum when the autocorrelation function is at a maximum. The second integral is defined as the *"range ambiguity function,"* $c(\tau)$:

$$c(\tau) = \int_{-\infty}^{\infty} s_a(t)s_a^*(t + \tau) \, dt \qquad (10\text{-}17)$$

The degree of target resolution is measured by $|c(\tau)|^2$. The maximum value of $|c(\tau)|$ occurs at $\tau = 0$, and if $|c(\tau)| = c(0)$ for any value of $\tau \neq 0$, then the two targets cannot be resolved. If $|c(\tau)|$ is close to $c(0)$ for some τ, then the two targets can be resolved only with difficulty. Equation (10-17) can be expressed in a form more convenient for computation by using the Correlation Theorem in Fourier Analysis (Burdic 1968),

$$c(\tau) = \mathfrak{I}^{-1}[S_a(f)S_a^*(f)] \qquad (10\text{-}18)$$

An example of a dolphin sonar click, with its autocorrelation function and corresponding envelope, is shown in Figure 10.3. The time resolution capability of a sonar signal can be defined by the *"time resolution constant"* of Woodward (1953), which is related to the range ambiguity function by the expression

$$\Delta\tau = \frac{\int_{-\infty}^{\infty} |c(\tau)|^2 \, d\tau}{c^2(0)} \qquad (10\text{-}19)$$

The relative resolution capabilities of different waveforms can be compared by using (10-19) to calculate the time resolution constant. The waveform having the smallest value for $\Delta\tau$ can be assumed to have the greatest potential for resolving two signals in time. From the Correlation

Theorem we derive (Bracewell 1978)

$$\Im[c(\tau)] = S_a^*(f)S_a(f) = 4|S(f)|^2 \quad (10\text{-}20)$$

From the Parseval Theorem, the denominator of (10-19) can be expressed as

$$c(0) = \int_{-\infty}^{\infty} |s_a(t)|^2\, dt = 2\int_{0}^{\infty} |S(f)|^2\, df \quad (10\text{-}21)$$

Substituting (10-20) and (10-21) into (10-19), we obtain for the Woodward time resolution constant

$$\Delta\tau = \frac{\int_0^{\infty} |S(f)|^4\, df}{[\int_0^{\infty} |S(f)|\, df]^2} \quad (10\text{-}22)$$

The time resolution constant can also be used to define another type of bandwidth referred to as the effective bandwidth,

$$\beta_e = \frac{1}{\Delta\tau} \quad (10\text{-}23)$$

The time resolution constants for various dolphin sonar signals were shown in Table 10.1. Typical values for *Tursiops* signals are between 12 and 15 μs, which translates to a resolvable range of approximately 1.0 to 1.1 cm between point targets. The time resolution problem can be visualized with the aid of Figure 10.4, which depicts the envelope of the autocorrelation function of echoes from two point targets separated by various times τ. Heptuna's signal shown in

Figure 10.3, with the properties listed in Table 10.1, was used to calculate the autocorrelation function. The normalized autocorrelation envelopes are not distinguishable for $\tau = 10$ μs, and barely distinguishable for a τ between 14 and 15 μs, which agrees well with a $\Delta\tau = 14.5$ μs. The relatively small time resolution constants for *Tursiops* probably play an important role in the animal's ability to make fine target discriminations.

10.1.4 Wideband Ambiguity Function

Up to this point, we have considered only stationary targets. Targets moving radially away from or toward a sonar will affect the time or range resolution characteristics of the signal because of Doppler effects. A narrow-band signal will experience a Doppler shift toward a higher frequency if the target is approaching, and toward a lower frequency if the target is moving away from the sonar source. Broadband sonar signals typical of dolphin sonar will not experience Doppler frequency shifts but will rather experience Doppler stretching or compression (Kelley and Wishner 1965). If the radial velocity of a point target is v, and the velocity of sound in water is c, the Doppler scale factor η is defined as

$$\eta = \frac{1 + \frac{v}{c}}{1 - \frac{v}{c}} \quad (10\text{-}24)$$

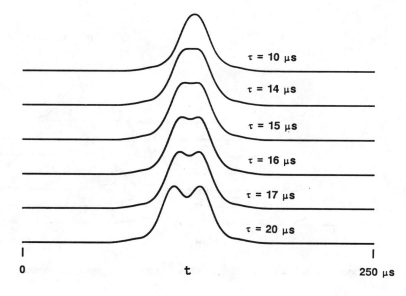

Figure 10.4. The normalized autocorrelation function envelopes of echoes from two point targets separated in time by the parameter τ. Heptuna's signal shown in Fig. 10.3, with a $\Delta\tau$ of 14.5 μs, was used to calculate the autocorrelation function.

$\tau = 10$ μs

$\tau = 14$ μs

$\tau = 15$ μs

$\tau = 16$ μs

$\tau = 17$ μs

$\tau = 20$ μs

0   t   250 μs

If we let $s_a(t)$ be the transmitted signal, the echo from a radially moving point target can then be expressed as $A \cdot s_a[\eta(t + \tau)]$, where τ is the time delay between the transmitted and received signal. Once again we can use the integrated square magnitude of the difference between the transmitted and received echo as a measure of the distinguishability between the two signals:

$$\varepsilon^2 = \int_{-\infty}^{\infty} |s_a(t) - As_a[\eta(t + \tau)]|^2 \, dt$$

$$= \int_{-\infty}^{\infty} |s_a(t)|^2 \, dt + A^2 \int_{-\infty}^{\infty} |s_a[\eta(t + \tau)]|^2 \, dt$$

$$- 2A^2 Re \int_{-\infty}^{\infty} s_a(t)s_a^*[\eta(t + \tau)] \, dt$$

$$(10\text{-}25)$$

The first two terms are the energy in the transmitted and reflected signals, respectively, which are constant for a given signal, target, and range. The difference, ε^2, will be at a minimum when the last integral is at a maximum. This integral represents the cross-correlation between the emitted signal and its delayed reflected echo from a target moving at a velocity v. The third integral, when normalized, is the wideband *ambiguity function* derived by Kelley and Wishner (1963):

$$\chi(\eta, \tau) = \sqrt{n} \int_{-\infty}^{\infty} S_a(t)S_a^*[\eta(t - \tau)] \, dt \quad (10\text{-}26)$$

The ambiguity function can be expressed in a more convenient form for computational purposes by using the Shift Theorem and the Parseval Theorem from Fourier Analysis (Bracewell 1978)

$$\chi(\eta, \tau) = \frac{1}{\sqrt{\eta}} \int_{-\infty}^{\infty} S_a(f)S_a^*(f)e^{j2\pi f\tau} \, df$$

$$= \frac{1}{\sqrt{\eta}} \mathfrak{I}^{-1}[S_a(f)s_a^*(f)]$$

$$(10\text{-}27)$$

Some of the properties of the wideband ambiguity function have been described by Altes (1973). A convenient technique to evaluate (10-27) using a digital computer has been described by Lin (1988). The envelope of the wideband ambiguity function for the *Tursiops truncatus* signal shown in Figure 10.3 is plotted in Figure 10.5 for different positive values of the Doppler scale factor η. The ambiguity function is symmetrical with respect to η so that $\chi(\eta, \tau)$ will be the same for $\pm \eta$ values. The target velocity for $\eta = 1.03$ is approximately 22 m/s. The wideband ambiguity function plotted in Figure 10.5 indicates that the range resolution capability of a typical dolphin sonar signal is not affected by the velocity of the target, so that the signal can be considered to be Doppler tolerant. The ambiguity diagram also indicates that the dolphin signal cannot be used to resolve the velocity of the target. Ambiguity function plots for the sonar signals of *Phocoena phocoena*, and *Delphinus delphis* have been pre-

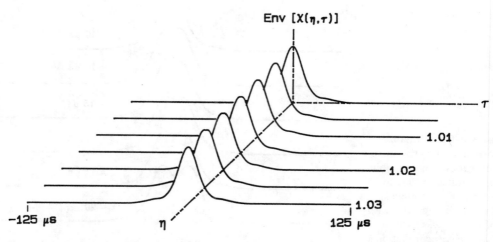

Figure 10.5. Wideband ambiguity diagram of the *Tursiops truncatus* signal found in Figure 10.3.

sented by Dziedzic et al. (1969, 1978). These signals were also Doppler tolerant with no velocity resolution capability.

The sonar signals of some bat species contain a constant frequency (CF) portion and have ambiguity plots that suggest very good velocity resolution capabilities. The sonar signals of *Rhinolophus ferrumequinum* were found by Beuter (1980) to have a velocity resolution capability as fine as 10 cm/s. Lin (1984, 1988) showed that the signal of *Myotis mystacinus* has good velocity resolution as well. Therefore, these bats may be able to estimate the radial velocity of their prey from a single echo. However, such capability may not be necessary for dolphins, since the speed of sound in water is approximately five times as high as in air. Therefore, echoes from prey at comparable ranges return to the dolphin much sooner than to a bat, allowing the dolphin to track the trajectory of the prey by receiving many consecutive echoes.

10.1.5 Time-Bandwidth Product

The *time-bandwidth* product of a sonar signal is another parameter that is often referred to in sonar and radar literature, and will be considered here for dolphin sonar signals. The mean-square bandwidth is defined in Eq (10-5), and in a similar manner, the *mean-square* duration of the signal waveform can be defined as (Bracewell, 1978)

$$\tau_d^2 = \frac{\int_{-\infty}^{\infty} (t - t_0)^2 |s(t)|^2 \, dt}{\int_{-\infty}^{\infty} |s(t)|^2 \, dt} \qquad (10\text{-}28)$$

where

$$t_0 = \frac{\int_{-\infty}^{\infty} t |s(t)|^2 \, dt}{\int_{-\infty}^{\infty} |s(t)|^2 \, dt} \qquad (10\text{-}29)$$

is the centroid of the time waveform. Since we are concerned only with the width of the time waveform and frequency spectrum, the analysis can be greatly simplified by relocating the centroid of the waveform and spectrum to the origin of their respective coordinate system (essentially letting t_0 and $f_0 = 0$). The produce of both mean-square function can now be expressed as

$$\tau_d^2 \beta^2 = \frac{\int_{-\infty}^{\infty} t^2 |s(t)|^2 \, dt \int_{-\infty}^{\infty} f^2 |S(f)|^2 \, df}{\int_{-\infty}^{\infty} |s(t)|^2 \, dt \int_{-\infty}^{\infty} |S(f)|^2 \, df} \qquad (10\text{-}30)$$

Equation (10-30) can be further simiplified in

a straight forward but involved process using the derivative theorem, Rayleigh's theorem and Schwartz's inequality found in Fourier analysis. The derivation will not be covered here; those interested in the details should refer to Bracewell (1978). After the appropriate manipulation of Eq (10-30), the time-bandwidth product can be expressed by the following inequality

$$\tau_d \beta \geq \frac{1}{4\pi} \qquad (10\text{-}31)$$

Equation (10-31) states that the time-bandwidth product must be greater than or equal to a minimum value of $1/(4\pi)$ or 0.08. The time-bandwidth product is often expressed as being ≥ 1 with the 4π factor cancelled out by defining the mean-square duration and bandwidth with a 4π constant appended to the right-hand side of Eqs (10-5) and (10-28).

Equation (10-31) is often expressed as the radar or sonar uncertainty principle deriving its name from Heisenberg's Uncertainty Principle of quantum mechanics. The inequality in Eq (10-31) places a constraint on the rms duration and bandwidth of a signal. For example, the range estimation accuracy of a signal expressed in Eq (10-11) is inversely proportional to its rms bandwidth, however the bandwidth cannot be made arbitrary large without regard to the rms duration. Gabor (1947) has shown that the only function that satisfies the equality in Eq (10-31) is a constant frequency sinusoidal pulse having a Gaussian envelope (see Eq 5-1). Gabor called this function the *elementary signal*, and suggested that it could be used as the basis for signal decomposition instead of the sinusoidal function in Fourier analysis.

The rms bandwidth, rms duration and time-bandwidth product of the signals depicted in Figs. 7.5 and 7.17 are given in Table 10.2. The time-bandwidth product for the *Tursiops* signals are closer to the theoretical lower limit than the signals for the smaller dolphins. However, even for *Tursiops'* signals, $\tau_d \beta$ is greater than the lower limit by 38 to 75%. These values do not support the notion of Kamminga and Beitma (1990) that dolphin sonar signals have $\tau_d \beta$ values that approach the lower theoretical limit to within 20%. As $\tau_d \beta$ approaches the lower limit, the number of possible signal shapes will become progressively

Table 10.2. Rms bandwidth, rms duration and time-bandwidth product of some dolphin sonar signals

Dolphin	β(kHz)	$\tau_d(\mu s)$	$\tau_d\beta$
T.T. (Ekiku)	21.5	5.5	0.12
T.T. (Ekahi)	28.1	5.1	0.14
T.T. (Heptuna)	21.5	5.4	0.12
T.T. (Sven)	21.4	5.1	0.11
Phocoena P.	16.6	19.2	0.32
Phocoenoides D.	17.0	27.9	0.47
Cephalorhynchus C.	15.8	22.7	0.36
Cephalorhynchus H.	25.9	26.7	0.69

smaller until $\tau_d\beta$ equals the lower limit, in which case, only the Gabor elementary signal expressed in Eq (5-1) is possible. However, this may be a moot point since considering the time-bandwidth product of the transmitted signals is not realistic or practical. We have seen in Section 8.1 that echoes arriving back to a dolphin will typically contain many highlights and will have a much longer duration than the transmitted signal. Therefore, the $\tau_d\beta$ product of echoes should be more important than the $\tau_d\beta$ product of the transmitted signal. If an echo has a duration equal to the broadband integration time for *Tursiops* (264 μs), then the time-bandwidth product can be as large as 7, a value almost 90 times larger than the lower limit of 0.08.

10.2 The Dolphin Modeled as an Energy Detector

The auditory threshold versus duration test and the critical ratio experiments of Johnson (1968a, b) with *Tursiops truncatus* indicate that the dolphin's inner ear functions like the human inner ear and that the animal integrates acoustic energy in the same way as humans. The study of Au et al. (1988) discussed in Section 8.4.1 indicates that for short duration, broadband, high frequency sonar signals, the auditory system of *Tursiops* behaves like an energy detector with an integration time of 264 μs. Green and Swet (1966) have shown that an energy detector is a good analog of the human auditory detection process. Therefore it seems reasonable to approach the dolphin auditory detection process as an energy detector.

10.2.1 Urkowitz Energy Detection Model

Urkowitz (1967) examined the detection of a deterministic signal in white Gaussian noise using an energy detector and derived expressions for $P(Y/n)$ and $P(Y/sn)$. In this section, we will present his derivation and apply his results to the dolphin data for target detection in masking noise from the previous section. In Urkowitz's model the signal, $y(t)$, is squared and then integrated over a time T. The noise is a zero-mean Gaussian random variable with a flat band-limited spectral density, N_0 (rms noise level per Hz), and a bandwidth of W Hz. The Gaussian probability density function of the noise can be expressed as

$$f_N(y_i) = \frac{1}{\sqrt{2\pi\sigma^2}} e^{\frac{-y_i^2}{2\sigma^2}} \qquad (10\text{-}32)$$

where the noise variance, σ^2 is equal to the rms noise level, so that

$$\sigma^2 = N_0 W = 2N_{02} W \qquad (10\text{-}33)$$

The parameter N_{02} is the two-sided noise spectral density and was used by Urkowitz (1967). The signal plus noise has a probability density function of

$$f_{SN}(y_i) = \frac{1}{\sqrt{2\pi\sigma^2}} e^{-\frac{(y_i-s)^2}{2\sigma^2}} \qquad (10\text{-}34)$$

The detection is a test of the following two hypotheses:

H_0: The input $y(t)$ is noise alone:
$\quad y(t) = n(t)$

H_1: The input $y(t)$ is signal plus noise:
$\quad y(t) = n(t) + s(t)$
$\quad E[n(t) + s(t)] = s(t)$

$E(y)$, the expected value of the random variable y, is zero for the noise since its mean is zero, and equals $s(t)$ for the signal plus noise. The output of the energy detector taken over an interval T is denoted as V and is taken as the test statistic, with

$$V = \frac{1}{N_{02}} \int_0^T y^2(t)\, dt \qquad (10\text{-}35)$$

The incoming signal is sampled at an interval of Δt, and from Shannon's Sampling Theorem, a bandlimited signal containing no frequency components above W Hz is uniquely determined by samples taken at intervals of $\Delta t = 1/(2W)$. According to Shannon's Sampling Theorem (Schwartz 1980), the sampled noise can thus be expressed as

$$n(t) = \sum_{i=1}^{2TW} n_i \frac{\sin[\pi(2WT - i)]}{\pi(2WT - i)} \quad (10\text{-}36)$$

where n_i is the i-th sample of the noise at time $\frac{i}{(2W)}$. The energy in the noise can be written as

$$\int_0^T n^2(t)\,dt = \int_0^T \sum_{i=1}^{2TW} n_i \frac{\sin[\pi(2WT - i)]}{\pi(2WT - i)}$$
$$\cdot \sum_{j=1}^{2TW} n_j \frac{\sin[(2WT - j)]}{\pi(2WT - j)}\,dt$$
$$(10\text{-}37)$$

However, the integral of (10-37) can be simplified since

$$\int_0^T \frac{\sin[\pi(2WT - i)]}{\pi(2WT - i)} \frac{\sin[\pi(2WT - j)]}{\pi(2WT - j)}\,dt$$
$$= \begin{cases} \frac{1}{2W} & \text{when } i = j \\ 0 & \text{when } i \neq j \end{cases} \quad (10\text{-}38)$$

Therefore, when the input to the energy detector is noise only, using (10-37) and (10-38), the test statistic of (10-35) can be expressed as

$$V = \sum_{i=1}^{2TW} b_i^2 \quad (10\text{-}39)$$

where

$$b_i = \frac{n_i}{\sqrt{2WN_{02}}} \quad (10\text{-}40)$$

Thus, V is the sum of the square of $2TW$ Gaussian random variables, each with a zero mean and unity variance. Such a random variable has a χ_i^2, or chi-square, distribution (Meyer 1965). Specifically, V has a chi-square distribution with $2TW$ degrees of freedom.

Now consider the situation when input to the energy detector consists of only the signal, $s(t)$. Using the Sampling Theorem, the signal can be expressed as

$$s(t) = \sum_{i=1}^{2TW} s_i \frac{\sin[\pi(2WT - i)]}{\pi(2WT - i)} \quad (10\text{-}41)$$

Following the same line of reasoning as with (10-33) and (10-34), we can write

$$\frac{1}{N_{02}} \int_0^T s^2(t)\,dt = \frac{1}{2W} \sum_{i=1}^{2WT} \beta_i^2 \quad (10\text{-}42)$$

where

$$\beta_i = \frac{s_i}{\sqrt{2WN_{02}}} \quad (10\text{-}43)$$

Using (10-39) and (10-42), the total input $y(t)$ with the signal plus noise present can be written as

$$y(t) = \sum_{i=1}^{2TW} (n_i + s_i) \frac{\sin[\pi(2WT - i)]}{\pi(2WT - i)} \quad (10\text{-}44)$$

Under hypothesis H_1, the test statistic V can now be written as

$$V = \frac{1}{N_{02}} \int_0^T y^2(t)\,dt = \sum_{i=1}^{2TW} (b_i^2 + \beta_i^2)^2 \quad (10\text{-}45)$$

The sum in (10-45) has a noncentral chi-square distribution with $2TW$ degrees of freedom and a noncentrality parameter λ given by

$$\lambda = \sum_{i=1}^{2TW} \beta_i^2 = \frac{1}{N_{02}} \int_0^T s^2(t)\,dt = \frac{E_s}{N_{02}} \quad (10\text{-}46)$$

where λ is merely the energy signal-to-noise ratio.

The probability of false alarm will be related to the area under the chi-square distribution curve in the manner discussed in Section 1.5. The probability of false alarm for a given threshold V_T derived by Urkowitz (1967) is given by

$$P(Y/n) = Pr\{V > V_T | H_0\} = Pr\{\chi_{2TW}^2 > V_T\} \quad (10\text{-}47)$$

For the same threshold V_T, the probability of detection is given by

$$P(Y/sn) = Pr\{V > V_T | H_1\} = Pr\{\chi_{2TW}^2(\lambda) > V_T\} \quad (10\text{-}48)$$

The symbol $\chi_{2TW}^2(\lambda)$ indicates a noncentral chi-square variable with $2TW$ degrees of freedom and the noncentrality parameter λ. The expression for the probability of false alarm can be rewritten as

$$P(Y/n) = 1 - Pr\{V_T \leq \chi_{2TW}^2\} \quad (10\text{-}49)$$

where Pr is the area under the chi-square distribution curve with $2TW$ degrees of freedom.

Similarly, the expression for the probability of detection can be rewritten as

$$P(Y/sn) = 1 - Pr\{V_T/G \leq \chi_D^2\} \quad (10\text{-}50)$$

where

$$D = \frac{(2TW + E/N_{02})}{(2TW + 2E/N_{02})} \quad (10\text{-}51)$$

$$G = \frac{(2TW + 2E/N_{02})}{(2TW + E/N_{02})} \quad (10\text{-}52)$$

Pr is now the area under the noncentral chi-square distribution with a modified number of degrees of freedom D and a threshold divisor G.

10.2.2 Application of the Urkowitz Model

These expressions derived by Urkowitz (1967) were applied by Au (1988) to the dolphin data for target detection in noise discussed in Section 8.3. In order to use the Urkowitz model, we need to assume that the dolphin behaves like an unbiased detector, which means that

$$P(Y/sn) = 1 - P(Y/n) \quad (10\text{-}53)$$

(see Fig. 1.10). The first step in applying the model is to choose a desired value of $2TW$ and calculate by an iterative method the V_T necessary to obtain different specific values of $P(Y/n)$ using (10-47) or

(10-49). From (10-53), once $P(Y/sn)$ is specified, $P(Y/sn)$ is also specified, so that the necessary E/N_{02} to obtain $P(Y/sn)$ for a given $2TW$ and V_T can be determined from (10-48) or (10-50). The values E/N_{02} necessary to achieve different specific $P(Y/sn)$ can be compared with the dolphin performance data. The process is continued for different $2TW$ values until the value of $P(Y/sn)$ as a function of E/N_{02} is found which best fits the dolphin data.

The result of such an iterative process using the Urkowitz model is plotted in Figure 10.6, along with the performance results of the three dolphin experiments on target detection in noise discussed in the previous section. Urkowitz's energy detector model agrees well with the dolphin's performance results, further supporting the notion that the dolphin behaves like an energy detector. If we insert the dolphin's integration time of 264 μs into $2TW = 20$, we get a bandwidth of 37.9 kHz (45.8 dB), for the energy detector. Au and Moore (1990) measured a critical ratio of approximately 18 kHz (42.6 dB) at 120 kHz for *Tursiops*, which is in relatively good agreement with the bandwidth for the energy detector model. The unbiased-detector assumption used to derive $P(Y/sn)$ is good for signal-to-noise ratio conditions that correspond to performance at or above the 75% correct response threshold, as can be seen in Figure 8.12.

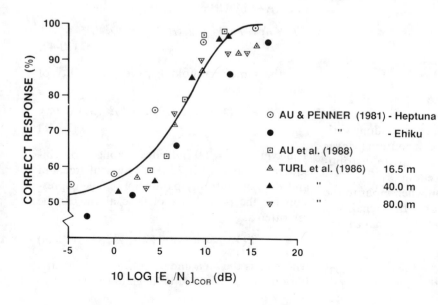

Figure 10.6. Dolphin performance results compared with results from the Urkowitz energy detector model for $2TW = 20$.

10.3 Signal Processing Models for Target Recognition

In this section we will consider methods by which broadband sonar signals like those used by dolphins can be processed in order to discriminate and recognize targets. Three different models will be considered, although there are no doubt other models that can be used to process broadband sonar signals. The three signal processing models share a common approach, their major difference being the manner in which features of targets are described and stored. The main steps involved in developing and using a signal processing model for target recognition are as follows:

1. Use a mathematical model to identify and extract pertinent features of targets from the echoes.
2. Collect echoes from known targets and store their features in a library.
3. Process echoes from an unknown target using the mathematical model developed in (1) and identify the target features. Then compare the features of the unknown target with features of known targets stored in the library. The library search and comparison process should lead to the identification or recognition of the unknown target if its features have been previously collected and stored in the library, or should indicate those known targets that are closest to the unknown target.

If we are only interested in performing a discrimination task to determine whether two targets A and B are different, steps 2 and 3 can be skipped. The features for both targets can be determined in step 1 and used directly to compare their echoes.

10.3.1 Energy Detection in a Filter Bank

The first signal processing model consists of characterizing a target by passing either its echo or its transfer function through a bank of contiguous filters and calculating the energy within each filter (Chestnut et al. 1979). If we let the frequency boundaries of the filters be f_0, f_1, $f_2, \ldots, f_N$, let $s(t)$ stand for the transmitted signal, and $y(t)$ for the echo, then the energy contained in the i-th filter can be expressed as

$$E_i = \int_{f_{i-1}}^{f_i} |Y(f)|^2 \, df \quad \text{(echo)} \quad (10\text{-}54)$$

$$E_i = \int_{f_{i-1}}^{f_i} \left| \frac{Y(f)}{S(f)} \right|^2 \, df \quad \text{(transfer function)} \quad (10\text{-}55)$$

where $S(f)$ and $Y(f)$ are the Fourier transforms of the transmitted signal and received echo, respectively. The collection of N energy terms, $E_1, E_2, \ldots, E_N$, will be the feature set for a target. The functions in the integrals of Eqs (10-54) and (10-55) are usually normalized to unity so that only features associated with the frequency spectrum will be computed and amplitude will not be a factor. The advantage of the transfer function form lies in the fact that one does not have to use the same signal to characterize the target for library storage and to examine unknown targets. Any well-characterized broadband signal can be used to measure the features of a target, and any other known signal can be used in the field. In order to apply (10-50), the same signal needs to be used both to measure the target features and to ping off targets in the field. The advantage of using only the echo consists of not having to perform a complex division of the received echo spectrum by the incident signal spectrum.

Two basic types of filters can be used, *constant-interval* and *constant-Q filters*. For constant-interval filters, the parameters that need to be specified are the lower and upper frequencies of interest, the bandwidth of each filter, and the number of filters. Since echo data will no doubt be processed with a computer and an FFT algorithm will be used to estimate calculate the frequency spectra in Eqs. (10-54) and (10-55), the parameters should be chosen to coincide with the FFT bin size and the boundaries of the FFT bins. If we let SR be the analog-to-digital sample rate and N the number of samples used to characterize the time domain echo, then the FFT bin size from (1-26) will be

$$\Delta f = \frac{SR}{N} = \frac{1}{N \Delta t} \quad (10\text{-}56)$$

where Δt is the time interval between samples. The boundaries of the FFT bins will be at frequencies corresponding to

$$f_i = (i - 1)\Delta f \qquad (10\text{-}57)$$

where $i = 1, 2, \ldots, N/2$. Therefore, it would be convenient to choose the lower and upper frequencies of the filter bank to coincide with (10-57), and to require that the filter interval be an integer multiple of Δf.

The second type of filter, the constant-Q filter, is closely related to the dolphin's auditory system since the critical ratio and critical bandwidth of *Tursiops truncatus* (Fig. 3.9) can be approximated by a bank of constant-Q filters. From (3-4), $Q = f_0/\Delta f$, where f_0 is the center frequency of the filter and Δf is its bandwidth. Let f_{i-1} be the lower frequency limit and f_i be the upper frequency limit of the i-th filter; then the center frequency will be the geometric mean between the frequency boundaries, or $f_0 = (f_i - f_{i-1})/2$, and the bandwidth $\Delta f = f_i - f_{i-1}$. Substituting these expressions into the definition of Q, the frequency boundaries of the i-th filter can be expressed as

$$f_i = \frac{2Q + 1}{2Q - 1} f_{i-1} \qquad (10\text{-}58)$$

Let f_1 be the lower frequency limit and f_h the upper frequency limit in the bank of constant filters; we can then express the ratio of the lower to upper frequency limit of each filter as

$$\frac{f_{l+1}}{f_1} = \frac{2Q + 1}{2Q - 1} \quad \text{1st filter} \qquad (10\text{-}59)$$

$$\frac{f_{l+2}}{f_{l+1}} = \frac{2Q + 1}{2Q - 1} \quad \text{2nd filter}$$

$$\vdots \qquad \vdots$$

$$\frac{f_u}{f_{u-1}} = \frac{2Q + 1}{2Q - 1} \quad N\text{-th filter}$$

Multiplying all frequency ratios with and canceling common terms, we get

$$\frac{f_u}{f_l} = \left[\frac{2Q + 1}{2Q - 1}\right]^N \qquad (10\text{-}60)$$

Solving for Q in (10-55), we arrive at

$$Q = \frac{1}{2} \frac{\left(\dfrac{f_u}{f_l}\right)^{\frac{1}{N}} + 1}{\left(\dfrac{f_u}{f_l}\right)^{\frac{1}{N}} - 1} \qquad (10\text{-}61)$$

There are three parameters which can be varied, f_1, f_u, and the number of filters, N. According to Chestnut et al. (1979) there is redundancy in the vectors of energy values used in the constant-Q filter bank, implying that the method is not as efficient in terms of the amount of information stored.

10.3.2 Measure of Feature Recognition

The features used to characterize targets, such as the energy in a bank of filters, constitute a multi-element vector with a certain pattern of distribution. In order to discriminate, recognize, or classify echoes, one must use some sort of pattern recognition scheme in which features of received echoes can be compared against archetypal feature vectors of echoes from known targets. The problem becomes one of recognizing the patterns of feature distribution, and determining a quantitative measure of the similarity or dissimilarity of one feature vector compared to a host of other feature vectors. There are many pattern recognition schemes that can be used, but, only one, the *Euclidean distance measure*, will be considered here. (Some other pattern recognition methods are discussed in Chestnut 1979). Let E_i be the features of an echo just received by a sonar and $E_i(k)$ be the features of a known target k stored in an archetype library, where $i = 1, 2, \ldots, N$; then the similarity between the feature vectors E_i and $E_i(k)$ can be expressed by the Euclidean distance

$$d_k = \sqrt{\sum_{i=1}^{N} [E_i - E_i(k)]^2} \qquad (10\text{-}62)$$

Features of the received echo are compared with corresponding features of all targets stored in the archetype library, and the target features to which they have the smallest Euclidean distance will be the most similar.

Chestnut et al. (1979) performed a sonar target

Table 10.3. Targets used in the sonar experiment of Chestnut et al. (1979)

Target	Diam.	Length	Wall
Hollow aluminum cylinder	3.81	17.78	0.48
Hollow aluminum cylinder	3.81	17.78	0.79
Hollow aluminum cylinder	3.81	17.78	0.95
Hollow aluminum cylinder	7.62	17.78	0.32
Hollow aluminum cylinder	7.62	17.78	0.64
Hollow aluminum cylinder	7.62	17.78	0.95
Hollow aluminum cylinder	7.62	17.78	1.58
Hollow aluminum cylinder	10.16	17.78	0.64
Hollow aluminum cylinder	6.36	17.78	0.64
Solid rock cylinder	3.81	17.78	—
Solid rock cylinder	6.35	17.78	—
Hollow PVC cylinder	7.62	17.78	1.27
Hollow aluminum sphere	7.62	—	—
Hollow aluminum sphere	15.14	—	—
Aluminum biconic	—	12.27	—

recognition experiment using the energy in a bank of filters and the Euclidean distance measure as one of the signal processing scheme. They also used an all-pole filter model to characterize echoes; however, only the results of the filter bank experiment will be discussed here. Echoes from the 16 simple geometrical targets listed in Table 10.3 were collected in a test pool by using a low frequency broadband signal with a peak frequency of 23 kHz as shown in Figure 10.7. A total of 240 echoes, 10 or 20 from each target, were collected. Four echoes from each target were used in the design set and the rest were used for the test set. The overall results of comparing the

echoes in the design set with the echoes in the test set as a function of the signal-to-noise ratio are displayed in Figure 10.8. The figure shows the probability of rejection (solid curves) and probability of misclassification (dashed curves) plotted as a function of signal-to-noise ratio. The results indicate that as the signal-to-noise ratio increases the probability of making an error decreases. They also indicate that progressively fewer errors were made as the number of filters increased to 30. The best results were obtained using 30 filters on the interval from 15 to 45 kHz. There were only slight differences between the uniformly spaced intervals and the constant-Q intervals.

10.3.3 Time Domain Highlight Features

Martin and Au (1988) describe a *time domain* feature extraction model which uses cues identified as salient for humans (Section 9.6). This model consists of extracting highlight separation and highlight amplitude ratios from echo envelopes to form the target feature sets. The first stage consists of passing a target echo through a matched filter with the transmitted signal as the reference. Let $s(t)$ be the transmitted signal and $y(t)$ be the received target echo; the output of the matched filter will be

$$x(t) = \mathfrak{I}^{-1}[Y(f)S^*(f)] \qquad (10\text{-}63)$$

where $S(f)$ and $Y(f)$ are the Fourier transforms of the time domain signals and $\mathfrak{I}^{-1}$ is the inverse

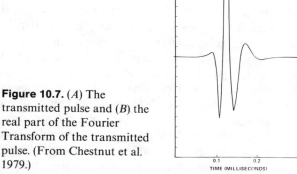

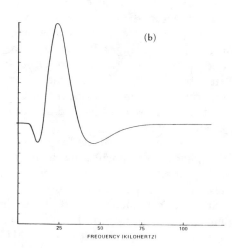

(a)

(b)

Figure 10.7. (*A*) The transmitted pulse and (*B*) the real part of the Fourier Transform of the transmitted pulse. (From Chestnut et al. 1979.)

TIME (MILLISECONDS)

FREQUENCY (KILOHERTZ)

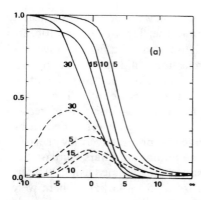 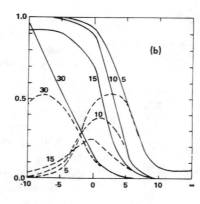

Figure 10.8. Probability of rejection (solid curves) and probability of misclassification (dashed curves) as a function of the signal-to-noise ratio for (*A*) constant filter width and (*B*) constant-*Q* model. The number next to the curves are the number of filters. (From Chestnut et al. 1979.)

Fourier Transform symbol. From (9-4), the envelope of the matched-filter output can be expressed as

$$Env[x(t)] = |x_a(t)| = \sqrt{x^2(t) + \hat{x}^2(t)} \quad (10\text{-}64)$$

where from (9-2) $\hat{x} = \mathfrak{J}^{-1}[-jX(f)]$ for $f > 0$. The resultant signal envelope is processed by a peak detector that determines the location of highlights by finding points where the slope of the envelope changes from positive to negative. Small amplitude extrema in the immediate neighborhood of a larger maximum are rejected. After obtaining a list of highlights for a given echo, the absolute maximum is assigned a time separation of zero and an amplitude of one. The other highlights are assigned negative or positive time separations according to position before or after the largest highlight. Amplitude ratios for each highlight are calculated with respect to the maximum. In the second stage, reference vectors are created from a set of echoes by aligning the features across the signals. Absolute maxima are aligned, so that every group of signals has a highlight occurring with an amplitude ratio of one and time separation zero. Other highlights are also aligned, and amplitude ratios and time separation become statistical quantities represented as means and standard deviations for the group of signals.

Martin and Au (1988) performed a sonar target recognition experiment using their time domain feature model. Target echoes were collected in a test pool using a simulated dolphin sonar signal with a peak frequency of 120 kHz similar to the incident signal shown in Figure 8.1. The echoes were digitized at a sample rate of 1 MHz. Reference and test data were represented by vectors of 25 features; each feature represented the time separation and relative amplitude of an echo highlight. Reference data were means of the feature vectors from 10 echoes. The time axis was partitioned into bins of 20 points or 20 μs each, and each bin was assigned an amplitude value determined by highlights in that portion of the echo. When any of the 25 time windows contained an echo highlight, that element was assigned the value of the highlight amplitude ratio relative to the maximum. If the partition did not contain a highlight, a value of zero was assigned. When more than one highlight was present in a partition, the largest amplitude was used. The Euclidean distance measure of equation (10-58) was used to compare echoes from the reference set with echoes from the test set.

Echoes from the cylindrical targets used in the material composition discrimination experiment of Hammer and Au (1980), along with a solid aluminum cylinder, were tested by Martin and Au (1988). The performances of the feature extraction and pattern recognition algorithms are shown as confusion matrices in Tables 10.4 and 10.5. Scores along the diagonal of the matrix represent correct responses; off-diagonal elements represent confusions. The signal processing scheme was 90% correct for material composition discrimination in Table 10.3 and 100% correct for internal structure discrimination in Table 10.4. Chance performance was 14.3% correct (1 in 7) for material composition and 25% correct (1 in 4) for internal structure discrimination. When echoes from two different glass cylinders were added to the data of Table 10.3 to make a 9-by-9 confusion matrix, performance dropped to 62% correct; chance dropped

Table 10.4. Confusion matrix for cylinder material composition discrimination using time domain features. Cyl-1 had an outer diameter of 3.61 cm; Cyl-2 and the solid aluminum cylinder had diameters of 7.62 cm. All of the cylinders had lengths of 17.78 cm. (From Martin and Au 1988.)

Test Target Echoes	Reference Target Echoes						
	Alum Cyl-1	Steel Cyl-1	Bronze Cyl-1	Alum Cyl-2	Steel Cyl-2	Bronze Cyl-2	Solid Alum Cyl
Alum Cyl-1	97%					3%	
Steel Cyl-1		100%					
Bronze Cyl-1			100%				
Alum Cyl-2				93%		7%	
Steel Cyl-2				37%	63%		
Bronze Cyl-2						100%	
Solid Alum Cyl							89%

Table 10.5. Confusion matrix for internal structure discrimination. All of the cylinders had outer diameters of 7.62 cm and lengths of 17.78 cm. (From Martin and Au 1988.)

Test Target Echoes	Reference Test Target			
	Alum Cyl-2 Airfilled	Coral Cyl-2 Solid	Alum Cyl-2 Waterfilled	Alum Cyl-2 Solid
Alum Cyl-2 Airfilled	100%			
Coral Cyl-2 Solid		100%		
Alum Cyl-2 Waterfilled			100%	
Alum Cyl-2 Solid				100%

to 11% correct (1 in 9). Many glass test echoes were incorrectly identified, and some echoes from other cylinders were wrongly identified as glass. Identification of echoes from the glass cylinders was also poor for both humans (Au and Martin 1989) and dolphins (Hammer and Au 1980; Schusterman et al. 1980). The study of Martin and Au (1988) was performed in a noiseless environment.

10.3.4 Spectrogram Correlation Model

A *spectrogram* is a time–frequency representation of a signal in which frequency spectra corresponding to different portions of the signal in the time domain are displayed. Therefore, a spectrogram represents the short-time spectral history of a signal (Altes 1980a) and is plotted in three-dimensional form like the ambiguity function shown in Figure 10.5, but with the velocity axis being replaced by a frequency axis. A spectrogram can be calculated by passing a signal through a time window, in a similar fashion as was done in calculating the different spectra in

Figure 4.14, by sliding a chi-square window past the signal and computing the frequency spectrum for different positions of the window in time. The data in Figure 4.14 could have been represented in typical spectrogram form by plotting the different spectra on a three-dimensional graph, with time and frequency axes in the horizontal plane and amplitude in the vertical plane. The features of a target can be determined by using (10-51) for each spectrum of the spectrogram, thereby creating a time–frequency description of the target with transfer function $H(f)$ and impulse response $h(t)$. Therefore, the time–frequency density function of the target can be expressed as (Rihaczek 1969)

$$e_{hh}(t, f) = h(t)H^*(f)e^{(-j2\pi ft)} \quad (10\text{-}65)$$

According to Altes, the time energy density function provides a useful description of a target because it contains information about both highlight and energy spectrum. Highlight ranges can be obtained from $e_{hh}(t,f)$ with the integral

$$|h(t)|^2 = \int_{-\infty}^{\infty} e_{hh}(t, f)\, df \quad (10\text{-}66)$$

Similarly, the frequency spectrum of the target impulse response is

$$|H(f)|^2 = \int_{-\infty}^{\infty} e_{hh}(t, f)\, dt \qquad (10\text{-}67)$$

The features of a spectrogram obtained from a sonar return can be denoted as E_{ij}, where i refers to the energy in the i-th frequency bin within the j-th time bin. The features from the k-th target in a library of archetypes can be expressed as $E_{ij}(k)$. The locally optimum detector is a spectrogram correlator (Altes 1980a) which computes the quantity

$$l(\underline{e}) = \sum_{i=1}^{N} \sum_{j=1}^{M} E_{ij} E_{ij}(k) \qquad (10\text{-}68)$$

Altes and Faust (1978) applied the spectrogram model to echoes from all of the targets listed in Table 10.3 as well as echoes from a 6-cm foam cube (0°, 45°), a 6 × 6-cm foam cylinder (0°, 90°), and a 6.35-cm O.D. × 17.8-cm long cylinder at incident angles from 0° to 90° in 5° increments. They used 16 frequency bins and 27 time bins to characterize the spectrogram of each target echo. Their results, comparing the 6.35 × 17.8-cm cylinder with all other targets, cylinders with non-cylinders, and metal with nonmetal, are shown in Figure 10.9. The probability of misclassification versus the signal-to-noise ratio is shown in Figure 10.9. The results indicate that the target classification scheme of Altes (1980a) involving spectrogram correlation can indeed be used to classify sonar echoes. However, this signal processing technique is extremely time consuming to implement since $N \times M$ features are required to characterize a target, compared with N features

a.   Pr[misclassification, 2.5 x 7" Al. Cyl. versus other targets]

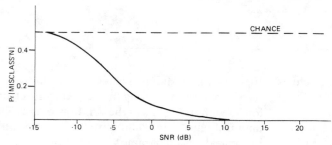

b.   Pr[misclassification, cylinder versus non-cylinder]

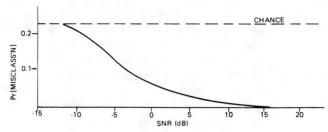

c.   Pr[misclassification, metal target versus non-metal]

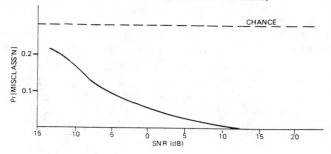

Figure 10.9. Classification performance of the spectrogram model in white noise. (From Altes 1980a.)

for the filter bank model and M features for the time domain model.

10.3.5 Comparative Evaluation of Target Recognition Models

The three signal processing models considered for target recognition all seem to have the capability to discriminate, recognize, and classify target echoes. The relative effectiveness of these three models as well as other mathematical models has not been adequately assessed under rigorous test conditions, and has not been compared with the dolphin sonar. All of the discriminations carried out by the mathematical models could also be performed relatively readily by the dolphin, with the exception of the hollow glass versus aluminum discrimination test performed by Martin and Au (1988). To identify suitable test targets for comparisons, dolphin sonar target recognition experiments need to be conducted in which the animal's performance varies from very good to chance as a target parameter is varied. These test targets can then be used to compare the performance of different signal processing models with the dolphin's performance. Initially, such tests should probably be performed with echoes collected in the stable, quiet environment of a test pool. Models that compare well with the dolphin's performance should then be further tested in a more realistic environment, and in real time. In a natural environment, factors such as target motion, transducer motion, ambient noise, and reverberation can have adverse effects on any sonar function.

10.4 Artificial Neural Networks and Target Recognition

We will now consider an area of research that has attracted considerable attention in the last several years, and has shown considerable promise in solving pattern recognition problems. Although research on *artificial neural networks* began in the late 1940s, interest in this area only began to mushroom in the mid-1980s as investigators began to achieve success in solving certain problems that had been essentially unsolvable before, or had not been solved satisfactorily or

reliably. Only a brief overview of artificial neural networks will be presented in this book. Those interested in a deeper presentation should refer to the many excellent books on the subject such as those written by Wasserman (1989), Hecht-Nielsen (1989), and Dayhoff (1990).

The human brain and nervous system regularly perform massive amounts of parallel processing, using an estimated 10^{11} neurons in perhaps 10^{15} interconnections over a relatively small volume (the size of the brain). Humans can solve complex pattern recognition problems on which computational methods and models have only achieved moderate and uninspiring success. Artificial neural networks are computational models that attempt to emulate the massive parallel processing nature of the nervous system and brain. The networks are built around mathematical analogs of interconnected neurons, with a different weight (emphasis) attached to each connection.

The *neuron* is the most basic cellular unit of the brain, and its mathematical model represents the most basic unit in an artificial neural network. A schematic of an artificial neuron, called a *processing element* (PE), is presented in Figure 10.10. In discussing the mathematical model, the biologically equivalent parts or functions of a real neuron will be given in parentheses. A PE has input connections (dendrites) through which it receives activations from other units, a summation function that combines the various inputs into a single activation, a transfer function that converts this summation into output activation, and an output connection (axon) by which a unit's output activation arrives as input to other units. Each input has a weight (synaptic strength) that modifies every input signal so that the summation by the j-th processing element can be thought of as a weighted sum:

$$I_j = \sum_{i=1}^{N} W_{ij} X_i \qquad (10\text{-}69)$$

The weighted sum is processed by a transfer function, $F(x)$, often referred to as a "*squashing function*," to produce an output. A variety of squashing functions exist to limit the output level of a PE. One popular squashing function is the "sigmoid" (meaning S-shaped) function, $F(x) = \frac{1}{(1+e^{-x})}$. The sigmoid function is plotted in Figure

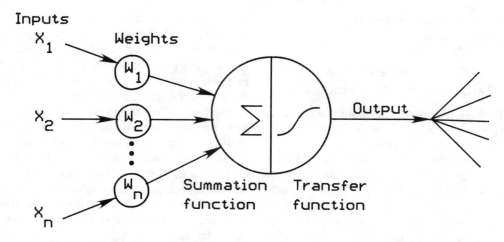

Figure 10.10. Schematic of a processing element in an artificial neural network.

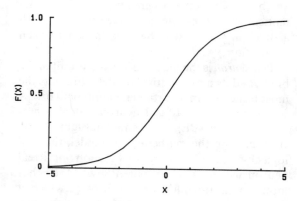

Figure 10.11. The sigmoid squashing or transfer function.

10.11. The output of the j-th PE with a sigmoid squashing function will be

$$O_j = F(I_j) = 1/(1 + e^{-Ij}) \qquad (10\text{-}70)$$

The squashing function serves to limit the output of the PE to within certain bounds. For large positive values of I, $F(I)$ in (10-70) will approach 1. For large negative values of I, $F(I)$ will approach 0. Therefore, the sigmoid squashing function will limit the output of a PE to values between 0 and 1. Another popular squashing function is the hyperbolic tangent or $\tanh(x)$ function, which will limit the output of a PE to between -1 and 1.

An artificial neural network consists of many interconnected PEs that operate in parallel, simulating a process in which the human brain may encode information. The processing elements are organized into a sequence of layers, with full connections between successive layers. There is an input layer where data are presented, an output layer which holds the response of the network to a given input, and "hidden layers" which are neither input nor output layers where the bulk of the processing is performed. An artificial neural network is not programmed to solve specific problems but is "taught" or trained to give acceptable answers to specific inputs. Training can be supervised or unsupervised. In supervised training each input is paired with a desired output, forming a training pair. Input data are given to the network and an output is calculated and compared with the desired target output. The difference, or error, is fed back through the network and the weight of each connection is adjusted according to a specific learning rule. The process is repeated continuously until the error is minimal. The resultant weights corresponding to a satisfactory solution constitute a learned response which the network can remember by storing the values of the weights.

In unsupervised training no desired outputs are required and the training set consists solely of input data. The Training process itself consists of modifying the network weights to produce outputs that are consistent, leading to internal data clustering.

10.4.1 Backpropagation Network

There are various ways in which processing elements can be configured to form a specific network. In this book, only two network architectures will be considered, the *backpropagation* and the *counterpropagation* networks. The backpropagation network is probably the most widely used neural network paradigm and has been applied successfully to a variety of problems. Werbos (1974) was the first to present a conceptual basis of backpropagation. Parker (1982) unknowingly reinvented the backpropagation concept. Rumelhart and McClelland (1986), unaware of both of these previous works, popularized the backpropagation network in their book *Parallel Distributed Processing*. A simplified diagram of a three-layer backpropagation network consisting of an input layer, an output layer, and one hidden layer is shown in Figure 10.12. Each PE is similar to the PE shown in Figure 10.10. Variations of this circuit would include additional hidden layers and different numbers of PEs in each layer. Initially all weights are assigned small random values and training begins by inputting a training vector set (x's) and computing the output using the random weights. The calculated output vector is compared with the desired target vector: the error between both vectors is calculated and appropriate correction factors are determined that would minimize the amount of error. The weights of the network are adjusted by progressively "backpropagating" the

error adjustment terms from the output toward the input, as shown schematically in Figure 10.12. The process is repeated by applying another set of input vectors, resulting in a new set of output and a new set of error terms, which will be backpropagated through the network again. This iterative process is continued indefinitely until the error falls within a desired limit. When training is terminated, the weights associated with each PE are saved and can be used later to classify data in a test set.

There are various learning rules by which error and associated weight adjustment can be calculated. However, the one most used with a backpropagation network is probably the delta learning rule. The delta rule is a simple and elegant learning heuristic that gives a network the ability to form and modify its own connections in ways that often approach optimal performance. In the backpropagation network, the error correction that will be backpropagated through the network must be done differently for the output and hidden layers. The difference between the target output T_j and the actual output O_j for the j-th PE in the output layer of Figure 10.10 is the output error. The error value, δ_j, that is backpropagated from the output layer is given by

$$\delta_j = (T_j - O_j)F'(I_j) \qquad (10\text{-}71)$$

where $F'(I_j)$ is the derivative of the squashing function and I_j is the weighted sum of the j-th PE in the output layer. The quantity $(T_j - O_j)$ is the amount of error, and $F'(I_j)$ scales the error to

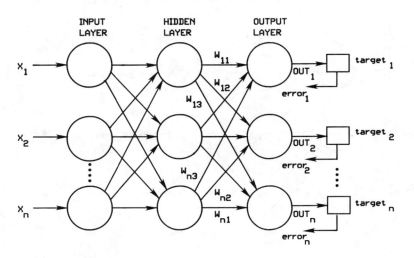

Figure 10.12. Diagram of a simplified three-layer backpropagation artificial neural network.

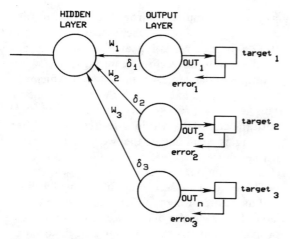

Figure 10.13. Simplified diagram showing error calculation for a PE in the hidden layer.

force a stronger correction when the sum I_j is near the rapid rise in the sigmoid curve (Fig. 10.11). The derivative of the sigmoid function $F'(I)$ is simply

$$F'(I) = F(I)(1 - F(I)) = O(1 - O) \quad (10\text{-}72)$$

where O is the output of the PE. The error for a PE in the hidden layer is more complicated; it can be illustrated with the diagram in Figure 10.13, showing a simplified situation involving the j-th PE in the hidden layer. The error will be the weighted sum of the errors associated with the PEs in the output layer multiplied by $F'(I)$

$$\delta_j = F'(I_j)(\delta_1 W_1 + \delta_2 W_2 + \delta_3 W_3) \quad (10\text{-}73)$$

Let η be a "learning rate" coefficient, then the

adjustment Δ to the weights of the PE can be expressed as

$$\Delta_j = \eta \delta O_j \quad (10\text{-}74)$$

If $w_j(n)$ is the weight before and $w_j(n + 1)$ is the weight after adjustment, the relationship between the weights before and after adjustment is

$$w_j(n + 1) = w_j(n) + \Delta_j \quad (10\text{-}75)$$

The delta rule can adjust weights for targets and desired outputs of either polarity, for both continuous and binary data.

10.4.2 Counterpropagation Network

A network that is useful in separating and learning categories is the counterpropagation network devised by Hecht-Nielsen (1987). For pattern classification problems, the counterpropagation network is a viable alternative to the backpropagation network with potentially faster learning and nearly equal effectiveness in solving problems. The network is a combination of two well-known algorithms: the self-organizing map of Kohonen (1988) and the Grossberg (1969) outstar. A diagram of a three-layer counterpropagation network consisting of an input layer, a Kohonen layer, and a Grossberg layer is shown in Figure 10.14. The PEs of the input layer perform no computation but serve a fan-out function by connecting the input vector to each neuron in the Kohonen layer. A set of weights is associated with each input to each Kohonen neuron (as shown in Fig. 10.14) so that the output

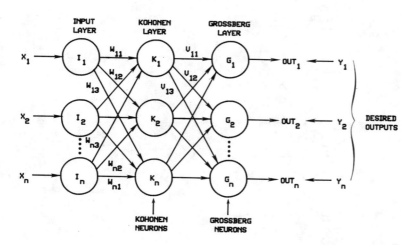

Figure 10.14. Diagram of a three-layer counterpropagation network.

of a Kohonen neuron will be the weighted sum of its input as expressed by equation (10-65). The Kohonen neuron with the largest output is set to 1 and the outputs of the other Kohonen neurons are set to 0. Therefore, the Kohonen layer serves as a "winner-takes-all" function and is often referred to as a competitive layer.

The Kohonen layer classifies the input vectors into similar groupings by adjusting the weights so that similar input vectors will activate the same Kohonen neuron. Training then is a self-organizing algorithm that operates in an un-supervised mode. The input vectors are usually normalized by dividing each component by the vector's length so that

$$x_i = \frac{x_i}{\sqrt{x_1^2 + x_2^2 + \cdots + x_n^2}} \quad (10\text{-}76)$$

This operation converts an input vector into a unit vector pointing in the same direction. Each normalized input vector can be thought of as being on the surface of a hypersphere of unit radius. The winning PE in the Kohonen layer will be the unit with weights that most closely matches the input vector and its weights are adjusted by the equation

$$W_{new} = W_{old} + \alpha(x - W_{old}) \quad (10\text{-}77)$$

where α is a training rate coefficient. Initially, the weights are randomized and normalized. As training progresses, each weight associated with the winning Kohonen neuron is changed by an amount proportional to the difference between its output and input values to minimize the difference between the weight and its input. The winner then becomes more likely to win the competition when the same or a similar input vector is presented to the network.

The output values of the Kohonen layer, k_1, $k_2, \ldots, k_n$, are applied to the Grossberg layer through a series of weights depicted as V_{11}, $V_{21}, \ldots, V_{n1}$. The output of the j-th Grossberg neuron can be expressed as

$$out_j = \sum_{i=1}^{n} k_i V_{ij} \quad (10\text{-}78)$$

where the vector V is the Grossberg layer weight matrix. Since only one PE in the Kohonen layer will have an output of 1 and the rest will be 0, the calculation for the Grossberg layer is rather simple. Training for the Grossberg layer is also relatively simple. After the output of the Grossberg neurons is calculated by (10-78) a given weight is adjusted only if it is connected to a Kohonen neuron having a nonzero output. The new weight for the j-th PE is related to the old weight by the equation

$$V_{ij,new} = V_{ij,old} + \beta(y_j - V_{ij,old})k_i \quad (10\text{-}79)$$

where k_i is the output of Kohonen neuron i and y_j is the component j of the vector of desired outputs. Initially β (training rate) should be set to approximately 0.1, and be gradually reduced as training progresses. Training in the Grossberg layer is supervised since the algorithm has a desired output to which it trains. The weights of the Grossberg layer will converge to the average values of the desired outputs whereas the weights of the Kohonen layer are trained to the average values of the inputs.

10.4.3 Application to Dolphin Sonar Discrimination

The principal use of neural networks in sonar signal processing involves recognition of the pattern distribution of features used to character-ize a target. The features obtained with the three signal processing models discussed in the previ-ous section can be used as input to a neural network. The performance of the network will depend on the features presented to it. In a sense, a neural network is a substitute for the Euclidean distance measure used in the previous section, or any other pattern recognition scheme or algorithm.

Roitblat et al. (1989) were the first to apply an artificial neural network to emulate a dolphin performing a sonar recognition task. Echoes from the target used in the delayed matching-to-sample experiment discussed in Section 9.6 were used in a counterpropagation network. The fre-quency spectra of echoes were divided into 20 equally spaced frequency bins between 63 and 162 kHz. Examples of echoes from the four targets used by Roitblat et al. (1989) are shown in Figure 9.35. Since the echoes were sampled at 1-μs time intervals and 1,024 points were col-lected per echo, the frequency spacing of the FFT calculation (from equation (1-26)) was

$$\Delta f = \frac{1}{N \Delta t} = 997 \text{ Hz} \qquad (10\text{-}80)$$

Each frequency bin for the neural network data was the averaged amplitude (on a linear scale) of five contiguous FFT bins, so the frequency spacing of the neural network data was 4.9 kHz. Ten echoes per target were used for the learning set and another ten echoes per target were used for the test set. The input layer of the counterpropagation network consisted of 20 input units corresponding to the 20 frequency bins of the data. The second layer contained 21 units and was used to normalize the inputs. The third layer was the Kohonen layer consisting of 8 units. The fourth layer was the output layer consisting of 4 units, each unit representing a specific target. All of the echoes from the training set (40 echoes, 10 echoes per target) were randomly fed into the network, and a total of 5,000 iterations of the data were used to train the network. When the 40 echoes from the test set were presented to the trained network, the network was able to classify the echoes with 100% accuracy. This compares with the dolphin's 94.5% accuracy in performing the matching-to-sample task (Section 9.6).

In the next phase of the study, Roitblat et al. (1989) used echoes from the sample targets collected while the dolphin echolocated on them. A directional hydrophone was placed adjacent to the observation aperture at the same depth as the center of the aperture. Echoes from the sample targets resulting from the dolphin's sonar signals were recorded on an instrumentation tape recorder operated at 152 cm/s (60 ips). Unlike the test pool echoes, these echoes exhibited fluctuations in amplitude and shape that caused the signal-to-noise ratio to vary from relatively high to very poor. These fluctuations probably resulted from variations of the animal-target-hydrophone alignment caused by wave- and wind-induced motions of the target and pen, and movement by the dolphin within the observation aperture. Ten echoes per target with high signal-to-noise ratios were digitized at a 1-MHz sample rate and used as the training set. Ten additional echoes per target, also with high signal-to-noise ratios, were digitized and used as the test set.

Only three of the original targets (large tube, sphere, and cone) were used. The natural echoes were used in the same counterpropagation network applied to the test pool echoes, except that the frequency range was changed to include frequencies between 40 and 138 kHz. The network, after training for 5,000 iterations, was able to correctly classify 29 of 30 echoes (96.7%). Another set of echoes was collected three months after the original set. After digitizing selected echoes that had high signal-to-noise ratios and using these echoes in the neural network, the results were similar to the results from the first set of natural echoes. The network was able to correctly classify 97% of the test echoes.

A backpropagation network consisting of 20 input units, an 8-unit hidden layer, and a 3-unit output layer was also trained for 5,000 iterations with the second set of natural echoes. This network was also able to correctly classify the echoes in the test set at an accuracy of 97%.

Moore et al. (1991) took a slightly different approach in emulating the dolphin with a neural network in the delayed matching-to-sample experiment. Instead of using only selected natural echoes having high signal-to-noise ratios, Moore et al. (1991) used sets of consecutive echoes. Therefore, the data included signals with a variety of signal-to-noise ratios. They used a backpropagation network that was modified to include an integration layer that preprocessed the data before they were introduced to a more standard backpropagation network. The frequency spectrum was divided into 30 bins of 1.95 kHz per bin from 31.3 to 146.5 kHz. The information in each frequency bin for each target was averaged from echo to echo before being applied to the backpropagation network.

The network was trained with six sets of ten successive echoes selected from the ends of randomly chosen echo trains. Two sets of echoes were chosen for each of three targets (large tube, sphere, and cone). The network was trained with 12,300 iterations before the chosen criterion, an RMS output error of 0.05, was achieved. The complete set of 1,335 sequential echoes was presented to the trained network and the network was allowed to classify each echo train. The results were determined in terms of "confidence ratio," defined as the ratio of the activation level

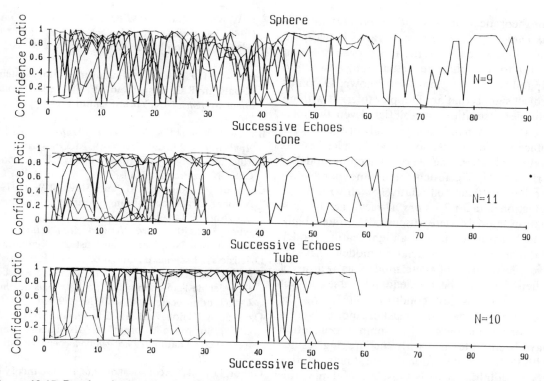

Figure 10.15. Results of generalization testing in terms of the confidence of the network in assigning the echo train to the proper category. (Reprinted with permission from Moore et al. 1991, © Pergamon Press Ltd.)

of the correct classification unit versus the total activation level of the three classification units in the output layer. The results of Moore et al. (1991) are shown in Figure 10.15, with the confidence ratio plotted as a function of the number of clicks used. Overall, the dolphin's performance was better than the network's, which achieved a correct classification performance between 90% and 93%. However, the network tended to make correct classifications with fewer echoes than the dolphin.

The data from Moore et al. (1991) were also applied to a standard backpropagation network without the preprocessing integration layer. The same number of input units, hidden layer units, and output units was used. In this situation, the network only correctly classified the targets with 63% accuracy. Therefore, preprocessing of the data by summing or integrating the signal in each frequency bin across the different echoes in the training set was very important for accurate target classification.

The artificial neural network studies of Roitblat et al. (1989) and Moore et al. (1991) show considerable promise in emulating the dolphin sonar process. However, these studies can only be considered preliminary incursions in applying artificial neural network technology to the field of dolphin sonar classification. The targets used were readily separable, as can be seen in the echoes shown in Figure 9.35. Eventually, the performance of neural networks must be compared with discrimination and classification tasks in which the dolphin exhibits differential performance and achieves a certain performance threshold as a function of a target variable, or as a function of the signal-to-noise ratio for the echo received.

10.5 Summary

A dolphin's ability to discriminate between targets is dependent to a large extent on the time or range resolution capability of its sonar signal.

The theoretical time resolution constant for typical *Tursiops truncatus* signals varies between 12 and 15 μs, which translates to a range resolution of 1.0 to 1.1 cm. The actual time resolution capability of a dolphin is not known. Dolphins project signals that are Doppler tolerant with no Doppler resolution capabilities. Even the relatively narrow-band sonar signals of the smaller cetaceans do not possess any Doppler resolution capabilities. The energy detection model of Urkowitz (1967), tested for a time–bandwidth product of 10, agreed relatively well with target detection data for *Tursiops*. Three target discrimination models were discussed: a filter bank model, a time domain highlight feature model, and a spectrogram correlation model. The relative effectiveness of these models as well as of others has not been adequately assessed yet under rigorous test conditions, including the presence of broadband masking noise. Neural network simulations of dolphin echolocation have been performed with encouraging results. Although the application of neural network theory to dolphin sonar is only in its infancy, this approach promises to be very useful.

References

Altes, R.A. (1973). Some invariance properties of the wideband ambiguity function. J. Acoust. Soc. Am. 53: 1154–1160.

Altes, R.A. (1980a). Models for echolocation. In: R.G. Busnel and J.F. Fish, eds., *Animal Sonar Systems*, New York: Plenum Press, pp. 625–671.

Altes, R.A. (1980b). Detection, estimation, and classification with spectrograms. J. Acoust. Soc. Am. 67: 1232–1246.

Altes, R.A., and Faust, W.J. (1978). Further development and new concepts for bionic sonar. Vol. 1: Software Processors. Naval Ocean Systems Center Techn. Rep. 404, San Diego, Cal.

Au, W.W.L. (1988). Detection and recognition models of dolphin sonar systems. In: P.E. Nachtigall and P.W.B. Moore, eds., *Animal Sonar: Processes and Performance*. New York: Plenum Press, pp. 753–768.

Au, W.W.L., and Martin, D.W. (1989). Insights into dolphin sonar discrimination capabilities from human listening experiment. J. Acoust. Soc. Am. 86: 1662–1670.

Au, W.W.L., and Moore, P.W.B. (1990). Critical ratio and critical bandwidth for the Atlantic bottlenose dolphin. J. Acoust. Soc. Am. 87: 1635–1638.

Au, W.W.L., Moore, P.W.B., and Pawloski, D.A. (1988). Detection of complex echoes in noise by an echolocating dolphin. J. Acoust. Soc. Am. 83: 662–668.

Beuter, K.J. (1980). Echo evaluation in auditory system of bats. In: R.G. Busnel and J.F. Fish, eds. *Animal Sonar Systems*. New York: Plenum Press, pp. 747–761.

Bracewell, R.N. (1978). *The Fourier Transform and Its Applications*, 2nd ed. New York: McGraw-Hill.

Burdic, W.S. (1968). *Radar Signal Analysis*. Englewood Cliffs, N. J.: Prentice-Hall.

Chestnut, P., Landsman, H., and Floyd, R.W. (1979). A sonar target recognition experiment. J. Acoust. Soc. Am. 66: 140–147.

Dayhoff, J. (1990). *Neural Network Architectures: An Introduction*. New York: Van Nostrand Reinhold.

Dziedzic, A., Escudie, B., Guillard, P., and Hellion, A. (1969). Mise en evidence de la tolerance a l'effet Doppler de l'emission sonar de *Delphinus delphis*. Cetace. odontocete. C.R. Acad. Sci. 279: 1313–1316.

Dziedzic, A., Chiollaz, M., Escudie, B., and Hellion, A. (1979). Sur quelques proprietes des signaux sonar frequence du dauphin *Phocoena phocoena*. Acustica 37: 258–266.

Gabor, D. (1947). Acoustical quanta and the theory of hearing. Nature 159, 591–594.

Green, D., and Swet, J. (1966). *Signal Detection Theory and Psychophysics*. Huntington, N. Y.: Krieger Pub. Co.

Grossberg, S. (1969). Some networks that can learn, remember and reproduce any number of complicated space-time patterns. J. Math. and Mech. 19: 53–91.

Hammer, C.E., Jr., and Au, W.W.L. (1980). Porpoise echo-recognition: an analysis of controlling target characteristics. J. Acoust. Soc. Am. 68: 1285–1293.

Hecht-Nielsen, R. (1987). Counterpropagation networks. In: M. Caudill and C. Butler, eds., *Proceedings of the IEEE First International Conference on Neural Networks*, Vol. 2. SOS Printing, San Diego, pp. 19–32.

Hecht-Nielsen, R. (1989). *Neurocomputing*. Reading, Mass.: Addison-Wesley.

Johnson, S.C. (1968a). Relation between absolute threshold and duration of tone pulse in the bottle-nosed porpoise. J. Acoust. Soc. Am. 43: 757–763.

Johnson, S.C. (1968b). Masked tonal thresholds in the bottlenosed porpoise. J. Acoust. Soc. Am. 44: 965–967.

Kamminga, C., and Beitsma, G.R. (1990). Investigations on cetacean sonar IX remarks on dominant sonar frequencies from *Tursiops truncatus*. Aquatic Mamm. 16.1: 14–20.

Kelly, E.J., and Wishner, R.P. (1965). Matched-filter

theory for high-velocity, accelerating targets. IEEE Trans. Microwave Theory Tech. 9: 56–69.

Kohonen, T. (1988). *Self-Organizing and Associative Memory*. New York: Springer-Verlag.

Lin, Z.-B. (1984). A method for computation of wideband ambiguity function and the numerical analysis of the bat's sonar signal. IEEE ICASSP Proc. 3: 47.11.1–4.

Lin, Z.-B. (1988). Wideband ambiguity function of broadband signals. J. Acoust. Am. 83: 2108–2116.

Martin, D.W., and Au, W.W.L. (1988). An automatic target recognition algorithm using time-domain features. In: P.E. Nachtigall and P.W.B. Moore, eds., *Animal Sonar: Processes and Performance*. New York: Plenum Press pp. 829–833.

Meyer, P.L. (1965). *Introductory Probability and Statistical Applications*. , Reading, Mass.: Addison Wesley.

Moore, P.W.B., Roitblat, H.L., Penner, R.H., and Nachtigall, P.E. (1991). Recognizing successive dolphin echoes with an integrator gateway Network. Neural Networks 4: 701–709.

Murchison, A.E. (1980). Detection range and range resolution of porpoise. In: R.G. Busnel and J.F. Fish, eds., *Animal Sonar Systems*. New York: Plenum Press, pp. 43–70.

Parker, D.B. (1982). Learning logic. Invention Report S81-64, File 1, Office of Technology Licensing, Stanford University, Stanford, Cal.

Rihaczek, A.W. (1969). *Principles of High-Resolution Radar*. New York: McGraw-Hill.

Roitblat, H.L., Moore, P.W.B., Nachtigall, P.E., Penner, R.H., and Au, W.W.L. (1989). Natural echolocation with an artificial neural network. Intern. J. Neural Net. 1: 239–248.

Rumelhart, D.E., and McClelland, J. L. (1986). *Parallel Distributed Processing*, Vol. 1. Cambridge, Mass.: MIT Press.

Schusterman, R.J., Kersting, D.A., and Au, W.W.L. (1980). Stimulus control of Echolocation Pulses in *Tursiops truncatus*. In: R.G. Busnel and J.F. Fish, eds., *Animal Sonar Systems*. New York: Plenum Press. pp. 981–982.

Schwartz, M. (1980). *Information Transmission, Modulation, and Noise*, 3rd ed. (New York: McGraw Hill.).

Urick, R.J. (1983). *Principles of Underwater Sound*, 3rd ed. New York: McGraw Hill.

Urkowitz, H. (1967). Energy detection of unknown deterministic signals. Proc. IEEE. 55: 523–531.

Wasserman, P.D. (1989). *Neural Computing Theory and Practice*. New York: Van Nostrand Reinhold.

Werbos, P.J. (1974). Beyond regression: new tools for prediction and analysis in the behavioral sciences. Masters thesis, Harvard University

Woodward, P.M. (1953). *Probability and Information Theory with Applications to Radar*, New York: Pergamon Press.

11

Comparison Between the Sonar of Bats and Dolphins

Another animal that possesses a sophisticated and highly developed sonar is the bat. A considerable amount of research is being performed on the auditory system of bats, much more than on dolphins. A conservative estimate would indicate that there are approximately three to four times more scientists involved with bat sonar and auditory research than with dolphins. There are many reasons for this imbalance; most are related to the high cost associated with a dolphin research facility. The expense of constructing a facility having large tanks filled with high-quality salt water, or installing a facility that includes animal holding pens and laboratories in an isolated but readily accessible lagoon, can be quite high. In comparison, a cage to house a colony of bats can be easily and inexpensively constructed and located in most laboratories even though bats require good control of room temperature and humidity. The amount of space needed per animal is considerably less for bats than for dolphins. The cost of long-term maintenance of dolphins is also considerably higher than that for bats. Dolphins are not as readily available as bats, and the expense of purchasing subjects can be very high. Cost aside, within the United States special permits issued by the National Marine Fisheries Service must be obtained in order to capture, acquire, or purchase marine mammals.

In this chapter we will compare the sonar of bats and dolphins, with emphasis on certain specific similarities and dissimilarities. The treatment of bat sonar will not be extensive since full treatment is beyond the scope of this book. Readers who want more detailed information on the sonar of bats can refer to the published proceedings of three excellent international NATO Advanced Study Institutes edited by Busnel (1967), Busnel and Fish (1980), and Nachtigall and Moore (1988). Recent advances in bat sonar research can also be found in a collection of reviews edited by Fenton et al. (1987).

There are many obvious differences—in fact, hardly any similarities—between bats and dolphins in general. Dolphins are air-breathing mammals which live in an aquatic environment whereas bats are flying mammals living in a terrestrial environment. Dolphins are considerably larger in size and weight than bats. A seemingly endless list of differences between the two classes of animals can be compiled, compared with only few similarities, among those being that both are mammals and echolocators. Although a common and important sonar function of both animals involves the capture of prey, there are large differences in the physical characteristics and behavior of prey types as well as in the environment they inhabit. Therefore it would not be surprising to find vast differences in the functioning, characteristics, and capabilities of the two sonar systems.

11.1 Comparison of Sonar Signals

11.1.1 Signal Characteristics

There are over 700 species of bats that use sonar to detect, localize, and capture prey. Most of these bats feed on flying insects; a few of the larger species are carnivorous and eat frogs, lizards, fishes, birds, and small mammals, while still others are omnivorous and eat everything from fruits to insects, frogs, and even other species of bats (Neuweiler 1990). Echolocation signals are brief sounds varying in duration from 0.3 to 200 ms and in frequency from 12 to 200 kHz (Neuweiler 1990). In contrast, we found in Chapter 7 that dolphins produce sonar signals that can be characterized as clicks; most dolphins produce broadband clicks, although some of the smaller species produce only narrow-band clicks. The structure of bat echolocation sounds is varied and diverse, being both species and situation specific (Pye 1980). In most species the sounds consist of either frequency-modulated (FM) components alone (instantaneous frequency varies with time) or a combination of a constant frequency (CF) component coupled with FM components. Sonar signals typically consist of the following elements or of combinations of them emitted as single or multiple harmonics: (1) downward FM sweep with linear or exponential time course (FM_{down}), (2) CF tone or shallowly modulated tonal element; (3) upward FM sweep with linear or curved time course (FM_{up}), which

only occurs in combination with other sound elements (Neuweiler 1990). FM-only sonar signals are brief in duration, varying from 0.5 to 10 ms. The sweep is usually downwards. CF signals are either short in duration, varying from 1 to 10 ms, or quite long, varying from 10 to 100 and sometimes to 200 ms.

Bat sonar signals are often analyzed with an instrument called sonograph. It produces visual spectrogram displays showing variations in the instantaneous frequency as a function of time. Relative amplitudes are indicated by the relative darkness of the traces. A schematic of a spectrogram display of orientation sounds emitted by four different bat species, each representing a different family, is shown in Figure 11.1. The signals of the big brown bat (*Eptesicus fuscus*) and the spear nose bat (*Phyllostomus hastatus*) are downward FM sweeps with higher harmonics present. The signal of the naked-backed bat (*Pteronotus suapurensis*) is a short-CF/FM and the signal of the greater horseshoe bat (*Rhinolophus ferrumequinum*) is a long-CF/FM tone. Some species emit a short upward FM tone followed by CF and downward-FM components (*Rhinolophus* signal); the initial upward FM sweep is not always present (Simmons et al. 1975). Examples of orientation signals in the time domain with their respective spectrograms are shown in Figure 11.2 for three species of bats. The signal of *Pteronotus parnelli* (mustache bat) is a long-CF/FM with a second harmonic. The signal of the *P. suapurensis* is a short-CF/FM with multiple harmonics. *Noctilio leporiunus* (fishing bat)

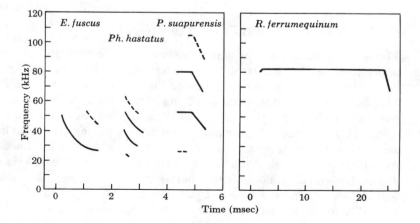

Figure 11.1. Representative orientation signals from four species of bats. Dashed lines indicate weaker harmonics. (From Simmons et al. 1975.)

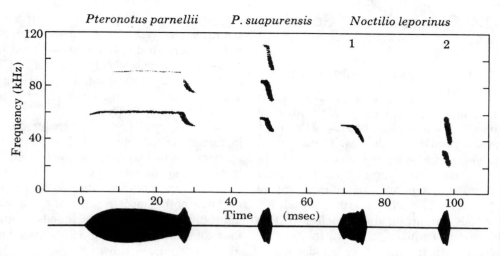

Figure 11.2. Orientation sonar signals of three species of bats shown in the time domain with the corresponding spectrograms above each time domain signal. (From Simmons et al. 1975.)

emits two types of sounds, a short-CF/FM and an FM with a second harmonic. The spectrogram displays in both Figures 11.1 and Figure 11.2 indicate that the bandwidth of the FM signals can be very wide, extending over an octave. Like dolphin echolocation signals, FM signals used by bats are Doppler tolerant (Altes and Titlebaum 1970) and are not affected significantly by either their own motion or that of prey. These FM signals cannot convey significant Doppler information to the bats. On the other hand, long-CF signals used by some bats are affected by the velocity of both bat and prey so that sonar echoes can contain Doppler information.

11.1.2 Signals During Pursuit of Prey

The sonar signals used by bats can best be understood in the context of prey detection, localization, tracking, and recognition. Figure 11.3 shows a typical interception of a flying insect based on stroboscopic photographs of Webster (1967). This sequence of flight maneuvers is fairly stereotyped for many species of bats, including *Myotis lucifugus* (little brown bat), *Eptesicus fuscus*, *Pipistrellus kuhli* (pipistrelle bat), *Rhinolophus ferrumequinum*, *Pteronotus parnellii*, and two other species of *Pteronotus* (Simmons 1989). According to Kick and Simmons (1984), there are

three distinct stages to the acoustic behavior of bats during the interception process; the search, approach, and terminal stage. These stages can be identified by the changes in sonar emission pattern during the interception process. Simmons (1987) presented a spectrogram record of the sonar sounds used by *Eptesicus fuscus* during a pursuit maneuver that is reproduced in Figure 11.4. It shows 34 signal emissions for a period extending from about 1.5 s before capture to the moment of capture. The first three emissions were FM sweeps over a narrow frequency range from about 28 to 22 kHz for the fundamental, or first, harmonic and from 56 to 44 kHz for the second harmonic. These were signals used to search for targets when flying in an open area (search phase), and were emitted at a rate of about 5 to 10 pulses per second. When it detected the prey, the bat reacted by emitting a distinctively new pattern of sounds. Beginning with emission #4, the FM sweep changed abruptly from shallow to steep, indicating that the insect was detected and pursuit had begun (approach phase). The signal bandwidth widened considerably, with the fundamental component sweeping from 50 to 60 kHz down to about 25 kHz, and the second harmonic sweeping from 100 kHz down to 50 kHz. During the approach phase (emission #4 to #10), both the signals and the intervals between emissions became progressive-

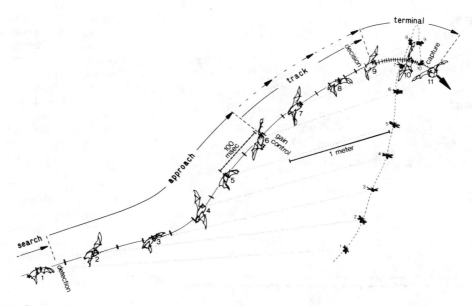

Figure 11.3. Successive positions of an insectivorous bat and a flying insect during pursuit and capture. The images of the bat and insect are separated by 100 ms. The distance between the bat and insect at any given time is shown by the dotted lines. The bat's sonar emissions are the short bars perpendicular to the bat's flight path. (From Kick and Simmons 1984.)

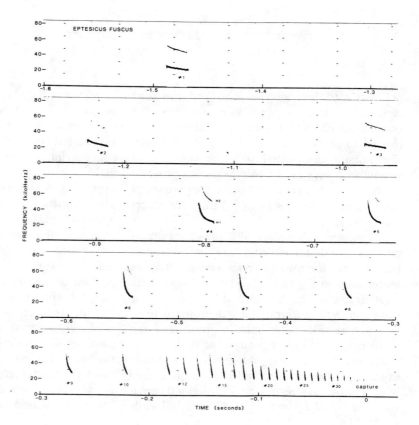

Figure 11.4. Continuous spectrogram record of a sequence of sonar signal emissions used by *Eptesicus fuscus* during a pursuit maneuver. (From Simmons 1987.)

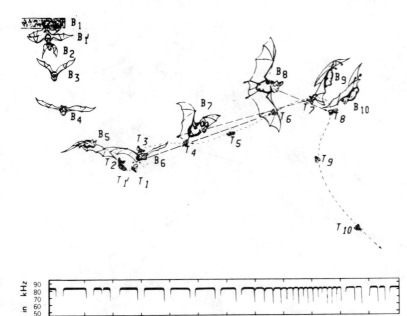

Figure 11.5. Multiexposure photograph of a horseshoe bat pursuing a moth (*top graph*), synchronized spectrogram record during the pursuit (*middle graph*), and frequency history of the CF portion of the bat's signals (*lower graph*). The frequency of the CF portion before take-off is $f_{Resting}$, and during flight it is $f_{emission}$. The frequencies, f_{E-Surr} and $f_{E-target}$ refer to the CF portion of the echoes from the stationary in front of the bat and from the insect, respectively. (From Trappe and Schnitzler 1982.)

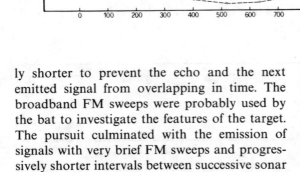

ly shorter to prevent the echo and the next emitted signal from overlapping in time. The broadband FM sweeps were probably used by the bat to investigate the features of the target. The pursuit culminated with the emission of signals with very brief FM sweeps and progressively shorter intervals between successive sonar emissions (emission #11 to 34) until the insect was captured.

Trappe and Schnitzler (1982) presented simultaneous visual and acoustic data on the pursuit behavior of a *Rhinolophus ferrumequinum* that are reproduced in Figure 11.5. The multi-exposure photograph of the bat and insect position is shown in the top graph, the synchronized spectrograph record in the middle graph, and the frequency of the CF portion of the signal in the lower graph. The first emission shown before exposure 1 (E1) corresponded to signals used while the bat was resting and scanning its environment. The frequency of the CF portion while

at rest was 83 kHz. The insect was probably detected by this emission, and the next emission occurred as the bat began its flight at exposure 1'. The bat was completely in the air when the emission immediately after E2 occurred. While in flight toward the insect, the duration of the CF portion of the signal progressively got shorter. The frequency of the CF portion during the approach phase (E2 to E5) decreased slowly to a minimum of about 81.5 kHz, and the interpulse intervals also decreased slowly. During the terminal phase (E6 to E9), the duration of the CF portion of the signal began to decrease rapidly, coupled with such rapid decrease in the intervals between emissions that successive pulses were emitted immediately after one another. The shorting of both the CF portion of the signal and the interval between emissions as the bat closed in on the moth seemed to be an attempt by the bat to prevent the echo from overlapping with the next emitted signal. During the terminal phase of pursuit, the CF portion of the echoes from the moth actually did overlap with the CF portion of the next signal; however, the FM_{down} portion of the echoes did not overlap with the FM_{down} portion of ensuing signals. The FM_{down} sweep of the signals during the terminal phase also increased by several kHz, presumably so that the bat could analyze the features of the target more closely. The moth escaped capture with a sharp turn between E7 and E8.

Little is known about how dolphins use their sonar in the detection and pursuit of prey. A few video records showing dolphins pursuing prey are available, but, simultaneously recorded video and acoustic data do not seem to exist. The narrow transmission beam coupled with distortions in the dolphin sonar signals when measured off the beam axis presents a difficult problem to overcome. Simultaneous acoustic and movie film recordings of a free-swimming bottlenose dolphin echolocating and moving toward stationary targets have been obtained by Evans and Powell (1967) and Johnson (1967). They found that as the dolphin swam toward the target, the click interval decreased continuously so that the target echo did not overlap with the next emitted click. However, unlike what we know about bats, there is no solid evidence of dolphins purposefully changing the spectral content and duration of their sonar signals when approaching prey or a stationary target (see Section 7.2 for detailed discussion).

11.1.3 Signal Levels

It is not meaningful to compare sound pressure levels measured in air and in water; acoustic intensity would be a more appropriate measure. The relationship between acoustic intensity and pressure was given by equation (1-3), which stated that intensity is equal to $p^2/\rho c$, where ρc is the characteristic impedance of the medium. At 20°C the density of air at sea level is 1.21 kg/m³ and the speed of sound is 344 m/s, so that the characteristic impedance of air is

$$(\rho c)_{air} = 416 \text{ Pa} \cdot \text{s/m} \qquad (11-1)$$

At 20°C, the density of sea water is 1,026 kg/m³ and the speed of sound is 1,500 m/s, so that the characteristic impedance is

$$(\rho c)_{sw} = 1.5 \times 10^6 \text{ Pa} \cdot \text{s/m} \qquad (11-2)$$

Hence, for a given acoustic pressure level, the intensity in sea water will be 3,606 times higher than in air. However, the sonar signals used by bats are much longer than the signals used by dolphins so that it is important to take into account the difference in signal duration. In order to compare the signal levels projected by bats and dolphins, we will calculate the energy flux density expressed in (11-3) for a signal of duration T, since it takes into account both signal duration and impedance of the media.

$$E = \frac{1}{\rho c} \int_0^T p^2(t) \, dt \qquad (11-3)$$

Unfortunately, this means that the integral in (11-3) must be evaluated.

Sound pressure level in airborne acoustics is usually expressed in terms of dB re 20 μPa and is designated as dB SPL. In order to convert the pressure amplitude to dB re 1 μPa, 26 dB must be added to the value expressed in dB SPL. Some investigators use peak equivalent SPL (dB peSPL), which is the rms sound pressure level of a continuous tone having the same amplitude as the bat signal (Møhl 1988). To convert from dB peSPL to dB SPL, one must add 9 dB to the dB peSPL value. Source levels of bat sonar signals are usually referenced to a distance 10 cm from the animal; if a 1-m reference distance is desired, 20 dB must be subtracted from the value at 10 cm.

The amplitude of bat sonar signals cannot be easily measured with free-flying bats because of the directional nature of the emitted sounds. Variability in amplitude of a bat signals is also very high, with the signals used in the search phase being the most intense and signals used in the terminal phase being the least intense. Accurate measurement of amplitudes seems to be limited to laboratory experiments in which the position of the bat's head can be controlled. Therefore, there are very few data on output amplitude levels for the various bat species. Troest and Møhl (1986) found (presumably during the detection phase) that the signals used by three *Eptesicus serotinus* varied between 94 and 104 dB peSPL. They also established for typical signals used by the bats that the energy flux

Table 11.1. Peak-to-peak source level and energy flux density for the sonar signals of some dolphins and one bat species

Species	SL (dB re 1 μPa)	E (joules/m^2)	E (db re 1 j/m^2)
Tursiops truncatus	220	8.3×10^{-3}	-20.8
Phocoena phocoena	162	4.0×10^{-8}	-74.0
Phocoenoides dalli	170	3.2×10^{-7}	-65.0
Cephalorhynchus commersonnii	160	2.3×10^{-8}	-76.3
Cephalorhynchus hectorii	151	4.0×10^{-9}	-84.0
Eptesicus serotinus (bat)	119	2.4×10^{-7}	-66.4

density was approximately equal to the sound pressure level in dB peSPL minus 124 dB. Therefore, a signal of amplitude 104 dB peSPL at 10 cm is equal to 84 dB peSPL (93 dB SPL) at 1 m and has an energy flux density of $84 - 124 = -40$ dB re 1 Pa^{2}s, which is equal to 1×10^{-4} Pa^{2}s. Dividing this value by the characteristic impedance of air ($\rho c = 416$), we obtain the source energy flux density used by the bats, which was 2.4×10^{-7} joules/m^2 or -66.2 dB re 1 joule/m^2. Kick (1982) reported a typical peak-to-peak output level of about 109 dB SPL at 10 cm for *Eptesicus fucus*. This corresponds to a pressure of 100 dB peSPL, which is only 4 dB lower than the levels reported by Troest and Møhl (1986) for *Eptesicus serotinus*. Therefore, the two species should have roughly the same source energy flux density.

The source energy flux density for dolphins can be calculated by using equation (7-19) and Table 7.3. Parameter *A* in (7-19) is the peak amplitude of a signal and is usually expressed in μPa, so that the units for *E* will be μPa^{2}s. Listed in Table 11.1 is the source energy flux density for the various dolphin species considered in Table 7.3 and the bat species used by Troest and Møhl (1986). The numbers indicate that the source energy flux density of a *Tursiops* sonar signal is about 45 dB or 3.5×10^4 times greater than the signal of *Eptesicus serotinus*. However, the energy flux density of the bat is comparable or even larger than that of most of the smaller dolphins, except for *Phocoenoides dalli*.

11.1.4 Transmission and Reception Beams

Sonar sounds of bats are produced in the larynx and vocal tract. Bats with a nose-leaf emit

sounds through their nostrils whereas bats without nose-leaf emit sound through their open mouths. Echoes and other sounds are received by bats through their external ears. Schnitzler and Grinnell (1977) measured the transmission directivity of a *Rhinolophus ferrumequinum* (a CF/FM bat) by placing a stationary microphone directly in front of the bat and having a second microphone mounted on a movable arm that could be directed toward the bat's head at any angle from $0°$ to $100°$ in any direction relative to the forward axis of the bat (the position of the bat's head was fixed). Both microphones were at a distance of 68 cm from the bat's head. The bat's signal consisted of a constant frequency portion of 84 kHz followed by a short terminal FM sweep. The difference between the two SPLs during the CF portion of sound emission gave the relative sound pressure level, SPL$_{rel}$, for each angle. The transmission beam (dashed line) for the vertical plane is shown in Figure 11.6 and for the horizontal plane in Figure 11.7. The maximum SPL was always measured by the fixed microphone directly in front of the bat. In the vertical plane the beam had a side lobe extending from about $-10°$ to $-68°$. The -3 dB level of the main beam occurred at $-8°$ and $+17°$, resulting in a vertical beamwidth of about $25°$. In the horizontal plane, the beam was symmetrical about the forward axis of the bat and the -3 dB points were at $\pm 15°$, resulting in a horizontal beamwidth of $30°$.

Simmons (1969b) measured the transmission beam of two species of bats, *Chilonycteris rubiginosa* and *Eptesicus fuscus*. The *Chilonycteris* emitted a signal with a CF component at 28 kHz which had a strong second harmonic, followed by a terminal FM$_{down}$ component, and the *Eptesicus* emitted a FM$_{down}$ signal sweeping from about 42 to 26 kHz. The average duration of the sounds

Figure 11.6. Various beam patterns in the vertical plane for a Greater Horseshoe Bat. The transmitting beam is denoted by the dashed line, receiving beam by the lines with dots and dashes and the combined transmitting and receiving beam by the solid line. (From Schnitzler and Grinnel, 1977).

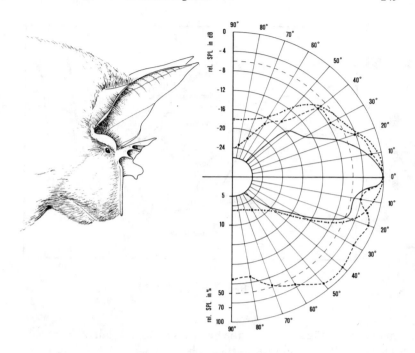

was 20 ms for the *Chilonycteris* and 1.8 ms for the *Eptesicus*. The beam patterns were measured with an array of 4 microphones at a frequency of 28 kHz for the *Chilonycteris* and at 30 kHz for the *Eptesicus*. The 3-dB beamwidth for the two bats was essentially the same, 45°. Both bats emitted their signals through their mouths and had wider beams than the nose-leaf bat *Rhinilophus*. The wider beamwidths measured by Simmons (1969b) may have been the result of his bats using lower frequencies. Another factor may have been the absence of nose-leaves in his bats.

The reception beam of a *Rhinilophus ferrumequinum* was measured by Grinnell and Schnitzler (1977). Whenever the bat, which was held stationary, emitted a signal, a phantom echo was projected back to it by a speaker located directly in front. The bat normally emitted signals consisting of a constant frequency component of 84 kHz followed by a short terminal FM sweep. A second, movable speaker was used to project noise to mask the echo. The noise level was gradually increased until the bat could no longer detect the phantom echo. The relative noise levels needed to totally mask the echo for the different speaker locations about the bat's head were used to determine the reception beamwidth. The reception beam denoted by the line of dots and dashes is depicted in Figure 11.6 for the vertical

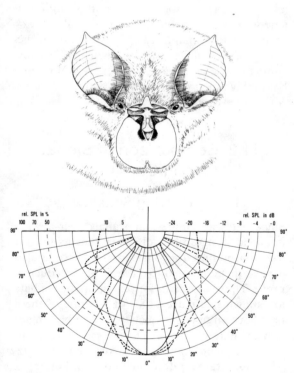

Figure 11.7. Various beam patterns in the horizontal plane for a Greater Horseshoe Bat. Beams are denoted as in Figure 11.6. (From Schnitzler and Grinnel 1977.)

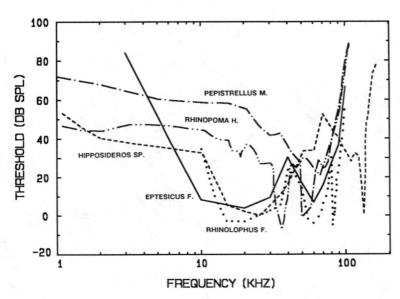

Figure 11.8. Audiogram for five species of echolocating bats. The curves were reproduced from data presented by Dalland (1965) for *Eptesicus f.*, by Long and Schnitzler (1975) for *Rhinolophus f.*, and by Neuweiler (1984) for *Pipistrellus m.*, *Rhinopoma h.*, and *Hipposideros* sp.

plane and in Figure 11.7 for the horizontal plane. The −3-dB points in the vertical plane were at angles of approximately −18° and +19°, so that the beamwidth was 37°. In the horizontal plane, the −3-dB points occurred at approximately ±23° so that the beamwidth was 46°. The behavioral measurements of Grinnell and Schnitzler (1977) agreed well with reception directivity measurements performed with electrophysiological techniques.

The beam patterns for the total echolocation system are denoted by the solid line in Figures 11.6 and 11.7. The combined beam had a 3-dB beamwidth of about 16° in the vertical plane and 20° in the horizontal plane. The side lobe of the transmission beam in the vertical plane was essentially flattened in the combined beam.

The various beam patterns for the horseshoe bat depicted in Figures 11.6 and 11.7 can be compared with corresponding beam patterns for the Atlantic bottlenose dolphin (Figs. 6.9 and 8.30). The dolphin transmission beamwidth was approximately 11° in both the vertical and horizontal planes. The reception beamwidth was approximately 18° in both planes. It is quite obvious that the various beam patterns for the dolphin are narrower than the corresponding beam patterns for the bat. This is not surprising, since the head of the bat is much smaller than that of the dolphin. However, the bat has two

fairly large pinnas that are very effective in providing good directional hearing capabilities, even with a small head size. The nose-leaf of the horseshoe bat is also effective in allowing for a narrower transmission beam than would be expected for the small head size of the bat. In the words of Schnitzler and Grinnell (1977), "the complex nose-leaves are not a caprice of nature but highly functional structures optimized in the course of evolution."

11.2 Comparison of Signal Detection Capabilities

11.2.1 Hearing Sensitivity

Since bats can emit sonar signals over a wide frequency range, especially when the second and third harmonics are considered, we should also expect them to hear over a broad frequency range. The audiograms for several species of bats are shown in Figure 11.8. All but one of these bats have an upper frequency limit of hearing of approximately 100 kHz. The *Hipposideros* has an upper frequency limit of approximately 150 kHz. The audiogram for dolphins shown in Figure 3.2 indicated comparable upper frequency limits, varying from about 100 to 150 kHz depending on the specific species. The lower limit of hearing for the bats in Figure 11.8 extends below 1 kHz. The *Eptesicus* curve suggests that the lower frequency limit is about 2 kHz. However, Poussin and Simmons (1982) studied the low frequency hearing of two *Eptesicus* from 200 Hz to 5 kHz and found that the audiogram had a sharp dip at 1 kHz with a sensitivity below 40 dB SPL. For frequencies below 1 kHz, down to 200 Hz, the bats' sensitivity decreased continuously reaching a level of about 87 dB SPL at 200 Hz. Dolphins seem to have similar low frequency hearing capabilities, also extending below 1 kHz.

The minimum sound pressure level that the bats represented in Figure 11.8 could detect varied between 0 dB SPL and −10 dB SPL. Acoustic pressures of 0 dB SPL and −10 dB SPL correspond to intensity levels of 10^{-12} and 10^{-13} watts/m². respectively. The audiogram for several species of dolphins in Figure 3.2 indicated that the typical maximum sensitivity level varied between 51 dB and 34 dB re 1 μPa. Sound pressure levels of 50 to 34 dB correspond to intensity levels of 8.4×10^{-14} to 1.7×10^{-15} watts/m². These data indicate that within the frequency range of best hearing, dolphins can hear lower intensity sounds than bats by 10 to 20 dB, depending on the specific species being compared.

11.2.2 Target Detection Range

Kick and Simmons (1984) measured the target detection capabilities of two *Eptesicus fuscus* using nylon spheres with diameters of 0.48 cm and 1.91 cm as targets. The bats were first pre-trained to detect each target at a close range of 20 cm. The distance from bat to target was gradually increased by increments of 25 cm until performance fell below the 75% correct response threshold. On each correct response trial the bats were rewarded with a piece of mealworm. The performance of one bat is shown in Fig. 11.9, with percent correct responses plotted against the target range. For the 1.91-cm sphere, the 75% correct response threshold range was 5.1 m with both bats. The averaged threshold range with the 0.48-cm sphere was 2.9 m. The experiment was conducted in a low-noise environment; the noise level was not measurable. Therefore, the bats were probably not masked by external noise but

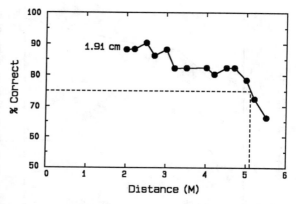

Figure 11.9. Target detection performance of one bat (*Eptesicus fuscus*) in target detection experiment with a 1.91-cm sphere. Each data point represents 50 trials. (Adapted from Kick 1982.)

by their own internal noise, and it was their absolute threshold for hearing their own sounds that was measured. Both bats typically emitted FM_{down} signals varying from 65 to 30 kHz, with strong first and second harmonics and a slight third harmonic.

In Section 8.3.1, we saw that the detection threshold range for *Tursiops truncatus* with a 2.54-cm diameter sphere was 73 m. This range is considerably longer than the one obtained for bats, especially in light of the fact that the dolphin was limited by external ambient noise. The difference in target strength between a 1.91-cm and a 2.54-cm sphere is not very large. The target strength of a sphere of radius a is given by equation (9-9) for $ka \gg 1$, where k is the wavenumber and is given as $2\pi/\lambda$; λ being the wavelength of the signal. For a sphere of radius $a_a = 0.96$ cm and a frequency of 30 kHz, $ka = 7.1$ in air. Correspondingly, for the sphere of radius $a_w = 1.27$ cm in water and a frequency of 120 kHz, $ka = 6.3$. Therefore, the target strength expression (9-9) should apply to both the air and underwater cases. The difference in target strength, ΔTS, will be

$$\Delta TS = 20 \log \frac{a_a}{a_w} \qquad (11\text{-}4)$$

The 1.91-cm sphere used with the bats had a target strength which was only about 2.5 dB lower than the target strength of the 2.54-cm sphere used with the dolphin.

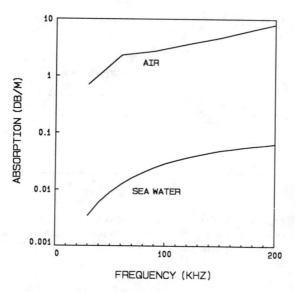

Figure 11.10. The attenuation of acoustic energy in air and water a function of frequency. The in-air curve was drawn from measurements made by Lawrence and Simmons (1982a). The seawater curve was calculated with eq. (1.12) using a temperature of 25°C.

One of the reasons for the bats to have such a small detection range is related to the high absorption of acoustic energy at ultrasonic frequencies in air. Lawrence and Simmons (1982a) measured the absorption of sounds as a function of frequency in their bat research laboratory and discussed the implications of high sound attenuation of echolocation by bats. Their measurements are plotted in Figure 11.10 along with a curve showing the attenuation of acoustic energy in seawater at a temperature of 25° Celsius. The attenuation in air is on the order of 40 dB/m higher than in seawater. The total two-way acoustic transmission loss (including spreading and absorption loss) experienced by the bats at a frequency of 30 kHz and a range of 5.1 m was approximately 78 dB. The transmission loss at a frequency of 60 kHz was approximately 91 dB. The spreading loss for the bats was calculated between a reference range of 10 cm and a target range of 51 cm. The dolphin experienced a comparable two-way loss of approximately 81 dB. The other reason for the dolphin's longer detection range is the larger energy flux density of the dolphin's signal compared with the bat's.

Møhl (1986) conducted an echo detection in noise experiment with a Pipistrelle bat (*Pipistrellus pipistrellus*). A phantom target electronic playback system was used to play back a previously recorded echolocation after a specific delay whenever the bat emitted a sonar signal. The delay was chosen so that the bat would receive the phantom echo 8 ms after it emitted a signal. The microphone to measure the bat's sonar signal and the playback speaker were located about 25 and 85 cm from the bat's platform, respectively. A constant amplitude broadband noise was played back with the phantom echo. The amplitude of the phantom echo was varied according to an up/down staircase procedure to determine the bat's detection threshold. Only when the bat responded correctly did it receive a mealworm reward. At threshold, the ratio of the total echo to the noise spectrum density was 50 dB. Møhl attributed this high E/N ratio to the effects of clutter caused by echoes produced by the speaker (front face diameter of 15 mm).

Troest and Møhl (1986) also measured the echo detection in noise capabilities of three serotine bats (*Eptesicus serotinus*), using the same phantom target system as Møhl (1986). However, the delay between the emission of a sonar signal and the reception of an artificial echo was reduced to 3.2 ms. The amplitude of the artificial echo was varied according to an up/down staircase procedure. The microphone and speaker were located at a distance of 22 cm and 88 cm, respectively, from the bat's platform. The average echo energy-to-noise ratio at threshold was found to be 36 dB. However, the presence of clutter from the speaker, which trailed the echo by 2 ms, may have affected the bats' performance.

The average echo energy-to-noise ratio at the detection threshold for four *Tursiops truncatus* from Figure 10.6 was approximately 7.5 dB. This signal-to-noise ratio is considerably smaller than the 36 and 50 dB measured in bat experiments at the University of Aarhus, Denmark. However, as mentioned, the results of the bat experiment were contaminated by clutter caused by the playback speakers. Nevertheless, the phantom echo in noise detection studies of Møhl (1986) and Troest and Møhl (1986) and the target detection experiments of Kick and Simmons (1984) strongly suggest that the target detection capability of *Tursiops truncatus* is considerably better than that of bats.

An interesting phenomenon studied by Kick and Simmons (1984) involves automatic gain control in the bat's receiver. The estimated echo amplitudes for echoes from the 0.48-cm and 1.9-cm spheres at threshold ranges of 2.9 m and 5.1 m, respectively, were in the region of 0 dB SPL. However, for target ranges less than about 1.1 m (6.4 ms delay from the emission of a signal to the reception of the echo), the echo amplitude at threshold actually increased at a rate of 11 to 12 dB per halving of range (Kick and Simmons 1984; Simmons et al. 1992). This implies that as a bat approaches an insect at ranges less than about 1 m, the increased threshold cancels out the increase in echo amplitude of 12 dB per halving of range, so that the echo amplitude is stabilized at the cochlea and the bat effectively has automatic gain control of the echoes. The elevated threshold for target ranges less than 1 m is probably the result of the bat's middle ear muscles contracting in synchrony with vocalization in order to protect the auditory system from damage from the relatively intense sounds being emitted. The middle ear muscles of the FM bats *Myotis lucifugus* and *Tadarida brasiliensis* contract at the moment of vocalization and relax over the course of up to 5 to 8 ms after the vocalization (Henson 1965; Suga and Jen 1975). The muscle contraction actually begins several milliseconds before the onset of vocalization (Suga and Jen 1975). Such an automatic gain control phenomenon has not been observed with dolphins.

11.3 Comparison of Target Discrimination Capabilities

11.3.1 Range Difference Discrimination

The target range resolution capability of *Eptesicus fuscus* was studied by Simmons (1969a) for his Ph.D. research. His results and conclusions have spawned a lively and interesting controversy that has yet to be settled after nearly 24 years. In one of his experiments involving four different species of bats (*Eptesicus fuscus, Phyllostomus hastatus, Pteronotus suapurensis,* and *Rhinolophus ferrumequinum;* Simmons 1973), the bats were required to lie on a starting platform and determine by echolocation which one of two triangular planar targets was closer to the platform. The targets were separated by an azimuth of 40°, with the origin being at the typical location of the bat's head on the starting platform. One of the targets was kept at a fixed distance from the bat's platform and the other at a closer, variable distance. The difference in distance between the nearer and farther targets was decreased from 10 cm down to zero in steps of 5 or 10 mm. The results of the range resolution experiment for *Eptesicus fuscus* are shown in Figure 11.11A, with the fixed target located at a range of either 30 or 60 cm. The two curves for the different absolute ranges were essentially identical. This

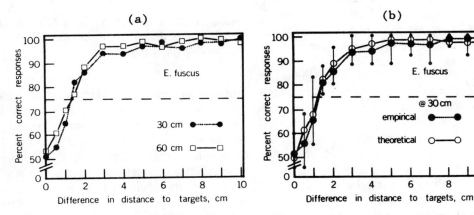

Figure 11.11. (*A*) Average target-range discrimination performance of *Eptesicus fuscus* at absolute ranges of 30 cm (for eight bats) and 60 cm (for three bats). (*B*) Average target-range discrimination performance of eight *Eptesicus* at an absolute range of 30 cm. The "empirical" curve shows the actual performance of the bats, and the "theoretical" curve shows the performance predicted from the envelope of the autocorrelation function of the bat's target-ranging sonar signal. (From Simmons 1973.)

Table 11.2. Target range discrimination by echolocating bats

Species	Absolute Range	Range Difference Threshold	$\Delta T = 2\,\Delta R/c$
Eptesicus fuscus[1]	30 cm	1.3 cm	77.8 μs
	60 cm	1.3 cm	77.8 μs
	240 cm	1.4 cm	83.8 μs
Myotis oxygnathus[2]	100 cm	0.8 cm	47.9 μs
	100 cm	2.3 cm	138.7 μs
Noctilio albiventris[3]	35 cm	1.3 cm	77.8 μs
Phyllostomus hastatus[1]	30 cm	1.3 cm	77.8 μs
	60 cm	1.2 cm	71.9 μs
	120 cm	1.2 cm	71.9 μs
Pipistrellus pipistrellus[4]	21–24 cm	1.5 cm	89.8 μs
Pteronotus gymnonotus[1]	30 cm	1.5 cm	89.8 μs
	60 cm	1.7 cm	101.8 μs
Rhinolophus ferrumequinum[1,2]	30 cm	2.8 cm	167.7 μs
	60 cm	4.1 cm	245.5 μs

[1] Simmons (1973); [2] Ayrapet'yants and Konstantinov (1974); [3] Roverud and Grinnell (1985); [4] Surlykke and Miller (1985)

led Simmons (1969a) to speculate that the bat may have a coherent or matched-filter receiver since the range discrimination capability of a coherent receiver does not depend on range for high signal-to-noise ratios. Another factor strengthening Simmons's conviction that bats process echoes like a matched filter were theoretical predictions of the bat's performance based on the envelope of the autocorrelation function of the emitted sonar signals as shown in Figure 11.11B. Equation (9-16) gives the theoretical prediction for a matched filter. Simmons (1969a, 1973) also used phantom electronic targets to study target range discrimination. The results for the phantom targets were essentially the same as for real targets.

The target range discrimination thresholds for seven species of bats are reproduced in Table 11.2 from data presented in Simmons and Grinnell (1988). The data indicate that target range discrimination thresholds for bats lie in the 1- to 2-cm region, except for *Rhinolophus*, where the threshold is 3 to 4 cm. Furthermore, the range discrimination threshold is relatively independent of target range, except for *Rhinolophus*. Once again, these data seem to indicate that bats may process echoes in a similar way as a cross-correlation or matched-filter receiver.

The corresponding target range discrimina-

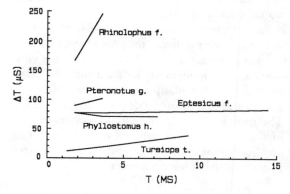

Figure 11.12. Graphs of ΔT as a function of the two-way travel time T for the bat data in Table 11.1 and the dolphin data in Figure 9.37.

tion performance for an echolocating Atlantic bottlenose dolphin was shown in Figure 9.38. The range difference thresholds at the 75% correct response level were 0.9, 1.5, and 3 cm for absolute target ranges of 1, 3, and 7 m. The time difference thresholds for dolphin and bat are plotted in Figure 11.12 as a function of the two-way travel time T for a signal to travel from the animal to the target and back. With the exception of *Rhinolophus f.*, the bats' time-difference thresholds were relatively constant as a function of T. The dolphin's threshold increased

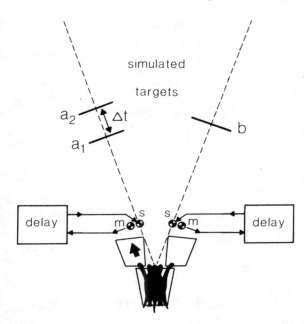

Figure 11.13. Diagram of the echo jitter experiment with electronic phantom targets. The bat was trained to detect the target that jittered in range from a_1 to a_2, and the size of the jitter interval, Δt. (From Simmons et al. 1990).

the next emission, back to t_1, and so on. Along the other arm of the Y a fixed delay time was used. The bat was trained to detect the echoes that jittered in range. Figure 11.14 shows the smallest detectable change in echo delay for *Eptesicus fuscus*, with percent correct response plotted as a function of the amount of echo jitter. The jitter time at the 75% correct response threshold was approximately 10 to 12 ns. Simmons et al. (1990) showed that the bats' performance could be predicted from the fine structure of the cross-correlation function of an echo and the emitted signal.

A comparable jitter echo experiment has not been performed with dolphins. However, in Chapter 8 it was shown that dolphins probably do not process echoes like a matched filter. Although the cross-correlation model of Simmons and his colleagues does seem to fit data obtained in range-difference and range-jitter experiments, there are some other considerations that must be addressed. First, a cross-correlation model implies that the emitted signal is exactly known. Yet it is not clear how a bat can obtain an accurate neural template of its own emitted signal. The bat's ears are probably in the near field of the projector (mouth or nostril) and are located behind the area where signals are emitted. If the template of the emitted signal is received via the external ears, then some sort of distortion should be expected. The emitted signals may be received through some other pathways within the bat's head. Nevertheless, whatever a bat hears of its emitted signal would probably not be exactly the same as the signal propagating toward the target. A good analogy of this line of argument comes from our human experience of hearing our own voice as we speak compared to hearing our voice through a tape recorder. The tape recording of our own voice sounds different than what we perceive as we speak. Some have argued that a bat can memorize its signal through experience in hearing echoes. However, our discussion of Figures 11.4 and 11.5 indicated that a bat changes its signal during pursuit, so its memory of a single or several signals would not be adequate. Furthermore, the results of Møhl (1986) seem to be inconsistent with a matched-filter model. In one of his tests, the playback echo was a time-reversed version of the emitted signals. Yet the

almost linearly with time. However, its threshold values were smaller than those of the bats, suggesting that the dolphin had the better time difference resolution capability. The broadband click signals used by dolphins have better time resolution capabilities than the longer bat signals, as can be seen by comparing the theoretical prediction curves for a dolphin (Fig. 9.38) and for a bat (Fig. 11.11).

In order to further study the range perception acuity of bats, Simmons (1973) and Simmons et al. (1990) performed an echo jitter experiment using phantom electronic targets with a setup that is depicted in Figure 11.13. The bat was trained to sit on a small Y-platform and emit its signals in the direction of the left and right branches. A microphone was located at the end of each arm of the Y to detect the bat's emissions. Two delay times, one representing target a_1 and the other target a_2, were used. The phantom echoes along one arm had delay times that alternated for successive sonar emissions, from t_1 (representing target a_1) on one emission to t_2 on

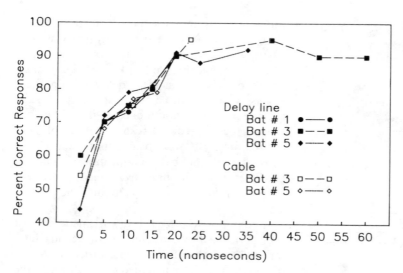

Figure 11.14. Results of the jitter experiment for three *Eptesicus fuscus*. 'Cable' refers to delays being obtained by sending the bat's emitted signal through a coaxial cable of appropriate length. (From Simmons et al. 1990).

bat's detection performance did not change. A cross-correlation receiver would require a 19 dB higher signal level to detect the time-reversed signal than the normal signal. Masters and Jacobs also performed an echo detection threshold experiment using phantom targets with normal and time-reversed model echoes. Like Møhl (1986), they obtained a similar detection threshold with both types of echoes. However, when they used the normal and time-reversed echoes in a range difference experiment, they found great variations in range difference threshold. Range difference threshold increased from about 1 cm with normal model echoes to 18 cm or more with reversed model echoes (Masters and Jacobs 1989). They suggested that range determination does involve some form of matched filtering on the part of the bat. For Masters and Jacobs, postulating "matched filtering" for the bat was not the same as suggesting that the bat had a coherent or semicoherent crosscorrelation receiver. They concluded that prior knowledge of an echo's time–frequency structure is crucial, and in this sense the bat's receiver is matched. Although the echo detection performance of bats is not fully consistent with the standard matched-filter model, that is the only model available to explain the bat's performance in the target range difference and jitter experiments. Perhaps the bat uses two different processing schemes, functioning separately but in parallel, to detect signals in noise and discriminate based on target range.

11.3.2 Specific Target Discrimination

Bats use their sonar to pursue prey and therefore must be able to discriminate nonprey from prey objects by analyzing the sonar returns. Bats that emit signals with a CF component may exploit the utility of CF signals in encoding Doppler information. FM bats will attack flying insects as well as a variety of other airborne objects they encounter (Simmons and Chen 1989). However, after some practice they will ignore inedible objects: it has been demonstrated that they can discriminate a mealworm from a host of plastic disks all simultaneously thrown into the air. These FM bats probably do not rely upon a simple set of acoustic features of echoes to discriminate inedible objects from flying insects.

Simmons and Chen (1989) measured the features of a nylon disk of 12.5 mm diameter and those of a mealworm by projecting impulsive sounds at these targets and measuring the echoes. The curvature of the circular disk resembled the typical curvature of mealworms. Figure 11.15 shows the echo waveform of the disk at four different orientations and that of the mealworm at five different orientations. Airborne acoustic signals will not penetrate most targets because of the large impedance difference between air and most materials. Therefore, echo highlights are the result of reflections from different parts of a target, as can be seen in Figure 11.15. In contrast, echo highlights from underwater targets can be

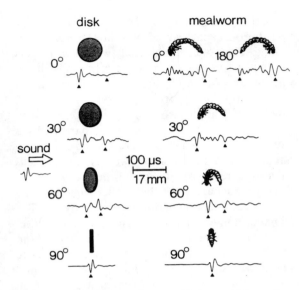

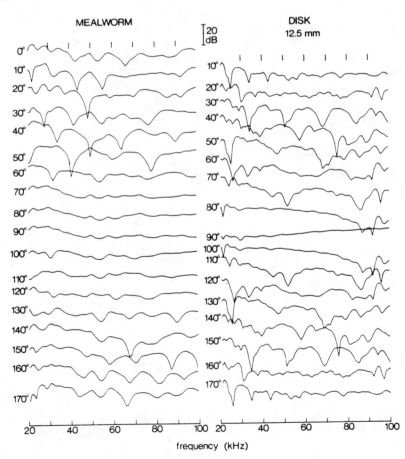

the result of acoustic energy penetrating a target and experiencing internal reflections. The multi-highlight echoes from the disk and the mealworm will cause the frequency domain representation of the echoes to have a rippled spectrum, as was discussed in Section 4.2. The frequency responses of the disk and mealworm at different orientations are shown in Figure 11.16. Simmons and Chen (1989) argued that differences in the spectral content, especially the differing location of the nulls, provided sufficient cues for bats to discriminate between the two objects.

Target cues used by dolphins were discussed in Chapter 8; among them were spectral ripples of echoes in the frequency domain. Therefore, bats and dolphins may use the same types of cues to discriminate between nonfluttering targets. However, it is not known for either bat or dolphin whether or not the echoes are in fact analyzed in the time or the frequency domain, or

Figure 11.15. Echoes from a 12.5-mm disk and a mealworm obtained with impulsive acoustic signals. (From Simmons and Chen 1989.)

Figure 11.16. Frequency response of the mealworm and the 12.5-mm disk for various orientations in the horizontal plane. (From Simmons and Chen 1989.)

both. Also, the possible generation of TSP within the auditory system of both animals by multi-highlight echoes cannot be ignored.

11.3.3 Vertical and Horizontal Localization

Lawrence and Simmons (1982b) examined the angular resolution capability of two *Eptesicsu fuscus* in the vertical plane using two arrays of horizontal rods, with two rods in each array, positioned along an arc. The rods in one array were separated in vertical angle by 6.5°, the rods in the other array by a greater vertical angle. The bats were required to indicate which array had the smaller angular separation between the rods. The results of the experiment are shown in Figure 11.17A. At the 75% correct response threshold, the bats could resolve angular separations of 3° to 3.5°. Lawrence and Simmons (1982b) demonstrated that the tragus of the bat's external ear played a major role in assisting the bats to perform the vertical localization task. They glued the tragus of each ear forward and down to the adjacent fur and found that the bats' performance fell significantly.

Simmons et al. (1983) conducted a similar experiment in order to investigate the angular resolution capability of bats in the horizontal plane. They once again used two set of targets each consisting of an array of vertical rods placed in an arc 44 cm from the bats' observation platform. First, an array of two rods separated by a small angle was presented together with another array of two rods, also separated by a small angle. The arrays were spaced randomly apart by an azimuth of 40 to 48°. Three *Eptesicus fuscus* were trained to locate the array of rods that were separated by the smaller angle, initially 6.5°, versus the pair of rods separated by 13°. The smaller angle was progressively increased in 1°-degree increments until the bats could no longer correctly identify it. Latter, an array of five rods was used. The results of the horizontal angle discrimination experiment are shown in Fig. 11.17B. The 75% correct response threshold occurred at an angular separation of 1.5°.

Bel'kovich et al. (1970), using a *Delphinus delphis*, performed a similar experiment as discussed for the bat using two arrays of two cylinders each. The cylinders were not positioned along an arc, however, which probably introduced a time delay cue caused by the targets being at slightly different distances from the

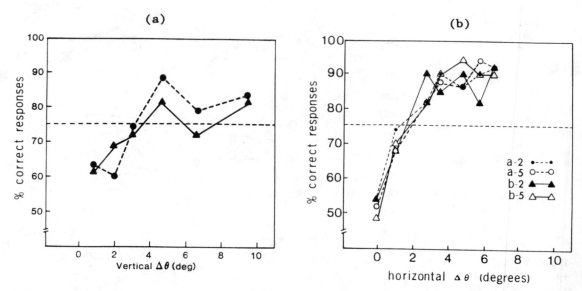

Figure 11.17. Results of angular resolution experiments with echolocating *Eptescius fuscus* (*A*) in the vertical plane (from Lawrence and Simmons 1982b), (*B*) in the horizontal plane with arrays of two and five rods (from Simmons et al. 1983).

dolphin, making the results of 0.028° suspect. Such a cue would not be present if the targets were positioned along an arc. However, the passive sound localization study discussed in Section 3.3.4 can provide a hint to the dolphin's sonar angle discrimination. Renaud and Popper (1975) measured a minimum audible angle of 0.7° in the vertical plane and 0.9° in the horizontal plane for *Tursiops* when the stimulus was a broadband click signal having a peak frequency of 64 kHz. Their data suggest that dolphins may have vertical and horizontal angular resolution capabilities slightly better than those of bats.

11.3.4 Target Size, Shape, and Composition Discrimination

In Section 9.2 we discussed the fact that a sonar's ability to recognize differences in target size really amounts to recognizing differences in target strength, which can be translated to differences in echo intensity. In discrimination experiments with cylinders of different diameters, *Noctilio leporinus* could recognize a 1-dB difference in target strength at the 75% correct response level; for *Myotis oxygnathus* the threshold was 2 dB, and for *Rhinolophus ferrumequinum* 4-5 dB (Konstantinov 1970; Ayrapet'yants and Konstantinov 1974). The capability of *Rhinolophus* was also tested with spherical targets, and discrimination thresholds of 2.4 to 4.4 dB were measured (Ayrapet'yants and Konstantinov 1974). *Rhinolophus* had a discrimination threshold of 2.8 to 3.2 dB when circular planar targets were used (Fleissner 1974). Simmons and Vernon (1971) found that a target strength difference of 1.5 to 3 dB was necessary for *Eptesicus fuscus* to discriminate triangular targets of different sizes. In Section 9.2 we saw that the target strength difference discrimination threshold for dolphins typically varies between 0.8 and 1.2 dB, with a high of about 2.8 dB in one study. Therefore, bats (with the exception of *Rhinolophus* and perhaps other bats that emit long CF signals) and dolphins have comparable target strength difference discrimination capabilities.

Dijkgraaf (1957) as reported by Schnitzler and Henson (1980) demonstrated that an echolocating *Plecotus auritus* could distinguish a disc from a cross of the same surface area. *Plecotus* could also discriminate between a cube, pyramid, and cylinder of equal volume by echolocation (Konstantinov and Akhmarova 1968). *Myotis oxygnathus* and *Rhinolophus ferrumequinum* were able to discriminate between a square with smooth edges and squares with serrated edges, and between a square with holes in it and concave or convex squares (Ayrapet'yants and Konstantinov 1974). Bradbury (1970) found that an echolocating *Vampyrum spectrum* could discriminate between a sphere and a prolate spheroid of similar target strength. The target shape discrimination capabilities of bats and dolphins cannot be directly compared since different types of targets were used. However, in Section 9.4 we saw that dolphins can discriminate between circular, square, and triangular disks, between spheres and cylinders, and between cylinders and cubes. The cues used in shape discrimination are probably similar for bats and dolphins: variation in echo intensity as the animals scan across a target, and fluctuation in echo intensity as a target is insonified from different aspects.

Dijkgraaf (1946) as reported by Schnitzler and Henson (1980) performed a material composition discrimination experiment in which *Eptesicus serotinus* demonstrated that it could discriminate between targets covered with glass and with velvet. Ayrapet'yants and Konstantinov (1974) found that *Myotis oxygnathus* and *Rhinolophus ferrumequinum* were able to discriminate between an aluminum plate and similar plates constructed of Plexiglas, plywood and textolite, but could not discriminate between plates made of brass and iron. These studies show that bats can discriminate between targets of different material composition only if the impedance difference between the targets is large. Glass has a much higher acoustic impedance than velvet. Similarly, aluminum has a much higher acoustic impedance than plywood and plastics such as plexiglas and textolite. Impedance differences could affect the amount of acoustic energy reflecting from targets. As to material discrimination in dolphins, we saw in Section 9.3.2 that they can discriminate between targets constructed of different metallic materials, implying that dolphins have a keener ability to discriminate based on material composition. However, this is not surprising since acoustic energy propagating in water will normally penetrate into a target and be reflected internally, thus providing additional cues to a dolphin. Acoustic energy in air will not normally penetrate into a target. Therefore, the difference in the reflection and scattering processes in air and water results in more cues being provided to dolphins than to bats.

11.3.5 Discrimination of Echo Highlights

Simmons et al. (1974) performed an experiment in which *Eptesicus fuscus* discriminated between a plate with holes of 8 mm depth and other plates with holes between 6.5 and 8 mm. The bats were able to discriminate based on depth differences of 0.6 to 0.9 mm. The primary cue in Simmons's experiment was the difference in time between the echo highlights from the front of the plate and the back surface of the holes. A depth difference of 0.6 to 0.9 mm can be translated into a time difference of 3.5 to 5.2 μs. The difference in time between highlights can be translated into a difference in ripple structure of the echoes in the frequency domain, as was discussed in Section 4.2. The experiments with stepped pyramids discussed in Section 9.3.5 showed that *Delphinus delphis* can discriminate between step sizes that correspond to about 4 μs difference in echo delay time. These data indicate that bats and dolphins have similar capabilities in discriminating delay time differences.

11.4 Doppler Compensation and Flutter Detection

We will now briefly discuss two capabilities of certain bats which emit CF signals—capabilities that dolphins do not possess. One is the ability to detect Doppler shifts caused by the velocity of the bat and its prey and the detection of target flutter (insect wing beat). The broadband echolocation click signals used by dolphins and the broadband FM signal used by many bats are Doppler tolerant and do not have the capability to encode velocity information. However, tonal CF signals used by rhinolophid bats, hipposiderid bats, and the mormoopid bat (*Pteronotus parnellii*) can encode velocity information (Schnitzler 1984). Let V_B be the velocity of a bat and let V_T

be the velocity of a prey; then the frequency of the CF portion of an echo returning to the flying bat will be (Schnitzler and Henson 1980)

$$f_{echo} = f_{trans}\left(1 + \frac{2V_R}{c}\right) \qquad (11\text{-}5)$$

where f_{echo} and f_{trans} are the frequencies of the echo and transmitted signal, respectively, and $V_R = V_B + V_T$ is the relative velocity between the bat and the target. The amount of Doppler frequency shift caused by the velocity of the bat and the target can be expressed as

$$\Delta f = f_{echo} - f_{trans} = f_{trans}\left(\frac{2V_R}{c}\right) \qquad (11\text{-}6)$$

Therefore, the Doppler shift is linearly related to emitted frequency and relative velocity between bat and target so that a bat's ability to determine this relative velocity will depend on its acuity in sensing the Doppler shift in the echo.

CF/FM bats that can sense Doppler shift in echoes compensate for the shift in frequency of the CF portion of echoes by lowering the emission frequency. This phenomenon is termed Doppler shift compensation. For a bat flying toward a stationary target or toward a prey flying away from it there will be a positive Doppler shift so that f_{echo} will be higher than f_{trans}, and the bat will lower its emission frequency in order to compensate for Doppler shift caused by its own flight velocity (Schnitzler 1984). The echo frequency is kept constant at the so-called reference frequency, a frequency about 200 Hz higher than the frequency emitted when at rest. Schnitzler (1973) conducted experiments with greater horseshoe bats flying in a wind tunnel and in a helium-oxygen gas mixture and found that the bats used a feedback control system to keep echo frequency constant. However, in the case of the greater horseshoe bats, compensation occurred only for the Doppler shift in the echoes from stationary surroundings and not for the additional Doppler shift caused by the flight of insects (Trappe and Schnitzler 1982). In this case, the frequency of the echoes from the surroundings was kept constant, near the reference frequency, and the frequency of the echoes produced by insects was above or below the reference frequency, depending on the insect's relative movements toward or away from the bat.

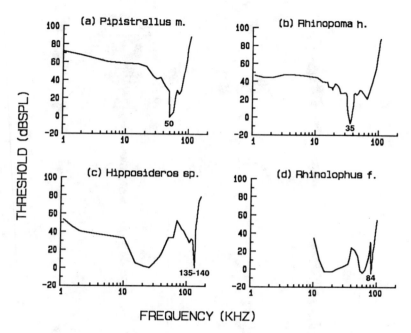

Figure 11.18. Audiograms showing specialization in the auditory systems of some CF/FM bats. Graphs (*A*), (*B*), and (*C*) were produced from data presented by Neuweiler (1984) and graph (*D*) was produced from data presented by Long and Schnitzler (1975). The numerical value below each audiogram indicates the frequency of the CF portion of the sounds emitted by the bats.

Doppler-sensing CF/FM bats have highly specialized auditory systems that aid them in sensing frequency fluctuations in echoes. There is usually a narrow frequency notch in their audiogram centered close to their reference frequency. In other words, these bats have narrow auditory filters that are tuned to the pure-tone echo frequency. Audiograms for some of them are shown in Figure 11.18. These specialized audiograms are also useful in detecting flutter in the amplitude of echoes caused by fluttering insects.

The other unique capability found in bats consists of a way in which certain bats detect and discriminate flying insects in the vicinity of bushes, trees, and other clutter: they are able to detect amplitude-modulated flutter in the echoes caused by the wing strokes of insects. Roeder (1963) found that the largest echo produced by moths flying at the same altitude as a bat occurred when the moths' flight path was perpendicular to the sound path and their wings were near the top of the stroke. The difference between the maximum and minimum echo was about 30 dB. Schnitzler et al. (1983) measured the echoes from *Autographa gamma* while the insect was simultaneously filmed with a 16-mm high-speed camera. Their results (Fig. 11.19) relate the amplitude-modulated echoes to the position of the flying insect. Schnitzler et al. (1983) also found that frequency modulations in echoes encoded the wingbeat frequency and contained species-specific information and information on the angular orientation of the insect. Bats that use wingbeat information will normally not pursue insects that stop beating their wings and enter into a free fall or insects that are stationary.

11.5 Summary

Differences in the types of signals used by bats and dolphins constitute one of the major differences between their sonars. Bats use a variety of signals which are species and situation specific. Bat sonar signals can be categorized into three general types: FM_{down}, CF/FM_{down}, and $FM_{up}/CF/FM_{down}$. The frequency of their signals varies from 12 to 200 kHz and the typical duration varies from 0.3 to 100 ms. In contrast, dolphins emit click signals; most dolphins emit broadband (about an octave bandwidth) signals with durations of about 70 to 100 μs, although signals from some of the smaller dolphin species are narrowband and have durations from about 125 to 250 μs. As a bat pursues its prey, the sonar signals change in their characteristics depending on the

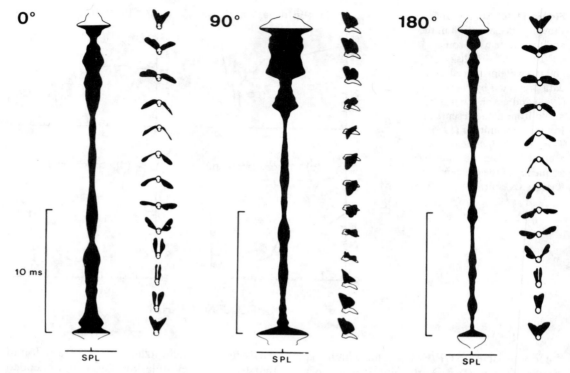

Figure 11.19. Amplitude-modulated echo of a flying insect oriented at three different angles. (From Schnitzler et al. 1983.)

specific phase of pursuit. The signals often become progressively shorter, the amplitude decreases, the FM sweep becomes broader, and the interpulse intervals become shorter as a bat goes into the approach and terminal phases of pursuit. Dolphins seem to control only the interclick interval and signal amplitude, decreasing both as a target is approached. Energy flux density calculations indicate that *Tursiops truncatus* emit signals with much higher energy levels than bats. However, bats may emit higher energy levels than some of the smaller dolphins. The transmission and reception beams of bats are generally wider than those of dolphins, which is not surprising considering the small size of a bat's head compared with a dolphin's. There is general agreement that bats produce sonar sounds in their larynx and vocal tract and emit them through their mouths or nostrils (for bats with a nose-leaf). Sounds and echoes are received through their ears. In contrast, there is considerable controversy as to how dolphins produce and receive sounds.

The capabilities of bats and dolphins with respect to sonar tasks such as the discrimination of differences in size, highlight delay time, and target shape are comparable. Bats have approximately as wide a hearing range as dolphins; however, in the frequency range of best hearing, dolphins have a greater sensitivity by about 10 to 20 dB. Dolphins perform better than bats on target detection tasks, on target range difference and material composition difference discrimination tasks. However, a good measure of the target detection capabilities of bats in noise, one that is not affected by clutter, still needs to be obtained. Given the available bat data which are contaminated by clutter, it appears that dolphins can detect targets in noise better than bats by at least 28 dB. This large difference in performance does not seem reasonable. Dolphins can also resolve acoustic propagation delay time differences better than bats for absolute target ranges that correspond to about 15 ms of two-way travel time. *Eptesicus fuscus* and *Phyllostomus hastatus*, which use short FM signals, can discriminate delay time differences best of all the bats observed. *Rhinolophus ferrumequinum*, which emit long CF/FM signals, seem to do worst and *Pteronotus suapurensis*, which use short CF/FM signals, seem to be in between the best and worst in discriminating delay time differences. Dolphins can recognize material composition differences of targets better than bats. Bats seem to be restricted to discriminating between materials that have large impedance differences, such as metal versus plastic or wood, and glass versus velvet. Dolphins, on the other hand, can discriminate between metallic targets of different composition. The penetration of underwater sounds into many targets gives an advantage to dolphins in discriminating by material composition.

Some bats that emit CF signals have a unique specialization in their hearing sensitivity that dolphins do not possess. These bats have audiograms with a highly tuned notch that is centered at their signal emission frequencies. This capability along with the long-CF nature of the signals allows these bats to detect Doppler shift caused by their own motion as well as by the motion of pursued prey. The highly tuned notch also allows these bats to easily detect fluttering prey and is useful in the rejection of clutter from stationary objects.

Unfortunately, this brief comparison of the sonar of bats and dolphins does not permit any discussion of the excellent electrophysiological research on bat brains by Nobuo Suga of St. Louis University and his present and former students and colleagues. Suga (1988, 1990) has spent many years mapping neural responses in the brains of bats, locating specific areas where neurons respond to certain facets of the sonar and echo signals. A discussion of his work deserves a separate volume.

References

Altes, R.A., and Titlebaum, E.L. (1970). Bat signals as optimally Doppler tolerant waveforms. J. Acoust. Soc. Am. 48: 1014–1020.

Ayrapet'yants, E., and Konstantinov, A.I. (1974). Echolocation in nature. Leningrad: Nauka. (English translation: Joint Publication Research Service, Arlington, Va.)

Bel'kovich, V.M., Borisov, V.I., and Gurevich, V.S. (1970). Angular resolution by echolocation by *Delphinus delphis*. Proc. 23rd Sci-Tech. Conf., Leningrad, pp. 66–67.

Bradbury, J. (1970). Target discrimination by the echolocating bat, *Vampyrum spectrum*. J. Exp. Zool. 173: 23–46.

Busnel, R.-G., ed. (1967). *Animal Sonar Systems: Biology and Bionics*. Laboratoire de Physiologie Acoustique, Jouy-en-Josas, France.

Busnel, R.-G., and Fish, J.F., eds. (1980). *Animal Sonar Systems*. New York: Plenum Press.

Dijkgraaf, S. (1946). Die Sinneswelt der Fledermäuse. Experientia 2: 438–448.

Dijkgraaf, S. (1957). Sinnesphysiologische Beobachtungen an Fledermäusen. Acta Physiol. Pharmacol. Neerlandica 6: 675–684.

Evans, W.W., and Powell, B.A. (1967). Discrimination of different metallic plates by an echolocating delphinid. In: R.-G. Busnel, ed., *Animal Sonar Systems: Biology and Bionics*. Laboratoire de Physiologie Acoustique, Jouy-en-Josas, France, pp. 363–382.

Fenton, M.B., Racey, R., and Rayner, J.M.V., eds. (1987). *Recent Advances in the Study of Bats*. London: Cambridge University Press.

Fleissner, N. (1974). Intensitätsunterscheidung bei Hufeisennasen (*Rhinolophus ferrumequinum*). Staatsexamensarbeit, Universität Frankfurt (Main).

Grinnell, A.D., and Schnitzler, H.-U. (1977). Directional sensitivity of echolocation in the horseshoe bat, *Rhinolophus ferrumequinum*. II. Behavioral directionality of hearing. J. Comp. Phys. A 116: 63–76.

Henson, O.W., Jr. (1965). The activity and function of the middle ear muscles in echolocating bats. J. Physiol. (London) 180: 871–887.

Johnson, C.S. (1967). Discussion. In: R.-G. Busnel, ed. *Animal Sonar Systems: Biology and Bionics*. Laboratoire de Physiologie Acoustique, Jouy-en-Josas, France, pp. 384–398.

Kick, S.A. (1982). Target-detection by the echolocating bat, *Eptesicus fuscus*. J. Comp. Physiol. 145: 431–435.

Kick, S.A., and Simmons, J.A. (1984). Automatic gain control in the bat's sonar receiver and the neuroethology of echolocation. J. Neurosci. 4: 2725–2737.

Konstantinov, A.I. (1970). A description of the ultrasonic location system of piscivorous bats. Tex. Dokl. 23-y Nauchno-Tekhn. Konf. LIAP, Leningrad, 51–52.

Konstantinov, A.I., and Akhmarova, N.I. (1968). Discrimination (analysis) of target by echolocation in *Myotis oxygnathus*. J. Biol. Sci. Moscow Univ. 4: 22–28.

Lawrence, B.D., and Simmons, J.A. (1982a). Measurements of atmospheric attenuation at ultrasonic frequencies and the significance for echolocation for bats. J. Acoust. Soc. Am. 71: 585–590.

Lawrence, B.D., and Simmons, J.A. (1982b). Echolocation in bats: the external ear and perception of the vertical positions of targets. Science 218: 481–483.

Long, G.R., and Schnitzler, H.-U. (1975). Behavioral audiogram from the bat *Rhinolophus ferrumequinum*. J. Comp. Physiol. A 100: 211–220.

Masters, W.M., and Jacobs, S.C. (1989). Target detection and range resolution by the big brown bat (*Eptesicus fuscus*) using normal and time-reversed model echoes. J. Comp. Physiol. A 166: 63–73.

Møhl, B. (1986). Detection by a pipistrelle bat of normal and reversed replica of its sonar pulses. Acustica 61: 75–82.

Møhl, B. (1988). Target detection by echolocating bats. In: P.E. Nachtigall and P.W.B. Moore, eds., *Animal Sonar: Processes and Performance*. New York: Plenum Press, pp. 435–450.

Nachtigall, P.E., and Moore, P.W.B., eds. (1988). *Animal Sonar: Processes and Performance*. New York: Plenum Press.

Neuweiler, G. (1984). Foraging, echolocation and audition in bats. Naturwissenschaften 71: 446–455.

Neuweiler, G. (1990). Auditory adaptations for prey capture in echolocating bats. Physiol. Rev. 70: 615–641.

Poussin, C., and Simmons, J.A. (1982). Low-frequency hearing sensitivity in the echolocating bat, *Eptesicus fuscus*. J. Acoust. Soc. Am. 72: 340–342.

Pye, J.D. (1980). Echolocation signals and echoes in air. In: R.-G. Busnel and J.F. Fish, eds., *Animal Sonar System*. New York: Plenum Press, pp. 309–353.

Renaud, D.L., and Popper, A.N. (1975). Sound localization by the bottlenose porpoise *Tursiops truncatus*. J. Exp. Biol. 63: 569–585.

Roeder, K.D. (1963). Echoes of ultrasonic pulses from flying moths. Biol. Bull. 124: 200–210.

Roverud, R.C., and Grinnell, A.D. (1985). Echolocation sound features processed to provide distance information in the CF/FM bat, *Noctillio albiventris*: evidence for a gated time window using both CF and FM components. J. Comp. Physiol. A 156: 447–456.

Schnitzler, H.-U. (1973). Control of Doppler shift compensation in the greater horseshoe bat, *Rhinolophus ferrumequinum*. J. Comp. Physiol. A 82: 79–92.

Schnitzler, H.-U. (1984). The performance of bat sonar systems. In: Varju and H.-U. Schnitzler eds., *Localization and Orientation in Biology and Engineering*. Berlin: Springer-Verlag, pp. 211–224.

Schnitzler, H.-U., and Grinnell, A.D. (1977). Directional sensitivity of echolocation in the horseshoe Bat, *Rhinolophus ferrumequinum*. I. Directionality of sound emission. J. Comp. Physiol. 116: 51–61.

Schnitzler, H.-U., and Henson, O.W., Jr. (1980). Performance of airborne animal sonar systems: I. Microchiroptera. In: R.-G. Busnel and J.F. Fish, eds., *Animal Sonar Systems*. New York: Plenum Press, pp. 109–181.

Schnitzler, H.-U., Menne, D., Kober, R., and Heblich, K. (1983). The acoustical image of fluttering insects in echolocating bats. In: F. Huber and H. Markl, eds., *Neuroethology and Behavioral Physiology*, Berlin: Springer-Verlag, pp. 235–250.

Simmons, J.A. (1969a). Depth perception by sonar in the bat *Eptesicus fuscus*. Ph.D. dissertation, Princeton University.

Simmons, J.A. (1969b). Acoustic radiation patterns for the echolocating bats *Chilonycteris rubigninosa* and *Eptesicus fuscus*. J. Acoust. Soc. Am. 46: 1054–1056.

Simmons, J.A. (1973). The resolution of target range by echolocating bats. J. Acoust. Soc. Am. 54: 157–173.

Simmons, J.A. (1987). Acoustic images of target range in the sonar of bats. Naval Research Review 39: 11–26.

Simmons, J.A. (1989). A view of the world through the bat's ear: the formation of acoustic images in echolocation. Cognition 33: 155–199.

Simmons, J.A., and Chen, L. (1989). The acoustic basis for target discrimination by FM echolocating bats. J. Acoust. Soc. Am. 86: 1333–1350.

Simmons, J.A., and Grinnell, A.D. (1988). The performance of echolocation: acoustic images perceived by echolocating bats. In: P.E. Nachtigall and P.W.B. Moore, eds., *Animal Sonar: Processes and Performance*. New York: Plenum Press, pp. 353–385.

Simmons, J.A., Lavender, W.A., Lavender, B.A., Doroshow, C.F., Kiefer, S.W., Livingston, R., Scallet, A.C., and Crowley, D.E. (1974). Target structure and echo spectral discrimination by echolocating bats. Science 186: 1130–1132.

Simmons, J.A., Howell, D.J., and Suga, N. (1975). Information content of bat sonar echoes. American Scientist 63: 204–215.

Simmons, J.A., Kick, S.A., Lawrence, B.D., Hale, C., Bard, C., and Escudie'. (1983). Acuity of horizontal angle discrimination by the echolocating bat, *Eptesicus fuscus*. J. Comp. Physiol. A 153: 321–330.

Simmons, J.A., Ferragamo, M., Moss, C.F., Stevenson, S.B., and Altes, R.A. (1990). Discrimination of jittered sonar echoes by the echolocating bat, *Eptesicus fuscus*: the shape of target images in echolocation. J. Comp. Physiol. A 167: 587–616.

Simmons, J.A., Moffat, A.J.M., and Masters, W.M. (1992). Sonar gain control and echo detection thresholds in the echolocating bat, *Eptesicus fuscus*, J. Acoust. Soc. Am. 91: 1150–1163.

Simmons, J.A., and Vernon, J.A. (1971). Echolocation: discrimination of targets by the bat *Eptesicus fuscus*. J. Exp. Zool. 176: 315–328.

Suga, N. (1988). Parallel-hierarchical processing of biosonar information in the mustached bat. In: P.E. Nachtigall and P.W.B. Moore, eds., *Animal Sonar: Processes and Performance*, New York: Plenum Press, pp. 149–159.

Suga, N. (1990). Biosonar and neural computation in bats. Sci. Am. 262 (June), 60–68.

Suga, N., and Jen, P.H.-S. (1975). Peripheral control of acoustic signals in the auditory system of echolocating bats. J. Exp. Biol. 62: 277–331

Surlykke, A., and Miller, L. A. (1985). The influence of arctiid moth clicks on bat echolocation: jamming or warning? J. Comp. Physiol. A 156: 831–843.

Trappe, M., and Schnitzler, H.-U. (1982). Doppler-shift compensation in catching horseshoe bats. Naturwissenschaften 69: 193–194.

Troest, N. and Møhl, B. (1986). The detection of phantom targets in noise by serotine bats: negative evidence for the coherent receiver. J. Comp. Physiol. A 159: 559–567.

Webster, F.A. (1967). Interception performance of echolocating bats in the presence of interference. In: *Animal Sonar Systems*, vol. I, R.G. Busnel, ed., Jouy-en-Josas; Laboratoire de Physiologie Acoustique, pp. 673–713.

12

Road Map for Future Research

In Chapters 2 through 10 we presented and discussed most of what is known of the sonar system of dolphins. While great progress has been made in the last quarter of a century in understanding the acoustic processes involved, there are still many gaps in our knowledge. We are still far from being able to construct realistic mathematical and analytical models of the various processes involved. One major goal in obtaining comprehensive understanding of the dolphin sonar is to be able to construct artificial sonars that can replicate the fine target recognition capabilities of these animals. In this final chapter, we will discuss various areas in which additional research is needed in order to fill the gaps in our understanding of the dolphin sonar system. The various items discussed will by no means be all-inclusive but will represent some of the areas that I feel are important. In performing such an inventory, I run the risk of not being sufficiently comprehensive and perhaps appearing short-sighted to those in disciplines other than my own. However, such an approach can be beneficial in bringing into focus some important issues, and promoting useful discussion and debate.

12.1 Mechanisms of Sound Reception and Hearing

There are many unknowns concerning the mechanisms of sound reception and hearing in dolphins. Evidence seems to point toward the notion that sound is received via the "acoustic window" of the mandible. However, some researchers still favor other reception sites, in particular the area around the external auditory meatus. Some pertinent questions concerning sound reception via the lower jaw include the following. If sound enters the dolphin's head via the pan bone of the mandible, how much loss of acoustic energy is associated with the process? How is the sound channeled to the external bulla, and what are the losses involved in that process? Norris (1968) first suggested that the special lipid in the region of the pan bone, which extends to the bulla, plays a major role in channeling received sounds to the bulla. What are the physical processes involved with this notion? How does acoustic energy actually enter the middle and inner ears, which are encased in a bony bulla? The physical processes involved with the coupling of sounds from outside the auditory bulla to the middle and inner ears are not clearly understood. What is the role of the ossicular chain (consisting of the malleus, incus, and stapes) in the perception of sounds? If sound enters via the external meatus, what are the physical mechanisms involved with its propagation along a thin fibrous tissue? Are there other acoustic paths associated with the head of a dolphin by which sound can finally reach the middle and inner ears, and what is the relative efficiency of the different acoustic paths? Finally, in regards to the reception of sound, how is the received beam formed and how does the received sound pressure level vary with source position in the near field? All these questions concerning the reception of sound by the dolphin need to be addressed in a way that allows for a broadband reception system sensitive to acoustic energy from several hundred Hertz to approximately 150 kHz.

There are also many areas of research that can be pursued to better understand the characteristics of hearing in dolphins. One of the more important questions concerns the shape of the auditory filter for different frequencies. The shape of the auditory filter will determine how well extraneous noise can be filtered, how broadband impulsive sounds are perceived, and how fine discriminations based on spectral differences of stimuli can be made. Another important subject concerns how dolphins perceive signals with rippled frequency spectra. Can dolphins perceive time separation pitch in a similar manner as humans when presented with signals having rippled frequency spectra? How well can a dolphin distinguish one rippled spectrum from another? Regarding the reception of acoustic signals with rippled spectra, is the processing done in the time or the frequency domain, or in both domains simultaneously? It would also be interesting to investigate what dolphins perceive when a rippled stimulus is composed of several delays in the time domain signals. How dolphins perceive complex sounds is another area that has not been addressed. In Chapter 3, I suggested that the minimum audible angle result for the vertical plane obtained by Renaud and Popper (1978) was puzzling. It would be worthwhile to investigate this issue further since passive localization of sound in three-dimensional space is an important capability for a dolphin.

Noninvasive electrophysiological techniques, particularly auditory evoked-potential techniques that measure the response of the early and late waves (Ridgway 1983) show considerable promise as tools for investigating the auditory systems of dolphins. However, these techniques need to progress beyond the restrictive manner in which they have been used in the past. Experiments have been performed with restrained dolphins held close to the water surface so that their blowholes and measuring electrodes were kept above the water surface (Ridgway 1980, 1983). The animals may even have been restrained in a small container. The acoustic environments in such experiments are virtually indescribable since there are many reflective surfaces (including the water surface) near the dolphin's head. Usually sound absorption materials are used, but the effectiveness of these materials over a broad frequency range is questionable. Electrophysical measurements should now be performed on trained dolphins with surface electrodes so that the acoustic environment can be more accurately described and the effects of unwanted reflections can be minimized. There is no doubt that better instrumentation, electrode design, and analytical techniques must be developed in order to gain the full benefits from electrophysiological experiments.

12.2 Mechanisms of Sound Production and Transmission

Turning now to the transmission side of the dolphin sonar system, there too are many areas in which research is needed to obtain a better understanding. The most obvious issue concerns the production of sonar and other acoustic signals. Although experimental evidence indicates that most acoustic signals are produced within the dolphin's head, the specific organs and specific physical mechanisms involved are not known. How are click sounds produced? How are the longer narrow-band, variable-frequency continuous tonal sounds commonly referred to as whistles produced? How does a dolphin vary the peak frequency of clicks and the frequency range of whistles? Are broadband clicks produced from multiple sources, and can multiple sound production sites be synchronized in such a manner as to cause internal steering of the outgoing beam? How is the acoustic energy coupled from the source into the melon for the various types of sounds? Precisely where on the dolphin's head does the major axis of the transmission beam for broadband click signals emanate? What is the source of energy for the production of high intensity clicks? How is the transmission beam formed in the head? These are but a small portion of a host of questions concerning sound production by dolphins. As some of these questions are addressed and answered, other questions will no doubt emerge.

Biosonar signal data obtained by scientists at

NOSC have involved experiments performed in either Kaneohe Bay or San Diego Bay, two bodies of water that have large populations of snapping shrimp. Three different species of dolphins, *Tursiops truncatus*, *Delphinapterus leucas*, and *Pseudorca crassidens*, have been studied. It is not clear how these species would perform in a much quieter environment, such as the open ocean or deep waters, where the effects of noise generated by bottom creatures would be minimized. Would these dolphins use high frequency clicks (peak frequencies > 100 kHz) in a target detection experiment performed in deep open ocean? High amplitude and high frequency clicks have been regularly observed for these dolphin species in Kaneohe and San Diego Bays. However, high frequency clicks have never been reported for these three species by other investigators, except by Poché et al. (1982) for a *Tursiops* in a tank lined with sound-absorbing material. Most of the signal data have been collected with animals close to the surface. What happens to the characteristics of the dolphin sonar signal as the animal dives to various depths? Are the air sacs, sound generation and acoustic propagational mechanisms affected as the external pressure increases with depth?

In Section 7.2.2, I presented data that suggest that dolphins probably do not purposefully adjust the spectral content of their sonar signals in order to optimize their target discrimination capabilities. I also suggested that although the data of Dziedzic and Alcuri (1977) seem to support the notion of spectral adaptation, it is difficult to interpret their data because of the geometry of their experiment. However, this issue needs to be better understand. One way of pursuing it further is to use suction cup hydrophones with long cables or a radio telemetry package attached to the animal. Although a suction cup hydrophone will detect signals in the near field, which will be slightly different from far-field signals, spectral changes, if present, will still be evident. The geometry between the animal and the hydrophone will not be a factor. If spectral adaptation does occur, it would be interesting to know under what conditions, at what point in a click train, and at which amplitude levels it occurs. It would also be interesting to examine whether spectral adaptation is the

result of internal beam steering or purposeful adjustment of spectral content by the dolphin.

A topic indirectly related to the mechanism of sound production has to do with the ontogeny of biosonar in dolphins. Determining when baby dolphins begin to emit sonar signals, in light of what is known about the physical development of structures within the animal's head, may provide insights to sound production mechanisms. The evolution of signal characteristics in babies may also provide important information as to the how the sonar process evolves in dolphins. Research by Brown and Grinnell (1980) indicates that with bats, in most instances there is correlation between the degree of morphological development at birth and the type of sound emitted.

12.3 Biosonar Capabilities and Mechanisms

In the area of sonar target detection, the various parameters of the dolphin's sonar system associated with the sonar equation for noise-limited situations have been determined for *Tursiops truncatus*. The dolphin's target detection capability has been modeled as an energy detector with an integration time of approximately 264 μs (Au 1990). Unfortunately, several of the parameters of the sonar equation are not known for other species. Data on target detection as a function of range have been obtained for *Pseudorca crassidens* (Thomas and Turl 1990). However, because the receiving directivity index is unknown for this species, the received signal-to-noise ratio as a function of target range cannot be calculated. The target detection data of Turl et al. (1987) for *Delphinapterus leucas*, suggesting that the beluga may have a detection threshold that is 8 to 13 dB lower than for *Tursiops*, are overly optimistic since the beluga would be able to perform better than an ideal detector. Alignment inaccuracies between the beluga, noise transducer and target probably contributed to the results, as was discussed in Chapter 8. However, the target detection in reverberation data for the beluga (Turl et al. 1991) suggested that this species may be more sensitive than *Tursiops* by 3.7 to 5.1 dB. It would not be surprising if the beluga's sensitivity for target detection in noise turned out to be better than that of *Tursiops*—at issue here is the size of the difference. Therefore, it is important to obtain a more accurate determination of the beluga's detection threshold in noise. Other important questions concerning the beluga sonar system should also be considered: How does the beluga obtain this processing gain? Does the different pattern of signal generation, such as the use of "packets" by the beluga, have a role in its superior detection sensitivity? What type of signal processing model would be appropriate for the beluga? Another piece of information, also important, is the beluga's receiving directivity index, which is needed to apply the sonar equation to the beluga's detection capability. Unfortunately, determining the receiving directivity index is a tedious process since the animal's receiving beam pattern must be measured at several different frequencies in both the horizontal and vertical planes (Au and Moore 1984). Perhaps electrophysiological techniques may be used to measure the received beam pattern quicker than possible in a behavioral study, or other techniques can be developed to measure the receiving directivity index directly.

Turning now from the detection of targets to discrimination, recognition, and classification of targets, we find that our understanding of the sonar processes utilized by dolphins is not as well developed. Some of the experiments discussed in Chapter 9 seem to suggest that time domain cues may play an important role in identifying certain targets. The signal analysis capability of the dolphin may be related to a capability to perceive time separation pitch. Experiments with stretched signals and human observers also discussed in Chapter 9 have highlighted the importance of time separation pitch in discerning multi-highlight echoes from targets. However, not many biosonar experiments have been performed to pinpoint target cues available to and perhaps used by dolphins to identify targets. Simulated electronic targets, similar to those used by Au et al. (1987), may be more useful than real targets because the characteristics of the echoes can be manipulated by the experimenter and thus finely and systematically varied. An area of research which may shed some light on the target discrimination and recognition process has to do with determining the time resolution capability of the dolphin. If we let Δt_1 be the time difference between highlights of a two-highlight echo and Δt_2 be the corresponding time difference for another two-highlight echo, how small can $\delta t = \Delta t_1 - \Delta t_2$ be for difference values of Δt_1, before the dolphin can no longer tell the difference between the two echoes? What happens to the dolphin's time resolution capability as the time separation Δt approaches the integration time constant of 264 μs for signal detection in noise? What are the effects of using signals in which the two highlights are of different amplitudes? What are the effects of noise on the time resolution capability of the dolphin? How will the dolphin's time resolution capability change as the signal becomes more complex, with a third highlight?

Simulated electronic targets can also be used to study how well dolphins can perceive spectral differences in broadband click signals. It is often assumed that dolphins can distinguish peak frequency differences in click signals. How small a peak frequency difference for different peak frequencies can be recognized? This issue may not be easily resolved since changing the peak frequency of a click signal may also affect its duration, and an experiment to measure a dolphin's capability to distinguish peak frequencies may actually be measuring the dolphin's ability to distinguish duration differences. However, if the peak frequency differences are not large, in the vicinity of 10%, the duration differences may not be an important cue. Another important issue associated with spectral discrimination involves the recognition of spectral differences for click signals having the same peak frequencies. Can a dolphin discriminate between two click signals having the same peak frequencies but slightly different spectra? The spectrum of a signal could be made systematically broader or narrower for frequencies above or below the peak frequency. How much do the spectra of two click signals need to differ before the dolphin can discriminate between them? The spectrum of a click signal could be altered by convolving the received digitized signals with various transfer functions to create a simulated electronic echo that is transmitted back to the dolphin.

Dolphin experiments involving target detection as well as target discrimination should be performed in the presence of different types of reverberation. In the clutter screen experiment of Au and Turl (1983), the targets were either in front of the clutter, a situation analogous to backward masking of auditory signals by noise in human hearing, or coincident with the clutter, analogous to simultaneous masking. It would be informative to determine how well the dolphin can detect and discriminate targets in the presence of different types of reverberation. However, a more appropriate source of reverberation than a screen of cork balls should be used to more closely simulate a natural environment. One natural source of reverberation is the ocean bottom: biosonar experiments can be performed in the presence of different types of bottoms ranging from soft muddy bottoms to coarse sandy bottoms. In any bottom reverberation experiment the amount of reflection from the bottom must be quantified so that the echo-to-reverberation ratio will be known. Closely associated with target detection and discrimination in the presence of reverberation is the dolphin's angular resolution capability. The smaller the angular resolution threshold, the better target echoes can be separated from reverberation.

Another area of research that would be worth pursuing involves the use of evoked-potential measurements to study the sonar process. Electrophysiological techniques have been used primarily in passive hearing studies and not with echolocation. Evoked-potential measurements may provide new insights on how the dolphin's brain receives and processes echo signals. For example, when sonar signals are emitted, the auditory system of a dolphin may be "sensitized" in a special way to receive low amplitude echoes from targets. It would be useful to determine whether the emission of sonar signals produces evoked potentials and prepares the auditory system for the reception of echoes. The various characteristics of evoked-potential responses to echoes can also be compared with properties of the received echoes, especially if simulated electronic targets are used to produce the echo stimuli. The properties of artificial echo stimuli can be varied in many ways, as has been discussed previously, and the evoked potentials these variations spawn can be studied to gain an understanding of how various acoustic signals are processed in the auditory pathways of the dolphin's system. The comments made in Section 12.1 concerning the use of trained animals and surface electrodes also apply to such studies of the active biosonar process using electrophysiological methods.

12.4 Signal Processing Models

Five different signal processing models were presented and discussed in Chapter 10, one for signal detection in noise and the others for target classification and recognition. Included in the latter four models were the backpropagation and counterpropagation neural networks. The energy detection model of Urkowitz (1967) seemed to fit the target detection data collected for *Tursiops* fairly well. However, other signal detection models, such as an envelope detection model, may also have close correspondence with the dolphin data and should be investigated. Whether the energy detection model is appropriate for the beluga is still an open question. The various signal processing models that classify and recognize targets should be tested against each other in noise. Joint experiments should be conceived in which the various discrimination capabilities of the dolphin sonar system are determined in noise in such a way that the dolphin's performance can be directly compared to the performance of different signal processing models. In such comparisons, the behavioral task must be designed to obtain a differential threshold for the animal. The only dolphin experiment that fits into this category is the wall thickness experiment of Au and Pawloski (1990) in which the dolphin's wall thickness discrimination capability was tested in noise. The importance of noise in such experiments cannot be overemphasized. I mentioned in Chapter 10 that some of the discrimination tasks can be performed quite well by various signal processing schemes if the signal-to-noise ratio is relatively high. It is only when the signal-to-noise ratio becomes small that the various models will begin to have difficulties with separating targets. This also applies to neural network models of the dolphin sonar system.

12.5 Natural and Dynamic Biosonar Behavior

Our understanding of the dolphin sonar system has come mainly through experiments using targets that are totally alien to these animals, in environments that may also be unnatural, and often requiring the animal to be constrained in a fixed location by a hoop or a bite plate. This approach is justifiable given the difficulty of working with free-ranging cetaceans. However, one has to wonder whether the various scanning and rocking motions that are associated with dolphins performing certain sonar tasks provide the animals with some kind of advantage. In the target detection experiment of Murchison (1980) using the experimental configuration shown in Figures 8.4B and 8.5, the dolphins often swam to the front of the pen, stuck their heads through the aperture and "rocked" about their longitudinal axis as they echolocated. Another favored maneuver consisted of the dolphins swimming parallel to the aperture with their bodies at a slight angle to the aperture and their heads canted almost 90°, presumably directed toward the target. What kind of information do the animals obtain by rocking about their longitudinal axis or by swimming almost perpendicular to the line-of-sight direction of the target? Scanning a target from slightly different locations may be helpful in nullifying the effects of bottom reverberation, but this premise needs to be tested.

Our perception of how dolphins utilize their sonar in the wild is based on extrapolation of knowledge obtained in "laboratory" experiments —we do not have the foggiest idea of how dolphins utilize their sonar in a natural environment. The experience of Jerry Diercks and William Evans as related in an article by Wood and Evans (1980) provides a sobering reminder that all is not what it seems. A *Tursiops truncatus*, wearing eyecups to block its vision was involved in a target discrimination experiment. The dolphin's sonar signals were monitored by an array of seven suction cup hydrophones mounted on its head. Echolocation signals were detected by the hydrophones during all discrimination trials. At the end of a session, a live fish was introduced into the tank, something the dolphin had not encountered in its several years of captivity. The dolphin did not emit any detectable sonar signals, but positioned itself adjacent to the fish and maintained its position with respect to the fish as the fish swam around the perimeter of the pool. The dolphin was able to catch the live fish and return it to the trainer without emitting a detectable sonar signal. The trainer then tossed the fish back into the pool. The dolphin repeated its behavior four times before the fish was eventually returned dead on the fifth retrieval. Instead of using its sonar to detect and localize the live fish as would be expected, the dolphin presumably used its hearing and passive localization capability to track and capture the fish.

There are many questions concerning how dolphins use their sonar in the wild: Do they usually use sonar for navigation purposes in enclosed bays, swamps, and in transiting open bodies of water? If so, how often are their sonars used? Do the animals usually use their sonar for the detection of predators and prey, or do they rely on their excellent hearing capabilities? If they use their sonar in the detection of predators and prey, what kinds of strategies are normally implemented? Is sonar used in long-range detection greater than approximately 100 m or for short ranges of 20 m or less? Dolphins normally exist in pods of a few individuals to several hundred animals. How is sonar used in a group of animals? Do the individuals in a pod use their sonar independently of each other, or does the leader of the pod usually assume the responsibility of probing the immediate environment, or is that duty relegated to another animal? It would not be surprising if dolphins relied on their sonar mainly at night or when in very turbid waters, and used their vision and hearing during the day. What kinds of sonar discrimination problems are dolphins typically faced with? Some of these questions have emerged because of the incidental catch of dolphins in monofilament driftnets. Target strength measurements of nets by Au and Jones (1991) indicate that dolphins should have little problem detecting monofilament driftnets at sufficiently long distances to avoid entanglement. However, since entanglement does occur frequently, field evidence suggests that dolphins may not utilize their sonar in the open ocean in ways that we would consider "normal" based on experimental results. To make progress in this area, miniature computer-controlled instrumentation packages are probably needed that can be attached to wild dolphins to monitor sonar activities and telemeter data via a satellite line.

The behavior of a group of Atlantic coastal spotted dolphins (*Stenella plagiodon*) and *Tursiops truncatus* in waters off the Bahamas has been studied by Herzing (1989, 1991). She has observed, and recorded on video tape, spotted dolphins hunting for fish prey that are buried in bottom sediment (Herzing, pers. comm.). These dolphins burrow into the bottom in order to capture their prey, usually lizardfish or razorfish. Do they use their active sonar capability to detect buried prey? What are the characteristics of their sonar signals? How deeply in the sediment can prey be detected? Spotted dolphins have been observed burrowing themselves in up to their pectoral fins, presumably to capture prey. *Tursiops* as well have been observed by Herzing hunting fish buried in the bottom sediment. However, *Tursiops* apparently forage in waters 15 to 20 m deep, making observation from the surface difficult. The detection of buried fish targets by dolphins is a subject area that has not been seriously addressed.

12.6 Concluding Remarks

The sonar system of dolphins has undergone millions of years of refinement under evolutionary pressures and has emerged as a sophisticated and highly sensitive sensory mechanism. It is superior to man-made sonar for short ranges (around 100 m) and in shallow waters containing bottom clutter and false targets. The dolphin's ability to discriminate and recognize features of targets with its sonar is a characteristic that man-made sonar systems do not possess. Although we have gained considerable knowledge of the capabilities and properties of the dolphin sonar, there is much we still need to understand before our knowledge can be of practical benefit. Nature has a way of optimizing various sensory faculties and functions in a compact and efficient manner that we are only slowly beginning to appreciate. However, as technology advances,

especially in the area of computer-controlled instrumentation, we will be able to use more channels with different types of sensors to monitor dolphins performing sonar scans, and hopefully gain a more thorough knowledge of their intriguing sensory mechanism. We have much to learn and gain from a deeper understanding of the dolphin sonar system.

References

Au, W.W.L. (1990). Target detection in noise by echolocating dolphins. In: J.A. Thomas and R. Kastelein, eds., *Sensory Abilities of Cetaceans*. New York: Plenum Press, pp. 203–216.

Au, W.W.L., and Jones, L. (1991). Acoustic reflectivity of nets and implications concerning the incident take of dolphins. Marine Mamm. Sci. 7: 258–273.

Au, W.W.L., and Moore, P.W.B. (1984). Receiving beam patterns and directivity indices of the Atlantic bottlenose dolphin *Tursiops truncatus*. J. Acoust. Soc. Am. 75: 255–262.

Au, W.W.L., Moore, P.W.B., and Martin, S.W. (1987). Phantom electronic target for dolphin sonar research, J. Acoust. Soc. Am. 82: 711–713.

Au, W.W.L., and Pawloski, D.A. (1992), Cylinder wall thickness discrimination by an echolocating dolphin, J. Comp. Physiol. A 172: 41–47.

Au, W.W.L., and Turl, C.W. (1983). Target detection in reverberation by an echolocating Atlantic bottlenose dolphin (*Tursiops truncatus*). J. Acoust. Soc. Am. 73: 1676–1681.

Brown, P.E., and Grinnell, A.D. (1980). Echolocation ontogeny in bats. In: R.G. Busnel and J.F. Fish, eds., *Animal Sonar Systems*. New York: Plenum Press, pp. 355–377.

Dziedzic, A., and Alcuri, G. (1977). Reconnaissance acoustique des formes et caracteristiques des signaux sonars chez *Tursiops truncatus*, famille des delphinides. C.R. Acad. Sc. Paris 285: Series D, 981–984.

Herzing, D.L. (1989). Social structure and underwater behavioral observations of Atlantic spotted dolphins, *Stenella plagidon*, in Bahamian waters: 1986–1989. Proc. 8th Biennial Conf. on the Biol. of Marine Mammm., Pacific Grove, Cal., Dec. 7–11, 1989.

Herzing, D.L. (1991). Underwater behavioral observations and sound correlations of free-ranging Atlantic spotted dolphin, *Stenella frontalis*, and bottlenose dolphin, *Tursiops truncatus*. Proc. 9th Biennial Conf. on the Biol. of Marine Mamm., Chicago, Ill., Dec. 5–9, 1991.

Murchison, A.E. (1980). Maximum detection range and range resolution in echolocating bottlenose porpoise (*Tursiops truncatus*). In: R.G. Busnel and J.F. Fish, eds., *Animal Sonar Systems*. New York: Plenum Press pp. 43–70.

Norris, K.S. (1968). The evolution of acoustic mechanisms in odontocete cetaceans. In: E. T. Drake, ed., *Evolution and Environment*. New Haven: Yale University Press, pp. 297–324.

Poché, L.B., Jr., Luker, L.D., and Rogers, P.H. (1982). Some observation of echolocation clicks from free-swimming dolphins in a tank. J. Acoust. Soc. Am. 71: 1036–1038.

Renaud, D.L., and Popper, A.N. (1978). Sound localization by the bottlenose porpoise *Tursiops truncatus*. J. Exp. Biol., 63: 569–585.

Ridgway, S.H. (1983). Dolphin hearing and sound production in health and illness. In: R.R. Fay and G. Gourevitch, eds., *Hearing and Other Senses: Presentations in Honor of E.G. Wever*. Groton, Conn.: Amphora Press, pp. 247–296.

Ridgway, S.H. (1980). Electrophysiological experiments on hearing. In: R.G. Busnel and J.F. Fish, eds., *Animal Sonar Systems*. New York: Plenum Press, pp. 483–493.

Thomas, J.A., and Turl, C.W. (1990). Echolocation characteristics and range detection by a false killer whale (*Pseudorca crassidens*). In: J.A. Thomas and R. Kastelein, eds., *Sensory Abilities of Cetaceans*, New York: Plenum Press, pp. 321–334.

Turl, C.W., Penner, R.H., and Au, W.W.L. (1987). Comparison of target detection capabilities of the beluga and bottlenose dolphin, J. Acoust. Soc. Am. 82: 1487–1491.

Turl, W.C., Skaar, D.J., and Au, W.W.L. (1991). The echolocation ability of the beluga (*Delphinapterus leucas*) to detect target in clutter. J. Acoust. Soc. Am. 89: 896–901.

Urkowitz, H. (1967). Energy detection of unknown deterministic signals. Proc. IEEE. 55: 523–531.

Wood, F.G., Jr., and Evans, W.E. (1980). Adaptiveness and ecology of echolocation in toothed whales. In: *Animal Sonar Systems*. R.G. Busnel and J.F. Fish, eds., New York: Plenum Press, pp. 381–425.

Index